FRACTAL ANALYSIS - APPLICATIONS IN PHYSICS, ENGINEERING AND TECHNOLOGY

Edited by **Fernando Brambila**

Fractal Analysis - Applications in Physics, Engineering and Technology
http://dx.doi.org/10.5772/65531
Edited by Fernando Brambila

Contributors

Alexander Potapov, Zeinab Eskandari, Asghar Keshtkar, Javad Ahmadi-Shokouh, Leila Ghanbari, Janusz Dudczyk, Jason Griggs, Wojciech Krzysztofik, Koji Nagata, Hiroki Suzuki, Shinsuke Mochizuki, Yasuhiko Sakai, Ksenia Potienko, Andrew Sedelnikov, Vlastimil Hotař, Oluranti Agboola, Maurice Onyango, Patricia Popoola, Opeyemi Alice Oyewo, Carlos Torres, Benito Fernando Martinez-Salgado, Rolando Rosas-Sampayo, Carlos Fuentes, Anthony Torres

Published by InTech
Janeza Trdine 9, 51000 Rijeka, Croatia

Publishing Process Manager Mirena Calmic
Technical Editor SPi Global
Cover InTech Design team

First published June, 2017
Printed in Croatia
Legal deposit, Croatia: National and University Library in Zagreb

Additional hard copies can be obtained from orders@intechopen.com

Fractal Analysis - Applications in Physics, Engineering and Technology, Edited by Fernando Brambila
p. cm.
Print ISBN 978-953-51-3191-5
Online ISBN 978-953-51-3192-2

For EU product safety concerns:
IN TECH d.o.o., Prolaz Marije Krucifikse Kozulić 3, 51000 Rijeka, Croatia,
info@intechopen.com or visit our website at intechopen.com.

Meet the editor

Fernando Brambila holds a PhD degree from UNAM, Mexico. His thesis on scattering theory was directed by Gunther Uhlmann at MIT. He obtained a postdoctoral position at ICTP, Italy, and has a diploma in Senior Technology Innovation Management at IPADE, Harvard. His research areas are mathematical analysis, partial differential equations, vectorial tomography, and hydraulic engineering. More recently, he has done research on fractional calculus and fractal geometry and applications with his collaborators K. Oleschko (UNAM), C. Fuentes (IMTA), and C. Chavez (UAQ). He is a full-time professor at the Mathematics Department of the School of Science at the National Autonomous University of Mexico, UNAM. He is a doctoral thesis advisor of F. Aceff, R. Mercado, J. Rico, B. Martinez, and C. Torres. Also, he is the former president of the Mexican Mathematical Society, and he is currently the president of AMITE (Mexican Association for Innovation in Educational Technology).

Contents

Preface

Fractal analysis has entered a new era. The applications to different areas of knowledge have been surprising. Let us begin with the fractional calculus-fractal geometry relationship, which allows for modeling with extreme precision of phenomena such as diffusion in porous media with fractional partial differential equations in fractal objects. Where the order of the equation is the same as the fractal dimension, this allows us to make calculations with enormous precision in diffusion phenomena—particularly in the petroleum industry, for new spillage prevention.

On the subject of fractional calculus, it is a problem proposed 300 years ago. In the I'Hopital letters, Leibniz and Bernoulli asked themselves how to define the fractional derivative and give it a physical and geometrical interpretation. Answers (but only about the definition) have been around since 1720, when Euler generalized the concept of *n!* for real and complex numbers with the Gamma function in order to give us the first answer to it. Then, in 1850, Riemann and Liouville gave us an answer about fractional integrals, so, too, others like Grunwald and Letnikov. But the physical and geometrical interpretation has only appeared in our days now that we understand that there is a relationship between fractional calculus and fractal geometry.

We have several partial answers to this 300-year-old problem:

the order of the fractional partial differential equations turned out to be related to the fractal dimension of the geometry where the phenomena take place. And the success of fractional calculus modeling physical phenomena reveals an underlying fractal nature of reality (S. Butera and M. Di Paola).

We have also made progress with the Nigmatullin-Rutman controversy—and the clarifications of Nigmatullin-Le Mehaute.

It's important to keep in mind Metzler, Glockle, and Nonnenmacher, who clarify that the parameters of the fractional partial differential equation that came from the anomalous diffusion are uniquely determined by the fractal Hausdorff dimension of the underlying object and the anomalous diffusion experiment.

An important note is that the fractional derivatives are not the weak Fourier-Laplace derivatives nor are the fractional operators pseudo-differential operators.

Fractal analysis is no longer just creating nice images, nor is it a branch of mathematics with little interaction with the other areas.

In the history of fractal geometry, important contributions were made toward seeing nature as it is, rather than as an approximation to classical geometry.

With the mathematics deduced from classical geometry, we could only model approximations of nature. Part of the history of fractal geometry was using them to create images of nature on a computer. For example, in 1978, Loren Carpenter achieved unprecedented images of mountains by using fractals, for a Boeing Commercial Aircraft. Carpenter based his work on the book *Fractals form, change and dimension*, by Benoit Mandelbrot. And so began the era of fractal images. Consider, for example, the lava scene from the movie Star Trek III —a masterpiece of fractal use.

This story takes us to 1999—the use of fractal antennas for cell phones. N. Cohen constructed the first fractal antenna for personal use; then, R. Hohlfeld and N. Cohen published an article in which they proved mathematically that in order to receive many frequencies, it is necessary to have fractal antennas.

More recently, while analyzing heart rate charts at Harvard, H. Goldberger realized that fractal analysis of these graphs allowed him to distinguish between a heart in good health and one that is not in good health.

It has also helped us detect changes and abnormalities in blood flow, which allows us to determine whether an organ, such as a kidney, has or will have cancerous tumors. Early cancer detection by use of fractal analysis, basically.

Power laws in nature are deduced from fractal dimension.

Fractals also allow us to know about the health of a forest. When we see a tree, we realize that a similar pattern is repeated all throughout the forest itself.

Now, we enter the modern part of fractal geometry. In this book, we have applications that Mandelbrot would have surely loved to see. Applications for the petroleum industry, numerical analysis, fractal antennas for cell phones, spacecraft, radars, image processing, measure, porosity, turbulence, scattering theory…

This book is divided into three sections, where the research chapters are presented in the following way.

The first part, called "Petroleum Industry, Numerical Analysis, and Fractal History," is about the use of fractal geometry and fractional calculus to model the difference in pressure with which the oil will come out. A history of fractional calculus and its relation accompany this section to the fractal dimension of the medium. That is because porosity can be measured using the fractal dimension, as shown by K. Oleschko. We also have in this section the numerical analysis for fractional partial differential equations, a new area of research.

Our second part is called "Industry, Antennas, Spacecraft, Radar, Images, and Measure." It presents applications of fractal analysis that are quite important today, such as fractal antennas and metamaterials that are used in all cell phones. Such topics include miniaturization and its use for efficiency in industry, spacecraft, microaccelearation, radars, image processing, industrial applications, time series, and precision measuring.

The third part is "Applications, Scattering, Porosity, and Turbulence." These classical subjects where mathematical modeling has used quantum mechanics, partial equations, and geometry nowadays have benefited from the use of fractal analysis. The applications of scattering theory of porosity and turbulence are of greater precision, for example.

It's impossible to picture today's research without fractal geometry.

Prof. Fernando Brambila
Mathematics Department, School of Sciences
National Autonomous University of Mexico, Mexico

Section 1

Petroleum Industry, Numerical Analysis and Fractal History

Chapter 1

Applications of Radial Basis Function Schemes to Fractional Partial Differential Equations

Carlos Alberto Torres Martínez and Carlos Fuentes

Additional information is available at the end of the chapter

http://dx.doi.org/10.5772/67892

Abstract

In modeling using diffusion equations, the relationship between fractal geometry and fractional calculus arises by modeling the conditions of the medium as a fractal whose fractional dimension determines the order of this equation. For this reason, it is very useful to have numerical methods that solve them and discretization of the domain is not determinant for the efficiency of the algorithm. In this work, it is proposed to show that meshless methods, in particular methods with radial basis functions (RBF), are an alternative to schemes in differences or structured meshes. We show that we can obtain numerical solutions to some fractional partial differential equations using collocation and RBF, over non equally distributed data.

Keywords: fractional partial differential equations (FPDE), meshless methods, radial basis functions (RBF), Caputo derivative, Riemann-Liouville derivative, Riesz derivative, diffusion-convection

1. Background

1.1. Radial basis function methodology

The Hardy-based radial-based functions (RBF) methodology [1] arises from the need to apply multivariate interpolation to cartography problems, with randomly dispersed data (also known as collocation nodes). Micchelli [2], Powell et al. [3] gave it a great boost by proving non-singularity theorems. Later, Kansa [4, 5] proposed to consider the analytical derivatives of the FBR to develop numerical schemes that deal with partial differential equations (PDE).

Regarding PDE over spaces of dimension greater than one, we generally opt for finite element (FE) type discretizations on meshes, structured or not; also pseudo spectral (PS) methods through base functions such as Fourier or Chebyshev. The high degree of computational efficiency in the procedure raises the cost in the regularity constraints on the form of the computational domain.

The FE methods involve decomposition of the domain; for example, in two dimensions rectangles are constructed with curvilinear mappings that allow the refinement of the mesh in critical areas. However, this type of implementation is complex and very close nodes are needed, mainly at the domain boundaries, which impairs stability conditions in time.

For this reason, we look for numerical techniques that do not depend to a great extent on data distribution. RBF collocation methods belong to the "meshfree" methods; that is, they only require scattered collocation nodes on the domain and the boundary. They are also an alternative for dealing with problems in larger dimensions and irregular domains. Hence, in recent decades, such methods have attracted the attention of researchers in order to solve partial differential equations. See for example: Chen et al. [6].

This type of technique approximates the solution by means of a linear combination of radial basis functions, which are globally defined and of exponential convergence, but they produce dense and badly conditioned interpolation matrices. On the other hand, some RBFs contain a shape parameter, a number that influences the precision of the numerical results [7, 8].

There are algorithms that also produce better conditioned interpolants, even when the shape parameter tends to zero, with a double precision arithmetic. Some of them are: *Contour-Padé* [9] and *RBF-QR* [10]. Related to the latter is the recent RBF-GA [11].

The RBF-QR method provides a numerically more stable alternative for converting basis that are very similar to each other, or almost linearly dependent on a set of scattered nodes, to a very well conditioned base that generates exactly the same space. This method was originally implemented for nodes on the surface of a sphere in Fornberg and Piret [10]. And recently for an arbitrary set of nodes in one, two, and three dimensions [12, 13].

1.2. Fractional calculus and differential equations: about some applications

The growing interest in fractional calculus has been motivated by applications of fractional equations in different fields of research.

One of the applications considered in this work is that of convection-diffusion equations, which appear not only in many applications of models in Physics and Chemistry [14–17], in problems of flow or heat transfer, but also in other fields such as financial [18, 19]. The solution of a convection-diffusion problem can be interpreted as a probability distribution of one or more underlying stochastic processes [18]. The classical diffusion equation (or heat equation) and its Gaussian solution existed long before Einstein established a connection with random walks.

The anomalous diffusion equations, on the other hand, were originally developed from stochastic walks models [20]. The parameters of these equations are uniquely determined by the fractal dimension of the underlying object [21].

Recent applications of these anomalous diffusion-convection models—those related to oil extraction and hydrological models for aquifers, food production and water distribution in large cities, are quite important [22]. Determining the behavior of the fluid inside the reservoir and the loss of permeability of the medium assists in the investigation of the mechanisms of oil migration. In this context, the geometric and physical interpretation of fractional derivatives in these differential equations is a "ultra slow" or "ultra fast" diffusion.

The relationship between fractal geometry and fractional calculus is given from the consideration that a particle moves in a porous medium with fractal structure [23]. Thus, when considering the derivative over that variable, a fractional derivative is obtained.

Fuentes et al. [24, 25] proposed a model that represents anomalous diffusion of petroleum (very fast or very slow) in three types of medium with different porosities: fractures, vugs and matrix. This model has a dimensionless form, which is seen as

$$(1-\omega_f-\omega_v)\frac{\partial^{\alpha_1} p_m}{\partial t^{\alpha_1}} = (1-\kappa_f-\kappa_v)\frac{1}{r}\frac{\partial}{\partial r}\left(r\frac{\partial^{\beta_1} p_m}{\partial r^{\beta_1}}\right) + \lambda_{mf}(p_f-p_m)+\lambda_{mv}(p_v-p_m), \quad (1)$$

$$\omega_f\frac{\partial^{\alpha_2} p_f}{\partial t^{\alpha_2}} = \kappa_f\frac{1}{r}\frac{\partial}{\partial r}\left(r\frac{\partial^{\beta_2} p_m}{\partial r^{\beta_2}}\right) - \lambda_{mf}(p_f-p_m)+\lambda_{fv}(p_v-p_f), \quad (2)$$

$$\omega_v\frac{\partial^{\alpha_3} p_v}{\partial t^{\alpha_3}} = \kappa_v\frac{1}{r}\frac{\partial}{\partial r}\left(r\frac{\partial^{\beta_3} p_m}{\partial r^{\beta_3}}\right) - \lambda_{mv}(p_v-p_m)+\lambda_{fv}(p_v-p_f). \quad (3)$$

This model must be solved for p's (pressure) at time t in the matrix of the medium p_m, the fractured medium p_f, and the vugular medium p_v.

κ represents each permeability tensor, which is assumed to be constant; r is the dimensionless variable related to the ratios of the wellbore. Values λ are dimensionless values and they are defined in terms of transfer coefficients at each interface, the radius of the wellbore, the dynamic viscosity of the fluid and the permeability tensor.

If time orders α's are less than one we say that the process is **subdiffusive** and if it is greater than one it is called **superdiffusive**. Spatial orders β's result from generalizing the classic Darcy law, which corresponds to $\beta = 1$.

The above equations, like the flow equation that considers the medium as a whole, are modified Bessel's equations.

Different porosities imply different fractal dimensions and therefore, fractional derivatives with different orders.

In Fractional Calculus there is not a single definition of derivative and between these definitions, in general, not equivalent to each other. Definitions such as Riemann-Liouville, Riesz, Caputo, Weyl or Grünwald-Letnikov are the most used to model anomalous diffusion, from a porous medium with fractal dimension.

But, fractional equations present serious numerical and mathematical difficulties in the context of diffusion equations. From the computational point of view, the challenge to overcome is to attenuate the numerical cost of the matrices that result in discretizing the problem. The proposal is to consider radial basis interpolators [26], taking into account that the domain geometry does not determine the efficiency of the algorithm and sounds like an immediate alternative for generalizing larger dimensions.

2. Introduction

The objective in this work is to deal with fractional differential operators, using radial basis functions (RBF) and optimizing discretization processes of such fractional operators, through QR matrix decomposition and to attenuate the bad condition due to the shape parameter. With RBF-QR, a very high precision and convergence are obtained without the need to increase a polynomial term to the interpolator [13]. As the computational cost increases, so does the range of problems to which such a technique can be applied.

3. Radial basis function method for interpolation

Definition 1. *Let s be a positive integer, a function $\varphi : \mathbb{R}^s \to \mathbb{R}$ is called* **radial** *whenever there is a function in one variable $\phi : [0, \infty) \to \mathbb{R}$ such that*

$$\varphi(\mathbf{x}) = \phi(r), \quad \text{where}\, r = \ ||\mathbf{x}|| \tag{4}$$

and $||\cdot||$ is a norm in $\mathbb{R}^s$ (usually the Euclidean standard norm). ***Table (1)*** *shows some examples, about the more used real functions $\phi(r)$.*

A standard interpolator in terms of radial basis functions, given the data u_k, of some real function u, in the corresponding **collocation nodes $\mathbf{x_k}$, $\mathbf{k}$** $= 1, \ldots, N$, has the form [13]

$$s_\varepsilon(\mathbf{x}) = \sum_{k=1}^{N} \lambda_k \phi(\varepsilon||\mathbf{x} - \mathbf{x_k}||) \tag{5}$$

where ϕ is one real variable function and the constant value ε is called **shape parameter**.

RBF name	RBF $\phi(r)$
Piecewiese smooth, global	
Polyharmonic spline	r^m, $m = 1, 3, 5, \ldots$
	$r^m \ln(r)$, $m = 2, 4, 6, \ldots$
Compact support ('Wendland')	$(1 - \varepsilon r)_+^m p(\varepsilon r)$, p polynomial.
Smooth, global	
Gaussian (GA)	$e^{-(\varepsilon r)^2}$
Multiquadric (MQ)	$\sqrt{1 + (\varepsilon r)^2}$
Inverse quadric (IQ)	$1/(1 + (\varepsilon r)^2)$
Inverse multiquadric (IMQ)	$1/\sqrt{1 + (\varepsilon r)^2}$
Bessel (BE) ($d = 1, 2, \ldots$)	$J_{d/2-1}(\varepsilon r)/(\varepsilon r)^{d/2-1}$

Table 1. Common elections for $\phi(r)$.

Note 1. *The radial basis function (FBR) we will take for the approach and examples is the* **Gaussian function**:

$$\phi(r) = e^{-r^2} \quad r \geq 0. \tag{6}$$

Because it is globally smooth and for which the RBF method can be applied together with QR matrix decomposition [26].

Notation 1. *Considering one dimensional case, we abbreviate* $\phi_k(x) := \phi(\varepsilon\|x - x_k\|)$. *Now the interpolator (5) is written as*

$$s_\varepsilon(x) \equiv \sum_{k=1}^{N} \lambda_k \phi_k(x). \tag{7}$$

The unknown coefficients λ_k can be determined by interpolation conditions

$$s_\varepsilon(x_i) = u_i, \quad i = 1, \ldots, N.$$

producing the linear equations system

$$\begin{array}{rcl} \lambda_1\phi_1(x_1) + \cdots + \lambda_N\phi_N(x_1) & = & u_1 \\ \vdots & \vdots & \vdots \\ \lambda_1\phi_1(x_N) + \cdots + \lambda_N\phi_N(x_N) & = & u_N \end{array}$$

which can be described in matrix form

$$\underbrace{\begin{bmatrix} \phi_1(x_1) & \cdots & \phi_N(x_1) \\ \vdots & \vdots & \vdots \\ \phi_1(x_N) & \cdots & \phi_N(x_N) \end{bmatrix}}_{A} \underbrace{\begin{bmatrix} \lambda_1 \\ \vdots \\ \lambda_N \end{bmatrix}}_{\lambda} = \underbrace{\begin{bmatrix} u_1 \\ \vdots \\ u_N \end{bmatrix}}_{u_X}.$$

Thus

$$A\lambda = u_X, \quad \text{where } A = \begin{bmatrix} \Phi(x_1)^T \\ \vdots \\ \Phi(x_N)^T \end{bmatrix} \quad \text{and} \quad \Phi(x) := \begin{bmatrix} \phi_1(x) \\ \vdots \\ \phi_N(x) \end{bmatrix}. \tag{8}$$

A is called **Gram matrix**.

4. RBF methodology for the discretization of differential operators

Suppose we want to apply a differential operator $\mathcal{L}$ to the function $u(x)$ in every **evaluation node** from the set $Z = \{z_1, \ldots, z_{N_e}\}$, given the values of the function at collocation nodes $X = \{x_1, \ldots, x_N\}$

$$\mathcal{L}u_Z \approx Du_X. \tag{9}$$

where $\mathcal{L}u_Z = \begin{bmatrix} \mathcal{L}u(z_1) \\ \vdots \\ \mathcal{L}u(z_{N_e}) \end{bmatrix}$ and $u_X = \begin{bmatrix} u(x_1) \\ \vdots \\ u(x_N) \end{bmatrix}$.

Assuming that $\mathcal{L}$ is linear, we apply it to the RBF interpolator, for each of the evaluation nodes, as an approximation to the values of $\mathcal{L}$ in u:

$$\begin{aligned} \mathcal{L}s_\varepsilon(z_i) &= \sum_{k=1}^{N} \lambda_k \mathcal{L}\phi_k(z_i), \quad i = 1, 2, \ldots, N_e; \\ \mathcal{L}(s_\varepsilon(z_1)) &= \lambda_1 \mathcal{L}\phi_1(z_1) + \cdots + \lambda_N \mathcal{L}\phi_N(z_1) \\ \vdots \quad &\vdots \quad\quad \vdots \\ \mathcal{L}(s_\varepsilon(z_{N_e})) &= \lambda_1 \mathcal{L}\phi_1(z_{N_e}) + \cdots + \lambda_N \mathcal{L}\phi_N(z_{N_e}) \end{aligned} \tag{10}$$

which in the matrix form is

$$\mathcal{L}s_\varepsilon = B\lambda \tag{11}$$

where

$$\mathcal{L}s_\varepsilon = \begin{bmatrix} \mathcal{L}(s_\varepsilon(z_1)) \\ \vdots \\ \mathcal{L}(s_\varepsilon(z_{N_e})) \end{bmatrix}, B = \begin{bmatrix} \mathcal{L}\phi_1(z_1) & \cdots & \mathcal{L}\phi_N(z_1) \\ \vdots & \vdots & \vdots \\ \mathcal{L}\phi_1(z_{N_e}) & \cdots & \mathcal{L}\phi_N(z_{N_e}) \end{bmatrix}.$$

The coefficient matrix $\lambda = \begin{bmatrix} \lambda_1 \\ \vdots \\ \lambda_N \end{bmatrix}$ is calculated by solving for λ in Eq. (8). In this way, from Eq. (11) we obtain

$$\mathcal{L}s_\varepsilon = (BA^{-1})u_X. \tag{12}$$

Thus, the matrix BA^{-1} is that matrix D which is mentioned in Eq. (9) and that approximates the values of the differential operator applied to the function.

5. Fractional calculus definitions

The definitions shown in this section are based on the texts of Podlubny [27] and Samko [28].

Generalizing the formula of **Cauchy**, we obtain

$$\mathcal{J}^n\left(f(t)\right) := \int \cdots \int_0^t f(\tau)d\tau = \frac{1}{(n-1)!}\int_0^t (t-\tau)^{n-1} f(\tau)d\tau$$

Definition 2. *The* **left-sided Riemann-Liouville fractional integral** *of order α of function $f(x)$ is defined as*

$$ {}_{a}^{RL}I_{x}^{\alpha}f(x) = \frac{1}{\Gamma(\alpha)}\int_{a}^{x}(x-\tau)^{\alpha-1}f(\tau)d\tau,\ x > a. \tag{13}$$

Definition 3. *The* **right-sided Riemann-Liouville fractional integral** *of order α of function $f(x)$ is defined as*

$$ {}_{x}^{RL}I_{b}^{\alpha}f(x) = \frac{1}{\Gamma(\alpha)}\int_{x}^{b}(\tau-x)^{\alpha-1}f(\tau)d\tau,\ x < b. \tag{14}$$

Based on the definition of fractional integral, the following fractional derivatives are constructed, which we will use in the numerical examples

Definition 4. *The* **left-sided Riemann-Liouville fractional derivative** *of order α of function $f(x)$ is defined as*

$$ {}_{a}^{RL}D_{x}^{\alpha}f(x) = \frac{1}{\Gamma(m-\alpha)}\frac{d^m}{dx^m}\int_{a}^{x}(x-\tau)^{m-\alpha-1}f(\tau)d\tau,\ x > a, \tag{15}$$

where $m = \lceil\alpha\rceil$.

Definition 5. *The* **right-sided Riemann-Liouville fractional derivative** *of order α of function $f(x)$ is defined as*

$$ {}_{x}^{RL}D_{b}^{\alpha}f(x) = \frac{(-1)^m}{\Gamma(m-\alpha)}\frac{d^m}{dx^m}\int_{x}^{b}(\tau-x)^{m-\alpha-1}f(\tau)d\tau,\ x < b, \tag{16}$$

where $m = \lceil\alpha\rceil$.

Definition 6. [28–30] *The* **Riesz fractional operator** *for α on a finite interval $0 \le x \le L$ is defined as*

$$ \frac{\partial^\alpha}{\partial|x|^\alpha}f(x,t) = -c_\alpha\left({}_{a}^{RL}D_{x}^{\alpha} + {}_{x}^{RL}D_{b}^{\alpha}\right)f(x,t), \tag{17}$$

where

$$ c_\alpha = \frac{1}{2\cos\left(\frac{\pi\alpha}{2}\right)},\ \alpha \neq 1, \tag{18}$$

$$ {}_{a}^{RL}D_{x}^{\alpha}f(x,t) = \frac{1}{\Gamma(m-\alpha)}\frac{d^m}{dx^m}\int_{a}^{x}(x-\tau)^{m-\alpha-1}f(\tau,t)d\tau, \tag{19}$$

$$ {}_{x}^{RL}D_{b}^{\alpha}f(x,t) = \frac{(-1)^m}{\Gamma(m-\alpha)}\frac{d^m}{dx^m}\int_{x}^{b}(\tau-x)^{m-\alpha-1}f(\tau,t)d\tau, \tag{20}$$

where $m = \lceil\alpha\rceil$.

Definition 7. *The* **left-sided Caputo fractional derivative** *of order α of function $f(x)$ is defined as*

$$ {}_{a}^{C}D_{x}^{\alpha}f(x) = \frac{1}{\Gamma(m-\alpha)} \int_{a}^{x} (x-\tau)^{m-\alpha-1} f^{(m)}(\tau) d\tau, \; x > a, \tag{21} $$

and $m = \lceil \alpha \rceil$.

The fact that the derivative of integer order appears within the integral in Definition 7, makes Caputo derivative the most suitable for dealing with initial conditions of the FPDEs.

Definition 8. *The* **right-sided Caputo fractional derivative** *of order* α *of function* $f(x)$ *is defined as*

$$ {}_{x}^{C}D_{b}^{\alpha}f(x) = \frac{(-1)^{m}}{\Gamma(m-\alpha)} \int_{x}^{b} (\tau-x)^{m-\alpha-1} f^{(m)}(\tau) d\tau, \; x < b, \tag{22} $$

and $m = \lceil \alpha \rceil$.

6. Numerical examples

6.1. Fractional partial differential equation with Riesz space fractional derivatives

The equation that we take in this part is the advection-diffusion equation (see [30, 6]).

$$ \begin{aligned} &\frac{\partial u(x,t)}{\partial t} = -K_{\alpha} \frac{\partial^{\alpha} u(x,t)}{\partial |x|^{\alpha}}, \; x \in [0,\pi], \; t \in (0,T], \\ &u(x,0) = u_0(x), \\ &u(0,t) = u(\pi,t) = 0, \end{aligned} \tag{23} $$

where u can be, for example, the concentration of a dissolute substance, K_{α} the dispersion coefficient and the fractional derivative of Riesz is given with fractional order $1 < \alpha \leq 2$.

Taking advantage of the fact that for RBFs we can consider non-equispaced collocation nodes, we divide the interval $[0, \pi]$ into N nodes, using Chebyshev distribution (see **Figure 1**):

$$ x_i = \frac{\pi}{2} \cos(\theta_i) + \frac{\pi}{2}, \quad \theta_i = \pi - i\frac{\pi}{N-1}, \quad i = 0, 1, \ldots, N-1. $$

The FPDE is solved by the method of lines based on the spatial trial spaces spanned by the Lagrange basis associated to RBFs. The Lagrange basis $L_1(x), \ldots, L_N(x)$ is generated by radial functions $\phi_j(x) = \phi(|x - x_j|)$, $j = 1, 2, \ldots, N$, taking into account the collocation nodes. This is done by solving the system

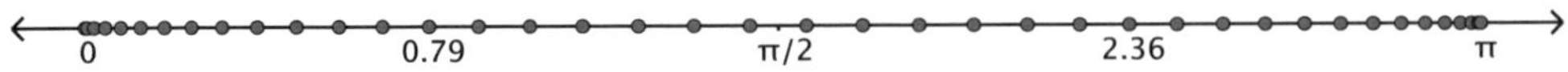

Figure 1. Chebyshev nodes distribution over $[0, \pi]$ interval.

$$L(x)^T = \Phi(x)^T A^{-1} \tag{24}$$

where

$$L(x)^T = [L_1(x) \quad \cdots \quad L_N(x)],\ \Phi(x)^T = [\phi_1(x) \quad \cdots \quad \phi_N(x)] \tag{25}$$

and the Gram matrix

$$A = \begin{bmatrix} \phi_1(x_1) & \cdots & \phi_N(x_1) \\ \vdots & \vdots & \vdots \\ \phi_1(x_N) & \cdots & \phi_N(x_N) \end{bmatrix}.$$

If $\mathcal{L}$ is a differential operator and RBF ϕ is sufficiently smooth, then the application of such operator to the Lagrange base is calculated through the relation

$$(\mathcal{L}L)(x) = (\mathcal{L}\phi)A^{-1}.$$

Because of Lagrange's standard conditions, the zero boundary conditions in $x_1 = 0$ and $x_N = \pi$ are de facto fulfilled if we use an approximation generated by the functions $L_2, \ldots, L_{N-1}$. This approximation is then represented as

$$u(x,t) = \sum_{j=2}^{N-1} \beta_j(t) L_j(x),$$

with unknown vector

$$\beta(t) = \begin{bmatrix} \beta_2(t) \\ \vdots \\ \beta_{N-1}(t) \end{bmatrix}.$$

Evaluating the interpolator at the PDE in each node x_i, we obtain

$$\sum_{j=2}^{N-1} \beta_j'(t) L_j(x_i) = -K_\alpha \sum_{j=2}^{N-1} \beta_j(t) \frac{\partial^\alpha}{\partial |x|^\alpha} L_j(x_i)$$

and initial conditions

$$\beta_j(0) = u_0(x_j),\ 2 \le j \le N-1.$$

From the latter two equations we obtain the following system of ordinary differential equations

$$\beta'(t) = -K_\alpha \left(\frac{\partial^\alpha}{\partial |\mathbf{x}|^\alpha} \mathbf{L} \right) \cdot \beta(t),\ \beta(0) = U_0,$$

where

$$\frac{\partial^\alpha}{\partial|\mathbf{x}|^\alpha}\mathbf{L} = \begin{bmatrix} \frac{\partial^\alpha}{\partial|x|^\alpha}L_2(x_2) & \cdots & \frac{\partial^\alpha}{\partial|x|^\alpha}L_{N-1}(x_2) \\ \vdots & \cdots & \vdots \\ \frac{\partial^\alpha}{\partial|x|^\alpha}L_2(x_N) & \cdots & \frac{\partial^\alpha}{\partial|x|^\alpha}L_{N-1}(x_N) \end{bmatrix} \text{ and } U_0 = \begin{bmatrix} u_0(x_2) \\ \vdots \\ u_0(x_{N-1}) \end{bmatrix}. \tag{26}$$

According to the equation data in articles [30, 6], the problem (23) is taken with parameters $\alpha = 1.8$, $K_\alpha = 0.25$ and $u_0(x) = x^2(\pi - x)$. For the numerical solution, Gaussian RBF is used with a shape parameter $\varepsilon = 0.8$, applying the scheme shown in Section 6.1. Graphs are shown in **Figure 2**.

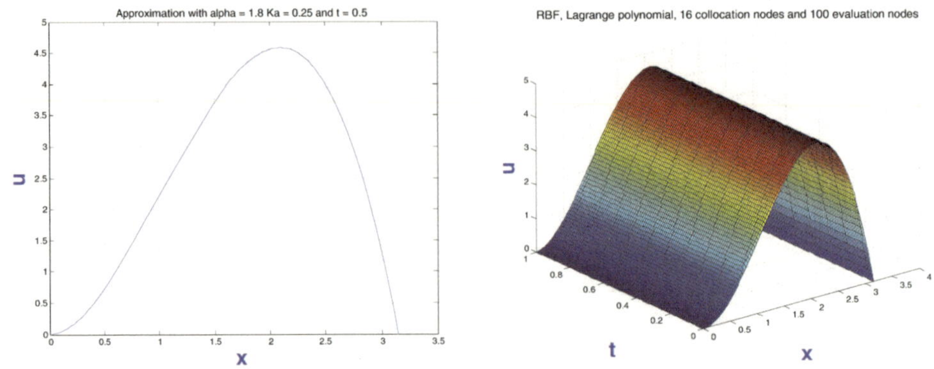

Figure 2. RBF approximation to solution, for Eq. (23), with $\alpha = 1.8$, $\varepsilon = 0.8$ and $K_\alpha = 0.25$.

6.2. Riemann-Liouville space-fractional diffusion equation

In this part we consider the problem introduced by Sousa [31] and taken up in Ref. [26] by Cécile and Hanert, the fractional diffusion equation in one dimension

$$\frac{\partial f(x,t)}{\partial t} = d(x)\,{}_{0}^{RL}D_x^\alpha f(x,t) + q(x,t),\ x \in [0,1] \text{ and } t > 0 \tag{27}$$

with ${}_{0}^{RL}D_x^\alpha$ Riemannn-Liouville derivative, $1 < \alpha \leq 2$,

$$d(x) = \frac{\Gamma(5-\alpha)}{24}x^\alpha \ y \ q(x,t) = -2e^{-t}x^4.$$

Initial conditions are considered

$$f(x,0) = x^4,\ x \in (a,b) \tag{28}$$

and Dirichlet boundary conditions

$$f(0,t) = 0,\ f(1,t) = e^{-t}. \tag{29}$$

The exact solution to Eq. (27) is

$$f(x,t) = e^{-t}x^4.$$

We apply RBF and the method of lines with Lagrange polynomials, in a similar way as it was applied in Section 6.1. With $N = 21$ Chebyshev collocation nodes and $Ne = 100$ evaluation nodes, various values for the order of the derivative α and for shape parameters.

The results obtained by Sousa [31] were obtained by applying implicit Crank-Nicholson schemes for time. The discretization of the fractional derivative was done using splines, with second-order precision. The results of both Sousa and RBF are shown in **Tables 2** and **3**, taking into account the error

$$||u_{exact} - u_{approximated}||_\infty \quad (30)$$

where $||\cdot||_\infty$ is the ℓ_∞ norm.

Global ℓ_∞ error (30) of time converged solution for three mesh resolutions at $t = 1$ for $\alpha = 1.2$, $\alpha = 1.4$ and $\Delta t = \Delta x$.

Δx	$\alpha = 1.2$	Rate	$\alpha = 1.4$	Rate
1/15	0.1275×10^{-2}		0.9070×10^{-3}	
1/20	0.7571×10^{-3}	1.8	0.5327×10^{-3}	1.8
1/25	0.5030×10^{-3}	1.8	0.3486×10^{-3}	1.9
1/30	0.3566×10^{-3}	1.9	0.2461×10^{-3}	1.9

Global ℓ_∞ error (30) of time converged solution for three mesh resolutions at $t = 1$ for $\alpha = 1.5$, $\alpha = 1.8$ and $\Delta t = \Delta x$.

Δx	$\alpha = 1.5$	Rate	$\alpha = 1.8$	Rate
1/15	0.7660×10^{-3}		0.4380×10^{-3}	
1/20	0.4493×10^{-3}	1.9	0.2540×10^{-3}	1.9
1/25	0.2929×10^{-3}	1.9	0.1649×10^{-3}	1.9
1/30	0.2067×10^{-3}	1.9	0.1150×10^{-3}	2.0

Table 2. Comparison of results for FPDE (27): Sousa results [31].

Global ℓ_∞ error (30) of time converged solutions at $t = 1$ for $\alpha = 1.2, 1.4, 1.5, 1.8$ and shape parameters $\varepsilon = 0.6, 1.2$.

$\varepsilon = 0.6$		
(r)1-2 Order	**Maximal error**	$\varepsilon = 1.2$
$\alpha = 1.2$	0.35574×10^{-4}	
$\alpha = 1.4$	0.46508×10^{-4}	
$\alpha = 1.5$	0.60498×10^{-4}	
$\alpha = 1.8$	0.16994×10^{-4}	0.18125×10^{-4}

Table 3. Comparison of results for FPDE (27): RBF results.

6.3. Caputo time fractional partial differential equations

In this part we consider examples of fractional partial differential equations (see [32]) of the type

$$\frac{\partial^\alpha u(x,t)}{\partial t^\alpha} + \delta \frac{\partial u(x,t)}{\partial x} + \gamma \frac{\partial^2 u(x,t)}{\partial x^2} = f(x,t), \tag{31}$$

where $t > 0$, $x \in [a, b]$, $0 < \alpha \leq 1$, δ and γ are real parameters, bounded initial condition $u(x,0) = u_0(x)$ and Dirichlet boundary conditions $u(a,t) = g_1(t)$ and $u(b,t) = g_2(t)$, $t \geq 0$. Caputo Derivative will be considered and rbf-qr routines from www.it.uu.se/research/scientific_computing/software/rbf_qr

6.3.1. *Example 1*

Putting $\delta = 1$, $\gamma = -1$ and $f(x,t) = \frac{2t^{2-\alpha}}{\Gamma(3-\alpha)} + 2x - 2$ in Eq. (31), we obtain a non-homogeneous, fractional and linear Burger equation

$$\frac{\partial^\alpha u(x,t)}{\partial t^\alpha} + \frac{\partial u(x,t)}{\partial x} - \frac{\partial^2 u(x,t)}{\partial x^2} = \frac{2t^{2-\alpha}}{\Gamma(3-\alpha)} + 2x - 2. \tag{32}$$

Using initial condition

$$u(x,0) = x^2, \tag{33}$$

Dirichlet boundary conditions

$$u(0,t) = t^2, \; u(1,t) = 1 + t^2. \tag{34}$$

The exact solution (see [33]) is

$$u(x,t) = x^2 + t^2, \tag{35}$$

This problem is solved using the method described in section 6.1 along with QR decomposition for spatial part and an implicit scheme for time, on domain $[0, 1]$, fractional derivative of order $\alpha = 0.5$; several shape parameters are considered and the one that gives the best approximation is taken. **Figure 3** shows a 3D image of the solution and maximal error for the solution at time $t = 1$.

In **Figure 4** we compare errors that result when choosing uniform collocation nodes (with a fixed step size) and Chebyshev, on definition interval $[0, 1]$. The election of Chebyshev nodes is due to attenuating instability, which manifests as oscillations in the graph, called the Gibbs phenomenon [34, 35].

[41 uniform collocation nodes and 100 evaluation nodes, maximal error for $t = 0.5$, $\alpha = 0.5$, $\varepsilon = 3.6$ is 0.83595×10^{-4}.] [41 Chebyshev collocation nodes and 100 evaluation nodes, maximal error for $t = 0.5$, $\alpha = 0.5$, $\varepsilon = 3.6$ is 0.14385×10^{-4}.]

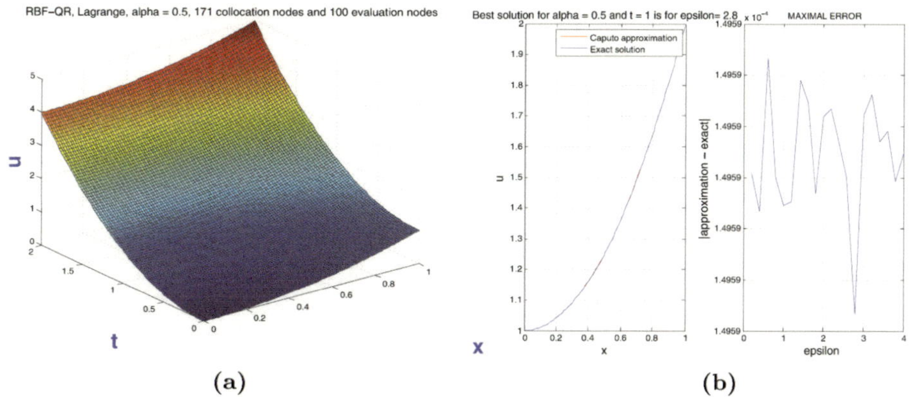

Figure 3. Numerical solution for Eq. (32) with $\alpha = 0.5$.

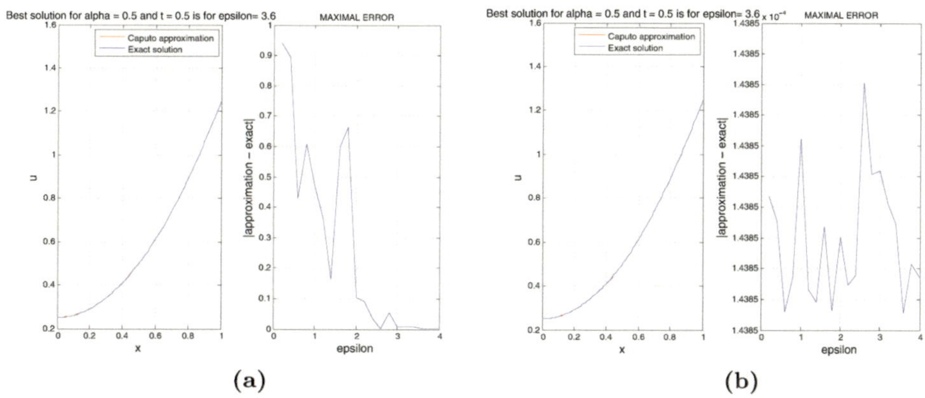

Figure 4. Comparison between results for Eq. (32) with $\alpha = 0.5$ and $t = 0.5$, due to election of nodes. (a) 41 uniform collocation nodes and 100 evaluation nodes, maximal error for $t = 0.5$, $\alpha = 0.5$, $\varepsilon = 3.6$ is 0.83595×10^{-4}. (b) 41 Chebyshev collocation nodes and 100 evaluation nodes, maximal error for $t = 0.5$, $\alpha = 0.5$, $\varepsilon = 3.6$ is 0.14385×10^{-4}.

6.3.2. Example 2

Putting $\delta = 1$, $\gamma = 0$ and $f(x,t) = \frac{t^{1-\alpha}}{\Gamma(2-\alpha)} \sin(x) + t\cos(x)$ in Eq. (31), we obtain

$$\frac{\partial^\alpha u(x,t)}{\partial t^\alpha} + \frac{\partial u(x,t)}{\partial x} = \frac{t^{1-\alpha}}{\Gamma(2-\alpha)} \sin(x) + t\cos(x). \tag{36}$$

Initial condition

$$u(x,0) = 0, \tag{37}$$

Next function is the exact solution (see [33]), which is used to set boundary conditions.

$$u(x,t) = t \sin x. \tag{38}$$

The problem is solved by Lagrange (like Section 6.1) and RBF-QR, for $\alpha = 0.6$, $N = 121$ collocation nodes on interval $[-3,3]$. In **Figure 5** we notice that when the number of uniform

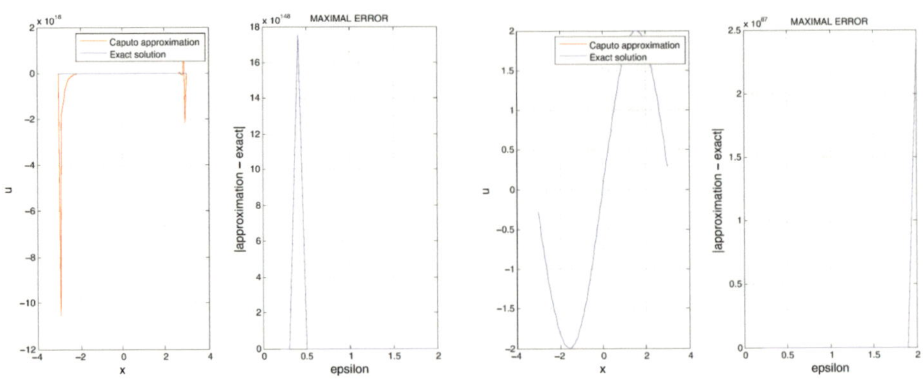

(a) (b)

Figure 5. Unstability due to node election, for Eq. (36). (a) 121 uniform collocation nodes and 101 evaluation nodes, maximal error for $t = 2$, $\alpha = 0.6$, $\varepsilon = 1.5$ is 0.10558×10^{19}. (b) 121 Chebyshev collocation nodes and 101 evaluation nodes, maximal error for $t = 2$, $\alpha = 0.6$, $\varepsilon = 0.2$ is 0.71054×10^{-15}.

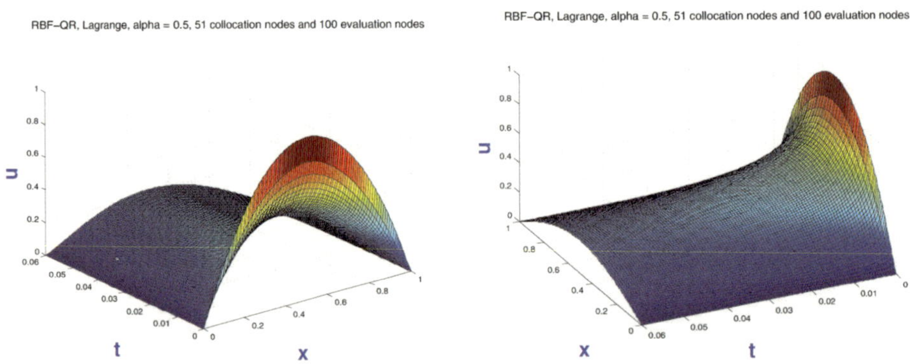

(a) (b)

Figure 6. Numerical solution for equation (39), order $\alpha = 0.5$ (superdiffusive phenomenon).

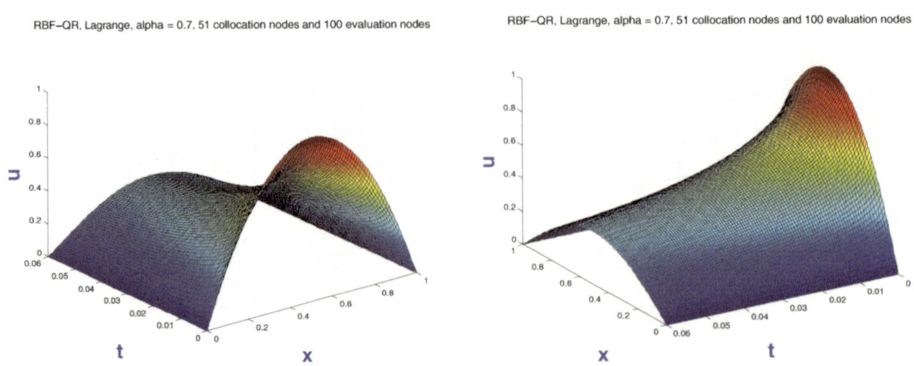

(a) (b)

Figure 7. Numerical solution for Eq. (39), order $\alpha = 0.7$ (subdiffusive phenomenon).

http://dx.doi.org/10.5772/67892

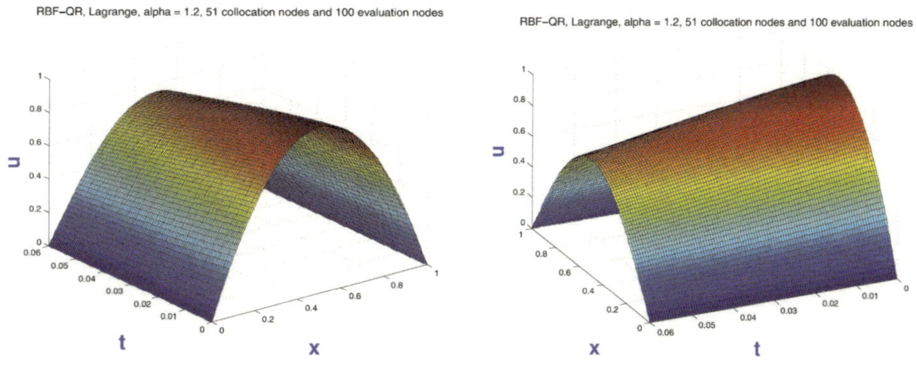

(a) (b)

Figure 8. Numerical solution for Eq. (39), order $\alpha = 1.2$ (superdiffusive phenomenon).

collocation nodes grows, there are large oscillations which cause unstability. For this reason, Chebyshev nodes are also preferred, as they attenuate this misbehavior.

6.3.3. Example 3

Putting $\delta = 0$, $\gamma = -1$ and $f(x, t) = 0$ in Eq. (31), we obtain

$$\frac{\partial^\alpha u(x,t)}{\partial t^\alpha} = \frac{\partial^2 u(x,t)}{\partial x^2}. \tag{39}$$

using initial condition

$$u(x,0) = 4x(1-x), \tag{40}$$

boundary conditions

$$u(0,t) = u(1,t) = 0. \tag{41}$$

The exact solution of this problem is not known, but is shown as an example of subdiffusive phenomenon which was discussed in the introduction. The problem is solved using the method described in section 6.1 along with QR decomposition for spatial part and an implicit scheme for time, on domain $[0, 1]$, fractional derivative of order $\alpha = 0.5$ and $\alpha = 0.7$, according to the results shown in Refs. [36, 32]. **Figures 6** and **7** show these results. **Figure 8** shows a what if situation when $\alpha = 1.2$ (superdiffusive phenomenon), mentioned in Section 1.2.

7. Discussion

The idea was to show that radial basis schemes are efficient and are on par with schemes like Finite Differences. They are an option to deal with multidimensional and irregular domain problems. The challenge is to adapt them to deal with diffusive problems, particularly with multidimensional systems of equations that consider that the medium does not have a single characteristic.

Fuentes et al. show in the work "The fractal models of saturated and unsaturated flows (capillary, diffusion and matrix) at micro, macro and mega scales", for the PEMEX oil company, how to model the pressure which the oil must leave in a wellbore, from a system of triple porosity and permeability of the fractured medium.

By converting the model into a dimensionless system, in terms of radial distances, non-mesh schemes such as RBFs sound like a viable option and it is a work that is still being addressed.

Author details

Carlos Alberto Torres Martínez[1]* and Carlos Fuentes[2]

*Address all correspondence to: inocencio3@gmail.com

1 Universidad Nacional Autónoma de México, Universidad Autónoma de la Ciudad de México (UACM), Mexico

2 Instituto Mexicano de Tecnología del Agua, Coordinación de Tecnología de Riego y Drenaje, Cuernavaca, Mexico

References

[1] Rolland L Hardy. Multiquadric equations of topography and other irregular surfaces. Journal of geophysical research, 76(8):1905–1915, 1971.

[2] Charles A Micchelli. Interpolation of scattered data: distance matrices and conditionally positive definite functions. In Approximation theory and spline functions pp. 143–145. Springer Netherlands. 1984.

[3] Michael JD Powell. The theory of radial basis function approximation in 1990. Department of Applied Mathematics and Theoretical Physics, University of Cambridge. pp. 105–209. 1990.

[4] Edward J Kansa. Multiquadrics—a scattered data approximation scheme with applications to computational fluid-dynamics—i surface approximations and partial derivative estimates. Computers & Mathematics with applications, 19(8):127–145, 1990.

[5] Edward J Kansa. Multiquadrics—a scattered data approximation scheme with applications to computational fluid-dynamics—ii solutions to parabolic, hyperbolic and elliptic partial differential equations. Computers & mathematics with applications, 19(8):147–161, 1990.

[6] Wen Chen, Zhuo-Jia Fu, and Ching-Shyang Chen. Recent advances in radial basis function collocation methods. Heidelberg: Springer, 2014.

[7] Marjan Uddin. On the selection of a good value of shape parameter in solving time-dependent partial differential equations using rbf approximation method. Applied Mathematical Modelling, 38(1):135–144, 2014.

[8] Shmuel Rippa. An algorithm for selecting a good value for the parameter c in radial basis function interpolation. Advances in Computational Mathematics, 11(2-3):193–210, 1999.

[9] Bengt Fornberg and Grady Wright. Stable computation of multiquadric interpolants for all values of the shape parameter. Computers & Mathematics with Applications, 48(5): 853–867, 2004.

[10] Bengt Fornberg and Cécile Piret. A stable algorithm for flat radial basis functions on a sphere. SIAM Journal on Scientific Computing, 30(1):60–80, 2007.

[11] Bengt Fornberg, Erik Lehto, and Collin Powell. Stable calculation of gaussian-based rbf-fd stencils. Computers & Mathematics with Applications, 65(4):627–637, 2013.

[12] Bengt Fornberg, Elisabeth Larsson, and Natasha Flyer. Stable computations with gaussian radial basis functions. SIAM Journal on Scientific Computing, 33(2):869–892, 2011.

[13] Elisabeth Larsson, Erik Lehto, Alfa Heryudono, and Bengt Fornberg. Stable computation of differentiation matrices and scattered node stencils based on gaussian radial basis functions. SIAM Journal on Scientific Computing, 35(4):A2096–A2119, 2013.

[14] Eli Barkai, Ralf Metzler, and Joseph Klafter. From continuous time random walks to the fractional Fokker-Planck equation. Physical Review E, 61(1):132, 2000.

[15] A Blumen, G Zumofen, and J Klafter. Transport aspects in anomalous diffusion: Lévy walks. Physical Review A, 40(7):3964, 1989.

[16] A Piryatinska, AI Saichev, and WA Woyczynski. Models of anomalous diffusion: the subdiffusive case. Physica A: Statistical Mechanics and its Applications, 349(3):375–420, 2005.

[17] AS Chaves. A fractional diffusion equation to describe lévy flights. Physics Letters A, 239 (1):13–16, 1998.

[18] Ali Safdari-Vaighani, Alfa Heryudono, and Elisabeth Larsson. A radial basis function partition of unity collocation method for convection–diffusion equations arising in financial applications. Journal of Scientific Computing, 64(2):341–367, 2015.

[19] L Sabatelli, S Keating, J Dudley, and P Richmond. Waiting time distributions in financial markets. The European Physical Journal B-Condensed Matter and Complex Systems, 27 (2):273–275, 2002.

[20] Mark M Meerschaert, Jeff Mortensen, and Stephen W Wheatcraft. Fractional vector calculus for fractional advection–dispersion. Physica A: Statistical Mechanics and its Applications, 367:181–190, 2006.

[21] Ralf Metzler, Walter G Glöckle, and Theo F Nonnenmacher. Fractional model equation for anomalous diffusion. Physica A: Statistical Mechanics and its Applications, 211(1):13–24, 1994.

[22] Erika Di Giuseppe, Monica Moroni, and Michele Caputo. Flux in porous media with memory: models and experiments. Transport in Porous Media, 83(3):479–500, 2010.

[23] K Oleschko, G Korvin, AS Balankin, RV Khachaturov, L Flores, B Figueroa, J Urrutia, and F Brambila. Fractal scattering of microwaves from soils. Physical Review Letters, 89 (18):188501, 2002.

[24] C Fuentes, F Brambila, M Vauclin, J-Y Parlange, and R Haverkamp. Fractal modeling of hydraulic conductivity in non-saturated soils. Hydraul. Eng. Mexico, 16 (2):119–137 (in spanish) 2001.

[25] C Fuentes. Approche fractale des transferts hydriques dans les sols non-saturés. These de doctorat de l'Université Joseph Fourier Grenoble I, Grenoble, France, 1992.

[26] Cécile Piret and Emmanuel Hanert. A radial basis functions method for fractional diffusion equations. Journal of Computational Physics, 238:71–81, 2013.

[27] Igor Podlubny. Fractional differential equations: an introduction to fractional derivatives, fractional differential equations, to methods of their solution and some of their applications, volume 198. Academic Press, USA. 1998.

[28] Stefan G Samko, Anatoly A Kilbas, and Oleg I Marichev. Fractional integrals and derivatives. Theory and Applications, Gordon and Breach, Yverdon, 1993, 1993.

[29] Rudolf Gorenflo and Francesco Mainardi. Random walk models for space-fractional diffusion processes. Fractional Calculus and Applied Analysis, 1(2):167–191, 1998.

[30] Qianqian Yang, Fawang Liu, and I Turner. Numerical methods for fractional partial differential equations with riesz space fractional derivatives. Applied Mathematical Modelling, 34(1):200–218, 2010.

[31] Erclia Sousa. Numerical approximations for fractional diffusion equations via splines. Computers & Mathematics with Applications, 62(3):938–944, 2011.

[32] Marjan Uddin and Sirajul Haq. Rbfs approximation method for time fractional partial differential equations. Communications in Nonlinear Science and Numerical Simulation, 16(11):4208–4214, 2011.

[33] Zaid Odibat and Shaher Momani. The variational iteration method: an efficient scheme for handling fractional partial differential equations in fluid mechanics. Computers & Mathematics with Applications, 58(11):2199–2208, 2009.

[34] David Gottlieb and Chi-Wang Shu. On the Gibbs phenomenon and its resolution. SIAM Review, 39(4):644–668, 1997.

[35] Tobin A Driscoll and Bengt Fornberg. A padé-based algorithm for overcoming the Gibbs phenomenon. Numerical Algorithms, 26(1):77–92, 2001.

[36] Igor Podlubny, Aleksei Chechkin, Tomas Skovranek, YangQuan Chen, and Blas M Vinagre Jara. Matrix approach to discrete fractional calculus ii: Partial fractional differential equations. Journal of Computational Physics, 228(8):3137–3153, 2009.

Chapter 2

Application of Fractional Calculus to Oil Industry

Benito F. Martínez-Salgado,
Rolando Rosas-Sampayo,
Anthony Torres-Hernández and Carlos Fuentes

Additional information is available at the end of the chapter

http://dx.doi.org/10.5772/intechopen.68571

Abstract

In this chapter, we present a discussion about the practical application of the fractal properties of the medium in the mathematical model through the use of fractional partial derivatives. We present one of the known models for the flow in saturated media and its generalization in fractional order derivatives. In the middle section, we present one of the main arguments that motivate the use of fractional derivatives in the porous media models, this is the Professor Nigmatullin's work. The final part describes the process for obtaining the coupled system of three equations for the monophase flow with triple porosity and triple permeability, briefly mentioning the method used for the first solutions of the system.

Keywords: fractional calculus, fractional derivatives, anomalous diffusion, porous media, fractal dimension

1. Introduction

The objects of nature rarely have a classical geometric form; in the particular case of oil reservoirs, the ground where the wells are found has been considered with Euclidean geometry; this is not sufficient in many cases to give good approximations in the mathematical models. Since its forms are closer to the fractal geometry, the knowledge of this can be useful to develop models that allow us to better manage the wells. This work presents an approach in fractional derivatives for the triple porosity and triple permeability monophasic saturated model, based on the one proposed by Camacho et al. [1, 2] and generalized partially by Fuentes et al. [3]. The main contribution is to consider the link between fractional equations and fractal geometry through the revision of Alexander-Orbach's conjecture [23], taken to the particular case.

2. Background of the approach of models of diffusion on fractal media

Fractional calculus was originated as a way to generalize classic calculus; however, it is more difficult to find a direct physical interpretation than in the classical version. When we consider an oil well as a fractal, it is important to choose which of its properties can be useful for elaborating a mathematical model [20, 21, 26, 27].

Alexander and Orbach [4] calculated the "spectral dimension (fracton)"; this parameter is associated to volume and fractal connectivity by being considered as an elastic fractal net of particles connected by harmonic strings. Thus, we consider the particle movement over this fractal and we find a relation of root mean square of an r aleatory walker dependent of time over the fractal, which is in accordance to the following relation:

$$\langle r^2(t)\rangle \approx t^{2/(2+\theta)}, \tag{1}$$

where r is in euclidean space. Alexander and Orbach defined $d_s = \frac{2d_f}{2+\theta}$ as the spectral dimension or fracton, where $d_w = 2+\theta$ is the dimension of the walk, θ gives us the dependence of the diffusion constant over the distance and d_f is the effective dimension [24, 25].

O'Shaughnessy and Procaccia [5] used the concepts of Alexander and Orbach to formulate their fractal diffusion equation:

$$\frac{\partial p(r,t)}{\partial t} = \frac{1}{r^{d_f-1}}\frac{\partial}{\partial r}\left[K_1 r^{-\theta} r^{d_f-1}\frac{\partial p(r,t)}{\partial r}\right],\ r>0,\ \mathrm{K_1 constant}, \tag{2}$$

with solution.

$$P(r,t) = \frac{2+\theta}{d_f\Gamma(d_f/(2+\theta))}\left[\frac{1}{(2+\theta)^2K_1t}\right]^{d_f/(2+\theta)} \exp\left[-\frac{r^{2+\theta}}{(2+\theta)^2K_1t}\right], \tag{3}$$

of which one finds a power law

$$\langle r^2(t)\rangle = \frac{\Gamma\left[\frac{d_f+2}{2+\theta}\right]}{\Gamma\left[\frac{d_f}{2+\theta}\right]}\left[(2+\theta)^2K_1t\right]^{2/(2+\theta)} = \langle r^2(1)\rangle r^{2/(2+\theta)}. \tag{4}$$

Metzler et al. [6] started with the characterization of an anomalous diffusion process 1. Here, they consider $d_w = \theta+2$ as the *anomalous diffusion exponent,* they are referencing the work of Havlin and Ben-Avraham [7] to calculate diffusion with a media (1). They obtain an "approach exponential":

$$P(r,t) \approx At^{-d_f/d_w}e^{-c(r/R)^u}, \tag{5}$$

valid in the asyntotic range $r/R \gg 1$, and $t\to\infty$ with R and r defined by

$$\begin{cases} R = \sqrt{\langle r^2(t)\rangle} \\ u = \dfrac{d_w}{d_w - 1} \end{cases}. \quad (6)$$

Thus, it is possible to obtain the solution of the fractional derivative diffusion equation:

$$\frac{\partial^{2/d_w}}{\partial t^{2/d_w}} P(r,t) = \frac{1}{r^{d_s-1}} \frac{\partial}{\partial r}\left(r^{d_s-1} \frac{\partial}{\partial r} P(r,t) \right) \quad (7)$$

where

$$\frac{\partial^{2/d_w}}{\partial t^{2/d_w}} P(r,t) = \frac{1}{\Gamma\left(1 - \frac{2}{d_w}\right)} \frac{\partial}{\partial t} \int_0^t d\tau \frac{P(r,\tau)}{(t-\tau)^{2/d_w}},\ 0 \leq \frac{2}{d_w} < 1. \quad (8)$$

3. Brief history of fractional calculus

In mathematics, one way to obtain new concept is to generalize by extending one definition or context for values not previously considered. For example, it is possible to generalize the power concept of x^n, for natural n values such as the concept of x, n times, to negative integers n, as the product of $\frac{1}{x}$, n times, then to n rational values such as $\sqrt[q]{x^p}$, if $n = \frac{p}{q}$, with positive p and q. In each step, the generalization modifies the concept a little, but it keeps the previous one as a particular case. This process can continue all the way to a complex n. In the same way as generalizations in differential and integral calculus have been made, in this case the generalization goes toward the n order of the $\frac{d^n y}{dx^n}$ derivative [22].

Leibniz: In a letter dated September 30, 1695, L'Hôpital, he has been inquired about the meaning of $\frac{d^n y}{dx^n}$, if $n = \frac{1}{2}$, in response he wrote: "You can see by that, sir, that one can express by an infinite series a quantity such as $d^{1/2}xy$ or $d^{1:2}xy$. Although infinite series and geometry are distant relations, infinite series admits only the use of exponents that are positive and negative integers, and does not, as yet, know the use of fractional exponents." Later in the same letter, Leibniz continues: "Thus, it follows that $d^{1/2}x$ will be equal to $x\sqrt{dx : x}$. This is an apparent paradox from which, one day, useful consequences will be drawn."

In his correspondence with Johan Bernoulli, Leibniz mentioned to him general order derivatives. In 1697, he established that differential calculus can be used to achieve these generalizations and used the $d^{1/2}$ notation to denote order $\frac{1}{2}$ derivative [22].

Euler: In 1730, Euler proposed derivatives as rate between functions and variables that can be expressed algebraically; the solution with this approach when the orders are not integers is the use of interpolations.

The (fractional) non-integer order derivative motivated Euler to introduce the Gamma function. Euler knew that he needed to generalize (or, as he said, interpolate), the $1 \cdot 2 \cdots n = n!$ product for non-integer n. He proposed an integral:

$$\prod_{k=1}^{n} n = \int_0^1 (-\log x)^n dx, \tag{9}$$

and used it to partially solve the Leibniz paradox. He also gave the basic fractional derivative (with modern notation $\Gamma(n+1) = n!$):

$$\frac{d^\alpha x^\beta}{dx^\alpha} = \frac{\Gamma(\beta+1)}{\Gamma(\beta-\alpha+1)} x^{\beta-\alpha}, \tag{10}$$

which is valid for non-integer α and β [22].

Laplace and Lacroix: Laplace also defined his fractional derivative via an integral. In 1819, Lacroix, applying the (10) formula and the Legendre symbol for Gamma function, was able to calculate the derivative with $y = x$ and $n = \frac{1}{2}$. He was also the first to use the term "fractional derivative." He thus achieved

$$\frac{d^{1/2}y}{dx^{1/2}} = \frac{2\sqrt{x}}{\sqrt{\pi}}. \tag{11}$$

Fourier: Joseph Fourier (1822), in his famous book *"The Analytical Theory of Heat"* making use of this expression of a function and an interpretation of the sines and cosines derivatives gave his definition of a fractional derivative:

$$f(x) = \frac{1}{2\pi} \int_{-\infty}^{\infty} f(\alpha) d\alpha \int_{-\infty}^{\infty} \cos p(x-\alpha) dp, \tag{12}$$

then

$$\frac{d^n}{dx^n} \cos p(x-a) = p^n \cos\left[p(x-\alpha) + \frac{1}{2} n\pi\right], \tag{13}$$

for an integer n. Formally replacing n with an arbitrary u, he obtained the generalization:

$$\frac{d^u}{dx^u} f(x) = \frac{1}{2\pi} \int_{-\infty}^{\infty} f(\alpha) d\alpha \int_{-\infty}^{\infty} p^u \cos\left[p(x-a) + \frac{1}{2} u\pi\right] dp. \tag{14}$$

Fourier thus establishes that the u number can be regarded as any quantity, positive or negative [8, 22].

Abel: In 1823, N. H. Abel published the solution of a problem presented by Hyugens in 1673: *The tautochrone problem*. Abel gave his solution in the form of an integral equation that is considered the first application of fractional calculus. The integral he worked with is

$$\int_0^x (x-t)^{-\frac{1}{2}} f(t) dt = k. \tag{15}$$

This integral is, except for the $1/\Gamma(\frac{1}{2})$ factor, a fractional integral of 1/2 order, Abel wrote the left part as $\pi\left[\frac{d^{-\frac{1}{2}}}{dx^{-\frac{1}{2}}}\right] f(x)$, thus he worked with both sides of the equation as

$$\sqrt{\pi}f(x) = \frac{d^{\frac{1}{2}}}{dx^{\frac{1}{2}}}k. \tag{16}$$

The first integral equation in history had been solved. Two facts may be observed: the regard for the sum of the orders, and that unlike in classical calculus, the derivative of a constant is not zero [8, 22].

Liouville: In 1832, Liouville made the first great study of fractional calculus. In his work, he considered $\left(\frac{d^{1/2}}{dx^{1/2}}\right)e^{2x}$. The first formula he obtained was the derivative of a function:

$$f(x) = \sum_{n=0}^{\infty} c_n e^{a_n x}, \Re(a_n) > 0, \tag{17}$$

from which he got

$$D^{\nu}f(x) = \sum_{n=0}^{\infty} x_n a_n^{\nu} e^{a_n x}. \tag{18}$$

that can be obtained using the extension

$$D^{\nu}e^{ax} = a^{\nu}e^{ax}, \tag{19}$$

for an arbitrary number ν. A second definition was achieved by Liouville from the defined integral:

$$I = \int_0^{\infty} u^{a-1}e^{-xu}du, \quad a > 0, \quad x > 0, \tag{20}$$

of which, after a change of variable and a suitable rewriting is obtained

$$D^{-\nu}x^{-a} = \frac{(-1)^{\nu}\Gamma(a+\nu)}{\Gamma(a)}x^{-a-\nu}, \quad a > 0. \tag{21}$$

Liouville also tackled the tautochrone problem and proposed differential equations of arbitrary order.

In 1832, he wrote about a generalization of Leibnitz's rule about the nth derivative of a product:

$$D^{\nu}f(x)g(x) = \sum_{n=0}^{\infty}\binom{\nu}{n}D^{n}f(x)D^{\nu-n}g(x), \tag{22}$$

where D^n is the ordinary n order differential operator, $D^{\nu-n}$ fractional operator, and $\binom{\nu}{n}$ the generalized binomial coefficient, expressed in terms of the Gamma function, $\frac{\Gamma(\nu+1)}{n!\Gamma(\nu-n+1)}$.

Liouville expanded the coefficients in Eq. (18) as

$$
\begin{aligned}
a_n^\nu &= \lim_{h\to 0}\frac{1}{h^\nu}(1-e^{-ha_n})^\nu, \quad a_n>0,\\
(-1)^\nu a_n^\nu &= \lim_{h\to 0}\frac{1}{h^\nu}(1-e^{ha_n})^\nu, \quad a_n<0.
\end{aligned} \tag{23}
$$

And inserted those equations in Eq. (18) to get

$$
\begin{aligned}
\frac{d^\nu}{dx^\nu}f(x) &= \lim_{h\to 0}\left\{\frac{1}{h^\nu}\sum_{n=0}^{\infty}\left[(-1)^m\binom{\nu}{n}f(x-mh)\right]\right\},\\
&= (-1)^\nu \lim_{h\to 0}\left\{\frac{1}{h^\nu}\sum_{n=0}^{\infty}\left[(-1)^m\binom{\nu}{n}f(x+mh)\right]\right\}.
\end{aligned} \tag{24}
$$

These formulas would be retaken by Grünwald in 1867.

Riemann: Riemann developed his Fractional Calculus theory when he was preparing his Ph.D. thesis, but his oeuvre was published posthumously around 1892. He searched for a generalization of Taylor's series, in which he defined the n-th differential coefficient of a $f(x)$ function as the h^n coefficient in the $f(x+h)$ expansion with integer powers of h. Thus, he generalizes this definition to non-integer powers and demands that

$$
f(x+h) = \sum_{n=-\infty}^{n=\infty} c-n+\alpha(\partial_x^{n+\tau}f)(x)h^{n+\alpha}, \tag{25}
$$

be valid for $n\in\mathbb{N}, a\in\mathbb{R}$. The $c_{n+\alpha}$ factor is determined by the $\partial^\beta(\partial^\alpha f)=\partial^{\beta+\alpha}f$ condition, and he found that it was $\frac{1}{\Gamma(n+\alpha+1)}$. Riemann then derived Eq. (25) expression for negative α:

$$
\partial^\alpha f = \frac{1}{\Gamma(-\alpha)}\int_k^x (x-t)^{-\alpha-1}f(t)dt + \sum_{n=1}^{\infty} K_n \frac{x^{-\alpha-n}}{\Gamma(-n-\alpha+1)}, \tag{26}
$$

where k, K_n are finite constants. Then, he extended the result to non-negative α.

Sonin and Letnikov: The Russian mathematicians N.Ya. Sonin (1868) and A.V. Letnikov (1868–1872) [29] made contributions taking as basis the formula for the nth derivative of the Cauchy integral formula given by

$$
D^n f(z) = \frac{n!}{2\pi i}\int_C \frac{f(\xi)}{(\xi-z)^{n+1}}d\xi. \tag{27}
$$

They worked using the contour integral method, with the contribution of Laurent (1884), they achieved the definition:

$$
{}_cD_x^{-\alpha}f(x) = \frac{1}{\Gamma(\alpha)}\int_c^x (x-t)^{\alpha-1}f(t)dt, \quad \Re(\alpha)>0. \tag{28}
$$

For an integration to an arbitrary order, when $x > c$ has the Riemann definition, but without a complementary definition, when $c = 0$ we get the shape known as Riemann-Liouville fractional integral:

$$ {}_0D_x^{-\alpha} f(x) = \frac{1}{\Gamma(\alpha)} \int_0^x (x-t)^{\alpha-1} f(t)dt, \quad \Re(\alpha) > 0. \tag{29} $$

Assigning c values in Eq. (19), we get different integrals of fractional order, which will be fundamental to define fractional derivatives.

If $c = -\infty$, we get

$$ {}_{-\infty}D_x^{-\alpha} f(x) = \frac{1}{\Gamma(\alpha)} \int_{-\infty}^x (x-t)^{\alpha-1} f(t)dt, \ \Re(\alpha) > 0. \tag{30} $$

Using integration properties, more definitions will be given.

Grünwald: Another contribution is that of Grünwald (1867) and Letnikov (1868). This extension of the classical derivative to fractional order is important because it lets us apply it in numerical approximations. They started with the definition of derivative as a limit given by Cauchy (1823):

$$ \frac{df}{dx} = \lim_{h\to\infty} \frac{[f(x) - f(x-h)]}{h}. \tag{31} $$

First generalizing for a nth integer derivative we get

$$ \begin{gathered} D^n f(x) = \lim_{h\to 0} \frac{\sum_{j=0}^{n} [(-1)^j \binom{n}{j} f(x-jh)]}{h^n}, \\ n \in \mathbb{N}, \quad \text{and} \quad f \in C^n[a,b], \ a < x < b. \end{gathered} \tag{32} $$

Grünwald generalizes Eq. (32) for an arbitrary q value, expressing it as

$$ D_a^q f(x) = \lim_{N\to\infty} h_N^{-q} \left[\sum_{j=0}^{N} (-1)^j \binom{-n}{j} f(x - jh_N) \right], q \in \mathbb{R}, \tag{33} $$

where the binomial coefficient is

$$ \binom{q}{j} = \frac{q(q-1)(q-2)\cdots(q-j+1)}{j!}, \tag{34} $$

also showing that

$$ (-1)^j \binom{q}{j} = \binom{j-q-1}{j} = \frac{\Gamma(j-q)}{\Gamma(-q)\Gamma(j+1)}. \tag{35} $$

and with those previous results, it is possible to establish this important property for $\alpha \in \mathbb{R}$ and $n \in \mathbb{N}$

$$\frac{d^n}{dx^n} D_a^\alpha f(x) = D_a^{n+\alpha} f(x). \tag{36}$$

In the twentieth and twenty-first centuries, more definitions will rise, but they will be given in terms within the Riemann-Liouville fractional integral and will be part of the Modern Fractional Calculus Theory, in all their fundamental definitions [22].

4. Fractional calculus

We will now present the assorted definitions and notations of fractional derivatives that will be used throughout this work. It is worth pointing out that this is necessary because such notation is currently standardized [18, 19].

4.1. Riemann-Liouville fractional derivative

The Riemann-Liouville derivative is the basis to define most fractional derivatives; it generalizes the Cauchy's formula for derivatives of high order. For an f function defined in a $[a, b]$ interval, a $\alpha \in \mathbb{C}$ value with $\Re(\alpha) > 0$ defines the left and right Riemann-Liouville integrals by

$$\left({}_a^{RL}I_x^\alpha f\right)(x) = \frac{1}{\Gamma(\alpha)} \int_a^x \frac{f(t)}{(x-t)^{1-\alpha}} dt, \quad (x > a), \tag{37}$$

$$\left({}_x^{RL}I_b^\alpha f\right)(x) = \frac{1}{\Gamma(\alpha)} \int_x^b \frac{f(t)}{(t-x)^{1-\alpha}} dt, \quad (x < b). \tag{38}$$

Following Riemann's notion of defining fractional derivatives as the integer order derivative of an fractional integral, we have the left and right derivative proposal as follows:

$$\left({}_a^{RL}D_x^\alpha f\right)(x) = \left(\frac{d}{dx}\right)^n \left(\left({}_a^{RL}I_x^{n-\alpha} f\right)(x)\right), x > a, \tag{39}$$

$$\left({}_x^{RL}D_b^\alpha f\right)(x) = \left(-\frac{d}{dx}\right)^n \left(\left({}_b^{RL}I_x^{n-\alpha} f\right)(x)\right), x < b, \tag{40}$$

with $n = -\lfloor \Re(\alpha) \rfloor$, i.e., $n = \lfloor \Re(\alpha) \rfloor + 1$.

As shown in Refs. [8–10], these operators generalize the usual derivation. In other words, when $\alpha \in \mathbb{N}_0$, then

$$\left({}_a^{RL}D_x^0 f\right)(x) = \left({}_x^{RL}D_b^0 f\right)(x) = f(x), \text{if } \alpha = 0, \tag{41}$$

$$\left({}_a^{RL}D_x^n f\right)(x) = f^{(n)}(x), \ \left({}_x^{RL}D_b^n f\right)(x) = (-1)^n f^{(n)}(x). \tag{42}$$

It is also possible to prove that the semigroup propriety about the order of integral operators (i.e., for $\alpha \in \mathbb{C}(\Re(\alpha > 0)), \beta \in \mathbb{C}(\Re(\beta) > 0))$ is achieved:

$$\left({}^{RL}_{a}I^{\alpha}_{x}\left({}^{RL}_{a}I^{\beta}_{x}f\right)\right)(x) = \left({}^{RL}_{a}I^{\alpha+\beta}_{x}f\right)(x), \tag{43}$$

$$\left({}^{RL}_{x}I^{\alpha}_{b}\left({}^{RL}_{x}I^{\beta}_{b}f\right)\right)(x) = \left({}^{RL}_{x}I^{\alpha+\beta}_{b}f\right)(x). \tag{44}$$

For the derivatives, we have

$$\left({}^{RL}_{a}D^{\alpha}_{x}\left({}^{RL}_{a}D^{\beta}_{x}f\right)\right)(x) = \left({}^{RL}_{a}D^{\alpha+\beta}_{x}f\right)(x) - \sum_{j=1}^{m}\left({}^{RL}_{a}D^{\beta-j}_{x}f\right)(a)\frac{(x-a)^{-j-\alpha}}{\Gamma(1-j-\alpha)}. \tag{45}$$

For $f(x) \in L_p(1 \leq p \leq \infty)$, the following relationships are valid:

$$\left({}^{RL}_{a}D^{\beta}_{x}\left({}^{RL}_{a}I^{\alpha}_{x}f\right)\right)(x) = {}^{RL}_{a}I^{\alpha-\beta}_{x}f(x), \tag{46}$$

$$\left({}^{RL}_{x}D^{\beta}_{b}\left({}^{RL}_{x}I^{\alpha}_{b}f\right)\right)(x) = {}^{RL}_{x}I^{\alpha-\beta}_{b}f(x). \tag{47}$$

If $\alpha = \beta$, we have the identity operator and the operators turn out to be inverted. On the other hand, if the order of the operators is inverted, it will have

$$\left({}^{RL}_{a}I^{\beta}_{x}\left({}^{RL}_{a}D^{\alpha}_{x}f\right)\right)(x) = f(x) - \sum_{j=1}^{n}\frac{f^{(n-j)}_{n-\alpha}(a)}{\Gamma(\alpha-j+1)}(x-\alpha)^{\alpha-j}, \tag{48}$$

where $\Re(\alpha) > 0, n = \lfloor\Re(\alpha)\rfloor + 1$ and $f_{n-\alpha}(x) = ({}^{RL}_{a}I^{n-\alpha}_{x}f)(x)$ in analogy for the right derivative.

All these properties can be used in the phenomena modeling and its solution; such models have shown to improve usual approaches. However, when using equations with Riemann-Liouville type fractional derivatives, the initial conditions cannot be interpreted physically; a clear example is that the derivative Riemann-Liouville of a constant is not zero, contrary to the impression that the derivatives gives a notion about the change that the function experiences when advancing in the time or to modify its position. This was the motivation for another definition that is better coupled with physical interpretations; this is the derivative of Caputo type.

4.2. Caputo fractional derivative

Michele Caputo [11] published a book in which he introduced a new derivative, which had been independently discovered by Gerasimov (1948). This derivative is quite important, because it allows for understanding initial conditions, and is used to model fractional time. In some texts, it is known as the Gerasimov-Caputo derivative.

Let $[a, b]$ be a finite interval of the real line $\mathbb{R}$, for $\alpha \in \mathbb{C}(\Re(\alpha) \geq 0)$. The left and right Caputo derivatives are defined as

$$\left({}_{a}^{C}D_{x}^{\alpha}y\right)(x)=\frac{1}{\Gamma(n-\alpha)}\int_{a}^{x}\frac{y^{(n)}dt}{(x-t)^{\alpha-n+1}}=\left({}_{a}^{RL}I_{x}^{n-\alpha}D^{n}y\right)(x), \tag{49}$$

$$\left({}_{x}^{C}D_{b}^{\alpha}y\right)(x)=\frac{(-1)^{n}}{\Gamma(n-\alpha)}\int_{x}^{b}\frac{y^{(n)}dt}{(t-x)^{\alpha-n+1}}=(-1)^{n}\left({}_{x}^{RL}I_{b}^{n-\alpha}D^{n}y\right)(x), \tag{50}$$

where $D=\frac{d}{dx}$ and $n=--\Re(\alpha)$, i.e., $n=\Re(\alpha)+1$ for $\alpha\notin\mathbb{N}_0$ and $n=\alpha$ for $\alpha\in\mathbb{N}_0$. And if $0<\Re(\alpha)<1$

$$\left({}_{a}^{C}D_{x}^{\alpha}y\right)(x)=\frac{1}{\Gamma(1-\alpha)}\int_{a}^{x}\frac{y'dt}{(x-t)^{\alpha}}=\left({}_{a}^{RL}I_{x}^{1-\alpha}Dy\right)(x), \tag{51}$$

$$\left({}_{x}^{C}D_{b}^{\alpha}y\right)(x)=-\frac{1}{\Gamma(n-\alpha)}\int_{x}^{b}\frac{y'dt}{(t-x)^{\alpha}}=-\left({}_{x}^{RL}I_{b}^{1-\alpha}Dy\right)(x). \tag{52}$$

The connection between Caputo and Riemann derivatives is given by the relations

$$\left({}_{a}^{C}D_{x}^{\alpha}y\right)(x)=\left({}_{a}^{RL}D_{x}^{\alpha}\left[y(t)-\sum_{k=0}^{n-1}\frac{y^{(k)}(a)}{k!}(t-a)^{k}\right]\right)(x), \tag{53}$$

$$\left({}_{x}^{C}D_{b}^{\alpha}y\right)(x)=\left({}_{x}^{RL}D_{b}^{\alpha}\left[y(t)-\sum_{k=0}^{n-1}\frac{y^{(k)}(b)}{k!}(b-t)^{k}\right]\right)(x). \tag{54}$$

In particular, if $0<\Re(\alpha)<1$, Eqs. (53) and (54) relation take the following shapes:

$$\left({}_{a}^{C}D_{x}^{\alpha}y\right)(x)=\left({}_{a}^{RL}D_{x}^{\alpha}[y(t)-y(a)]\right)(x), \tag{55}$$

$$\left({}_{x}^{C}D_{b}^{\alpha}y\right)(x)=\left({}_{x}^{RL}D_{b}^{\alpha}[y(t)-y(b)]\right)(x). \tag{56}$$

For $\alpha = n$, then the Caputo derivatives match classical derivatives except for the sign of the right derivative.

However, for $k=0,1,\ldots,n-1$, we have

$$\left({}_{a}^{C}D_{x}^{\alpha}(t-a)^{k}\right)(x)=0,\quad \left({}_{x}^{C}D_{b}^{\alpha}(b-t)^{k}\right)(x)=0, \tag{57}$$

in particular,

$$\left({}_{a}^{C}D_{x}^{\alpha}1\right)(x)=0,\quad \left({}_{x}^{C}D_{b}^{\alpha}1\right)(x)=0. \tag{58}$$

On the other hand, if $\Re(\alpha)>0$ and $\lambda>0$, then

$$\left({}_{a}^{C}D_{x}^{\alpha}e^{\lambda t}\right)(x)\neq\lambda^{\alpha}e^{\lambda t},\quad \text{for } \alpha\in\mathbb{R}. \tag{59}$$

The Caputo derivatives behave like inverted operators for the left Riemann-Liouville fractional integrals ${}_{a}^{RL}I_{x}^{\alpha}$ and ${}_{x}^{RL}I_{b}^{\alpha}$, if $\Re(\alpha)>0$ and $y(x)\in C[a,b]$

$$\left({}_{a}^{C}D_{x}^{\alpha}\left({}_{a}^{RL}I_{x}^{\alpha}y\right)\right)(x)=y(x),\quad \left({}_{x}^{C}D_{b}^{\alpha}\left({}_{x}^{RL}I_{b}^{\alpha}y\right)\right)(x)=y(x). \tag{60}$$

On the other hand, if $\Re(\alpha)>0$ and $n=-\lfloor-\Re(\alpha)\rfloor$, then for good conditions for $y(x)$

$$\left({}_{a}^{RL}I_{x}^{\alpha}\left({}_{a}^{C}D_{x}^{\alpha}y\right)\right)(x)=y(x)-\sum_{k=0}^{n-1}\frac{y^{(k)}(a)}{k!}(x-a)^{k}, \tag{61}$$

$$\left({}_{x}^{RL}I_{b}^{\alpha}\left({}_{x}^{C}D_{b}^{\alpha}y\right)\right)(x)=y(x)-\sum_{k=0}^{n-1}\frac{(-1)^{k}y^{(k)}(b)}{k!}(b-x)^{k}. \tag{62}$$

In particular if, $0<\Re(\alpha)\leq 1$, then

$$\left({}_{a}^{RL}I_{x}^{\alpha}\left({}_{a}^{C}D_{x}^{\alpha}y\right)\right)(x)=y(x)-y(a), \tag{63}$$

$$\left({}_{x}^{RL}I_{b}^{\alpha}\left({}_{x}^{C}D_{b}^{\alpha}y\right)\right)(x)=y(x)-y(b). \tag{64}$$

In his early articles and several after that, Caputo used a Laplace transformed of the Caputo fractional derivative, which is given by

$$(\mathcal{L}\{{}_{0}^{C}D_{x}^{\alpha}y\})(s)=s^{\alpha}(\mathcal{L}y)(s)-\sum_{k=0}^{n-1}s^{\alpha-k-1}(D^{k}y)(0). \tag{65}$$

When $0<\alpha\leq 1$, then

$$(\mathcal{L}\{{}_{0}^{C}D_{x}^{\alpha}y\})(s)=s^{\alpha}(\mathcal{L}y)(s)-s^{\alpha-1}y(0). \tag{66}$$

These derivatives can be defined over the whole real axis resulting in the expressions:

$$({}^{C}D_{x}^{\alpha}y)(x)=\frac{1}{\Gamma(n-\alpha)}\int_{-\infty}^{x}\frac{y^{(n)}(t)dt}{(x-t)^{\alpha-n+1}}, \tag{67}$$

$$({}_{x}^{C}D^{\alpha}y)(x)=\frac{(-1)^{n}}{\Gamma(n-\alpha)}\int_{x}^{\infty}\frac{y^{(n)}(t)dt}{(t-x)^{\alpha-n+1}}, \tag{68}$$

with $x\in\mathbb{R}$.

5. Fractal geometry and fractional calculus

The phenomenon of anomalous diffusion is mathematically modeled by a fractional partial differential equation. The parameters of this equation are uniquely determined by the fractal dimension of the underlying object.

There are some results that show the relationship between fractals and fractional operators [24]; two of the most important that motivated the particular study of the equations to determine the pressure deficit in oil wells are highlighted below.

5.1. Cantor's Bars and fractional integral

In 1992, Nigmatullin [12] presents one of the most distinguished contributions to the search of the concrete relationship between the fractal dimension of a porous medium and the order of the fractional derivative to model a phenomena through such a medium; in this, he achieves the evolution of a physical system of a Cantor set type.

In his research, Nigmatullin proposes a relationship between the fractal dimension of a Cantor type set and the order of a fractional integral of the Riemann-Liouville type. The systems he considers are named phenomena with "memory." The use of fractional derivatives given by assuming a transference function $J(t)$ in relationship to a rectifiable function $f(t)$ through the convolution operator with a distribution $K^*(t)$ establishes that

$$J(t) = K^\star(t) * f(t) = \int_0^t K^\star(t-\tau) f(\tau) d\tau. \tag{69}$$

Where the distribution to apply (see Refs. [13, 14]) is a so-called "Cantor's Bars" $K_{T,\nu}(t)$, supported in the [0, T] interval, with a fractal dimension $\nu = \ln(2)/\ln(1/\xi)$, with $\xi \in [0, 1/2]$ being the compression factor, normalized in L^1.

Through the result of distribution values, he establishes the relation:

$$J(t) = \langle K_{T,\nu}(t) \rangle * f(t) = B(\nu) T^{-\nu} {}_0^{RL}D_x^{-\nu}[f(t)] = \frac{B(\nu)T^{-\nu}}{\Gamma(\nu)} \left(t^{\nu-1} * f(t) \right), \tag{70}$$

$$\langle K_{T,\nu}(t) \rangle = \int_{-1/2}^{1/2} K_{T,\nu}(t\xi^{-x})(2\xi)^{-x}\, dx = \frac{B(\nu)}{\Gamma(\nu)} \left(\frac{t}{T} \right)^{\nu-1}, \tag{71}$$

$$B(\nu) = \int_{-1/2}^{1/2} q_\nu(z + x\ln\xi)\, dx. \tag{72}$$

Thus, assuming a porous medium with a ν fractal dimension, we establish a fractional derivative of $-\nu$ order.

The initial results were strongly questioned by different authors, including Roman Rutman (see Refs. [15, 16]), who asserts that the relation is so artificial. However, recent works suggest that Nigmatullin's statements are not far from reality, but it is necessary to reduce the set of functions and that of fractals for which the necessary convergence is fulfilled.

6. Fractional calculus for modeling oil pressure

In this section, the Equation Continuity which follows from the law of conservation of mass is established. Darcy's law is used to relate fluid motion to pressure and gravitational gradients.

The combination of the Continuity Equation and Darcy's Law leads to a heat-conducting differential equation in mathematical physics describing the transfer of the fluid. We obtain a system formed by three partial differential equations, one for each fluid. This multiphase system must be solved considering the relevant boundary and initial conditions [30].

In the particular case of naturally fractured reservoirs (see Refs. [1, 2]), usually it is possible to discern three porosity types: matrix, fracture, and vugs; with this conception, it is accepted that the three porosities have associated a solid phase, and with this both Continuity Equation and Darcy's law can be expressed for each fluid in each geometrical media. If we only consider oil (monophasic) in a isotropic and saturated media, we can obtain a three equations system; for this, we begin with standard continuity equation and standard Darcy's law, respectively (see Ref. [17]):

$$\frac{\partial(\rho\theta)}{\partial t} + \nabla \cdot p(\rho q) = \rho \Upsilon, \qquad q = -\frac{1}{\mu}k(p)(\nabla p - \rho g \nabla D), \tag{73}$$

where θ is the volumetric content of fluid; $q = (q_1, q_2, q_3)$ is the Darcy flux, with its spatial components (x ,y, z), t is the time; ρ is the density of the fluid; μ is the dynamic viscosity of the fluid; g gravitational acceleration; Υ is a source term and represents a volume provided per volume unit of porous media in the time unity; p is the pressure; D is the depth as a function of spatial coordinates, usually identified to the vertical coordinate z; k is the permeability tensor of the partially saturated porous media and it depends on the pressure. The relations $\theta(p)$ and $k(p)$ are the fluid-dynamics characteristics of the media.

General fluid transfer equation results combining the formulas in Eq. (73):

$$\frac{\partial(\rho\theta)}{\partial t} = \nabla \cdot p[\frac{\rho}{\mu}k(p)(\nabla p - \rho g \nabla D)] + \rho \Upsilon. \tag{74}$$

This differential equation contains two dependent variables, namely the humidity content and fluid pressure, but they are related. For this reason, the saturation $S(p)$ is defined so that

$$\theta(p) = \phi(p)S(p) \tag{75}$$

where ϕ is the total porosity of the medium, and the specific capacity defined by

$$C(p) = \frac{d(\rho\phi S)}{dp} = \phi S \frac{d\rho}{dp} + \rho S \frac{d\phi}{dp} + \rho\phi \frac{dS}{dp}, \tag{76}$$

in consequence

$$\frac{\partial(\rho\theta)}{\partial t} = C(p)\frac{\partial p}{\partial t}, \tag{77}$$

6.1. Triadic media

The porous media is considered to be formed by three porous media: the matrix, fractured media, and vuggy media. The total volume of the porous media (V_T) is equal to the sum of the total volume of the matrix (V_m), of the total volume of the fractured medium (V_F) and of the total volume of the vuggy media (V_G). In other words

$$V_T = V_M + V_F + V_G, \tag{78}$$

each of the porous media contains solids and voids so that

$$V_M = V_{MS} + V_{MV} \tag{79}$$

$$V_F = V_{FS} + V_{FV} \tag{80}$$

$$V_G = V_{GS} + V_{GV}. \tag{81}$$

The porous medium as everything contains solids and voids, with the following relations:

$$V_T = V_{TS} + V_{TV},\ V_{TS} = V_{MS} + V_{FS} + V_{GS},\ V_{TV} = V_{MV} + V_{FV} + V_{GV}. \tag{82}$$

The volume fraction occupied by the matrix is defined as (ν_M), the volume fraction occupied by the fractured media as ν_F, and the fraction that occupies the vuggy media as (ν_G) relative to the total volume of the porous medium given by

$$\nu_M = \frac{V_M}{V_T}, \nu_f = \frac{V_F}{V_T}, \nu_G = \frac{V_G}{V_T}, \qquad \nu_M + \nu_F + \nu_G = 1. \tag{83}$$

The porosity of the porous media (ϕ), in matrix (ϕ_M), fracture media (ϕ_F) and vuggy media (ϕ_G) are defined by

$$\phi = \frac{V_{TV}}{V_T},\ \phi_M = \frac{V_{MV}}{V_M},\ \phi_F = \frac{V_{FV}}{V_F},\ \phi_G = \frac{V_{GV}}{V_G}. \tag{84}$$

From the above equations, we deduce the relation between the porosities:

$$\phi = \nu_M \phi_M + \nu_F \phi_F + \nu_G \phi_G \tag{85}$$

When the empty space contains fluid partially, the total volumetric content of the fluid (θ) as the total fluid volume (V_{TW}) with respect to the total volume of the porous medium is $(V_T) : \theta = V_{TW}/V_T$. In an analogous way, the volumetric content of fluid in the matrix is defined $\theta_M = V_{MW}/V_M$, in the fractured media $\theta_F = V_{FW}/V_F$ and vuggy media $\theta_G = V_{GW}/V_G$. It follows that

$$\theta = \nu_M \theta_M + \nu_F \theta_F + \nu_G \theta_G \tag{86}$$

which is reduced to Eq. (85) when the porous medium is fully saturated with fluid. It is satisfied: $0 < \theta < \phi, 0 < \theta_M < \phi_M, 0 < \theta_F < \phi_F, 0 < \theta_G < \phi_G$. The relation between the total volumetric flow of the fluid per unit area in the porous medium (q), the volumetric flow per unit area in the matrix (q_M), the volumetric flow per unit area in the fractured medium q_F, and the volumetric flow per unit area in the vuggy media (q_G) is analogous to Eq. (86), namely

$$q = \nu_M q_M + \nu_F q_F + \nu_G q_G \tag{87}$$

The continuity equations for the matrix, the fractured medium, and the vuggy media considering Eq. (86) acquire the form

$$\frac{\partial(\rho\theta_i)}{\partial t} + \nabla \cdot p(\rho q_i) = \rho \Upsilon_i,\ i = M,\ F,\ G. \tag{88}$$

Darcy's law for the matrix, the fractured medium, and the vuggy media, takes the form

$$q_i = -\frac{1}{\mu} k_i(p_i)(\nabla p_i - \rho g \nabla D), i = M,\ F,\ G \tag{89}$$

The equation of continuity of the porous medium, Eq. (73), is deduced from the sum of Eq. (88) previously multiplied by $\nu_m,\ \nu_F,\ \nu_G$, respectively, if the source terms are related by

$$\Upsilon = \nu_M \Upsilon_M + \nu_F \Upsilon_F + \nu_G \Upsilon_G \tag{90}$$

from Eqs. (87) and (89), the following relationships are deduced:

$$k(p) = \nu_M k_M(p_M) + \nu_F k_F(p_F) + \nu_G k_G(p_G) \tag{91}$$

$$\Phi(p) = \nu_M \Phi_M(p_M) + \nu_F \Phi_F(p_M) + \nu_G \Phi_G(p_G) \tag{92}$$

where Φ represents the potential of Kirchoff which is generically defined as

$$\Phi(p) = \int_{-\infty}^{p} k(u) du \tag{93}$$

If there is no fluid gain or loss in the porous medium, then $\Upsilon = 0$ and in consequence:

$$\nu_M \Upsilon_M = \Upsilon_{MF} + \Upsilon_{MG} \tag{94}$$

$$\nu_F \Upsilon_F = -\Upsilon_{MF} + \Upsilon_{FG} \tag{95}$$

$$\upsilon_G \Upsilon_G = -\Upsilon_{MG} - \Upsilon_{FG} \tag{96}$$

where Υ_{MF} is the input of fluid that receives the matrix from the fractured medium, Υ_{MG} is the fluid input that receives the matrix of the vuggy media, and Υ_{FG} is the contribution of fluid that receives the fractured medium from the vuggy media.

The system of differential equations is defined as follows:

$$\frac{\partial(\rho\theta_M)}{\partial t} = \nabla \cdot p[\frac{\rho}{\mu} k_M(p_M)(\nabla p_M - \rho g \nabla D)] + \frac{\rho}{\nu_M}(\Upsilon_{MF} + \Upsilon_{MG}) \tag{97}$$

$$\frac{\partial(\rho\theta_F)}{\partial t} = \nabla \cdot p[\frac{\rho}{\mu} k_F(p_F)(\nabla p_F - \rho g \nabla D)] - \frac{\rho}{\nu_F}(\Upsilon_{MF} + \Upsilon_{FG}) \tag{98}$$

$$\frac{\partial(\rho\theta_G)}{\partial t} = \nabla \cdot p[\frac{\rho}{\mu}k_G(p_G)(\nabla p_G - \rho g \nabla D)] - \frac{\rho}{\nu_G}(\Upsilon_{MG} + \Upsilon_{FG}). \tag{99}$$

The contributions of fluid in each porous medium are modeled with the following relations:

$$\Upsilon_{MF} = a_{MF}(p_F - p_M), \tag{100}$$

$$\Upsilon_{MG} = a_{MG}(p_G - p_M), \tag{101}$$

$$\Upsilon_{FG} = a_{FG}(p_G - p_F), \tag{102}$$

where a_{MF}, a_{MG}, and a_{FG} are transfer coefficients at each interface, which may depend on the pressures on the adjacent media.

6.2. Monophasic flow saturated in triadic media

In the case of the monophasic flow saturated in triadic means, the continuity equations in each porous medium can be written as follows:

$$\frac{\partial(\rho\phi_i)}{\partial t} + \nabla \cdot p(\rho q_i) = \rho\Upsilon_i,\ i = M, F, G. \tag{103}$$

Darcy's law for each porous media takes the form

$$q_i = -\frac{1}{\mu}k_i(\nabla p_i - \rho g \nabla D), i = M, F, G. \tag{104}$$

The substitution of Darcy's law in the continuity equation leads to the following equations:

$$\frac{\partial(\rho\phi_M)}{\partial t} = \nabla \cdot p[\frac{\rho}{\mu}k_M(\nabla p_M - \rho g \nabla D)] + \frac{\rho}{\nu_M}(\Upsilon_{MF} + \Upsilon_{MG}), \tag{105}$$

$$\frac{\partial(\rho\phi_F)}{\partial t} = \nabla \cdot p[\frac{\rho}{\mu}k_F(\nabla p_F - \rho g \nabla D)] - \frac{\rho}{\nu_F}(\Upsilon_{MF} - \Upsilon_{FG}), \tag{106}$$

$$\frac{\partial(\rho\phi_G)}{\partial t} = \nabla \cdot p[\frac{\rho}{\mu}k_G(\nabla p_G - \rho g \nabla D)] - \frac{\rho}{\nu_G}(\Upsilon_{MF} + \Upsilon_{FG}). \tag{107}$$

When the fluid is considered at constant density and viscosity and the means of constant permeability, with $D = z$, we have

$$\phi_M c_M \frac{\partial p_M}{\partial t} = \frac{k_M}{\mu}\Delta p_M + \frac{1}{\nu_M}(\Upsilon_{MF} + \Upsilon_{MG}),\ c_M = \frac{1}{\phi_M}\frac{\partial \phi_M}{\partial p_M}, \tag{108}$$

$$\phi_F c_F \frac{\partial p_F}{\partial t} = \frac{k_F}{\mu}\Delta p_F - \frac{1}{\nu_F}(\Upsilon_{MF} - \Upsilon_{FG}),\ c_F = \frac{1}{\phi_F}\frac{\partial \phi_F}{\partial p_F}, \tag{109}$$

http://dx.doi.org/10.5772/intechopen.68571

$$\phi_G c_G \frac{\partial p_G}{\partial t} = \frac{k_G}{\mu} \Delta p_G - \frac{1}{\nu_G} (\Upsilon_{MF} - \Upsilon_{FG}),\ c_G = \frac{1}{\phi_G} \frac{\partial \phi_G}{\partial p_G} \tag{110}$$

6.3. Triple porosity and triple permeability model

The porosity of each medium has been defined as the volume of the space occupied by the medium. However, the porosity can be defined as the volume of empty space in each medium with respect to the volume of the total space occupied by the porous medium as a whole. These new porosities will be denoted with subscripts in lowercase letters and clearly have

$$\phi_m = \nu_M \phi_M,\ \phi_f = \nu_F \phi_F, \phi_g = \nu_G \phi_G \tag{111}$$

$$\phi = \phi_m + \phi_f + \phi_g \tag{112}$$

In an analogous way, the corresponding Darcy´s flow can be defined in each medium:

$$q_m = \nu_M q_M,\ q_f = \nu_F q_F, q_g = \nu_G q_G \tag{113}$$

$$q = q_m + q_f + q_g \tag{114}$$

Eq. (113) implies that the permeability of the Darcy's law in each medium is defined as

$$k_m = \nu_M k_M,\ k_f = \nu_F k_F,\ k_g = \nu_G k_G \tag{115}$$

The nest system by Eqs. (108)–(110), by congruently changing the subscripts in uppercase by lowercase in the pressures, in terms of compressibility, is written as follows:

$$\phi_m c_m \frac{\partial p_m}{\partial t} = \frac{k_m}{\mu} \Delta p_m + (\Upsilon_{mf} + \Upsilon_{mg}), c_m = \frac{1}{\phi_m} \frac{\partial \phi_m}{\partial p_m}, \tag{116}$$

$$\phi_f c_f \frac{\partial p_f}{\partial t} = \frac{k_f}{\mu} \Delta p_f - (\Upsilon_{mf} - \Upsilon_{fg}), c_f = \frac{1}{\phi_f} \frac{\partial \phi_f}{\partial p_f}, \tag{117}$$

$$\phi_g c_g \frac{\partial p_g}{\partial t} = \frac{k_g}{\mu} \Delta p_g - (\Upsilon_{mg} + \Upsilon_{mg}), c_g = \frac{1}{\phi_g} \frac{\partial \phi_g}{\partial p_g} \tag{118}$$

with $p_m \equiv p_M, p_f \equiv p_M, p_g \equiv p_G, \Upsilon_{mf} \equiv \Upsilon_{MF}, \Upsilon_{mg} \equiv \Upsilon_{MG}, \Upsilon_{fg} \equiv \Upsilon_{FG}, c_m \equiv c_M, c_f \equiv c_F, c_g \equiv c_G$.

The substitution of Eqs. (116)–(118) in Eqs. (100)–(102) leads to the system of differential equations that finalize the pressure in the matrix, fractured media, and vuggy media:

$$\phi_m c_m \frac{\partial p_m}{\partial t} = \frac{k_m}{\mu} \Delta p_m + a_{mf}(p_f - p_m) + a_{mg}(p_g - p_m), \tag{119}$$

$$\phi_f c_f \frac{\partial p_f}{\partial t} = \frac{k_f}{\mu} \Delta p_f - a_{mf}(p_f - p_m) + a_{fg}(p_g - p_f), \tag{120}$$

$$\phi_g c_g \frac{\partial p_g}{\partial t} = \frac{k_g}{\mu} \Delta p_g - a_{mg}(p_g - p_m) - a_{fg}(p_g - p_f), \tag{121}$$

in which this system constitutes a triple porosity and triple permeability model. In polar coordinates, the system reduces to

$$\phi_m c_m \frac{\partial p_m}{\partial t} = \frac{k_m}{\mu} \frac{1}{r} \frac{\partial}{\partial r}(r \frac{\partial p_m}{\partial r}) + a_{mf}(p_f - p_m) + a_{mg}(p_g - p_m), \tag{122}$$

$$\phi_f c_f \frac{\partial p_f}{\partial t} = \frac{k_f}{\mu} \frac{1}{r} \frac{\partial}{\partial r}(r \frac{\partial p_f}{\partial r}) - a_{mf}(p_f - p_m) + a_{fg}(p_g - p_f), \tag{123}$$

$$\phi_g c_g \frac{\partial p_g}{\partial t} = \frac{k_g}{\mu} \frac{1}{r} \frac{\partial}{\partial r}(r \frac{\partial p_g}{\partial r}) - a_{mg}(p_g - p_m) - a_{fg}(p_g - p_f) \tag{124}$$

6.4. Dimensionless variables

Now we will give a process of dimensionlessness to better manage the variables. This is a technique commonly used to make the parameters or variables in an equation having no units, bring to a range the possible values of a variable or constant in order that its value is known, and in this way, more manipulable.

The system of Eqs. (122)–(124) takes the following form after making the changes mentioned in the previous paragraph:

$$(1 - \omega_f - \omega_v) \frac{\partial p_{Dm}}{\partial t_D} = (1 - \kappa_f - \kappa_v) \frac{1}{r_D} \frac{\partial}{\partial r_D}(r_D \frac{\partial p_{Dm}}{\partial r_D}) + \lambda_{mf}(p_{Df} - p_{Dm}) + \lambda_{mv}(p_{Dv} - p_{Dm}) \tag{125}$$

$$\omega_f \frac{\partial p_{Df}}{\partial t_D} = \kappa_f \frac{1}{r_D} \frac{\partial}{\partial r_D}(r_D \frac{\partial p_{Df}}{\partial r_D}) - \lambda_{mf}(p_{Df} - p_{Dm}) + \lambda_{fv}(p_{Dv} - p_{Df}) \tag{126}$$

$$\omega_v \frac{\partial p_{Dv}}{\partial t_D} = \kappa_v \frac{1}{r_D} \frac{\partial}{\partial r_D}(r_D \frac{\partial p_{Dv}}{\partial r_D}) - \lambda_{mv}(p_{Dv} - p_{Dm}) - \lambda_{fv}(p_{Dv} - p_{Df}) \tag{127}$$

where

$$\omega_f = \frac{\phi_f c_f}{\phi_m c_m + \phi_f c_f + \phi_v c_v}, \omega_g = \frac{\phi_v c_v}{\phi_m c_m + \phi_f c_f + \phi_v c_v}, r_D = \frac{r}{r_w} \tag{128}$$

$$\kappa_f = \frac{k_f}{k_m + k_f + k_v}, \kappa_g = \frac{k_v}{k_m + k_f + k_v} \tag{129}$$

$$\lambda_{mf} = \frac{a_{mf} \mu r_w^2}{k_m + k_f + k_v}, \lambda_{mv} = \frac{a_{mv} \mu r_w^2}{k_m + k_f + k_v}, \lambda_{fv} = \frac{a_{fv} \mu r_w^2}{k_m + k_f + k_v} \tag{130}$$

$$p_{Dj} = \frac{2\pi h(k_m + k_f + k_v)(p_i - p_j)}{Q_0 B_0 \mu}, t_D = \frac{t(k_m + k_f + k_v)}{\mu r_w^2(\phi_m c_m + \phi_f c_f + \phi_v c_v)} \tag{131}$$

Eqs. (128)–(131) represent dimensionless variables so they have no units. The boundary conditions to which the previous model is subjected are

$$\lim_{r_D \to 1} r_D (1 - \kappa_f - \kappa_v) \frac{\partial p_{Dm}}{\partial r_D} + r_D \kappa_v \frac{\partial p_{Dv}}{\partial r_D} = -1 \tag{132}$$

$$p_w(t) = p_{Dm}(r_D, t)|_{r_D=1} = p_{Df}(r_D, t)|_{r_D=1} = p_{Dv}(r_D, t)|_{r_D=1} = -1 \tag{133}$$

Substituting derivatives $\frac{\partial p_{Di}}{\partial t_D}$ by Caputo fractional derivatives $\frac{\partial^{\alpha_i} p_{Dj}}{t_D^{\alpha_i}}$ with $0 < \alpha_i \leqq 1$, and $\frac{\partial}{\partial r_D}\left(r_D \frac{\partial p_{Dj}}{\partial r_D}\right)$ by Riemann-Liouville complementary derivative, i. e., with infinite limit of integration, (also called Weyl derivative) $\frac{\partial^{\gamma_i}}{\partial r_D^{\gamma_i}}\left(r_D \frac{\partial^{\beta_i} p_{Dj}}{\partial r_D^{\beta_i}}\right)$, with $1 < \gamma_i + \beta_i \leq 2$, $i = 1, 2, 3$; $j = v, f, m$; $\alpha_i, \beta_i, \gamma_i$ rational numbers.

The choice of the derivatives, Caputo and Riemann-Liouville (Weyl), obeys the nature of the problem and the ease with which they can be manipulated.

The monophase flow model with triple porosity and triple permeability is expressed as follows: For the matrix

$$(1 - \omega_f - \omega_v) \frac{\partial^{\alpha_1} p_m}{\partial t^{\alpha_1}} = (1 - \kappa_f - \kappa_v) \frac{1}{r} \frac{\partial^{\gamma_1}}{\partial r^{\gamma_1}} \left(r \frac{\partial^{\beta_1} p_m}{\partial r^{\beta_1}} \right) + \lambda_{mf}(p_f - p_m) + \lambda_{mv}(p_v - p_m), \tag{134}$$

for fracture media

$$\omega_f \frac{\partial^{\alpha_2} p_f}{\partial t^{\alpha_2}} = \kappa_f \frac{1}{r} \frac{\partial^{\gamma_2}}{\partial r^{\gamma_2}} \left(r \frac{\partial^{\beta_2} p_f}{\partial r^{\beta_2}} \right) - \lambda_{mf}(p_f - p_m) + \lambda_{fv}(p_v - p_f), \tag{135}$$

for vuggs

$$\omega_v \frac{\partial^{\alpha_3} p_v}{\partial t^{\alpha_3}} = \kappa_v \frac{1}{r} \frac{\partial^{\gamma_3}}{\partial r^{\gamma_3}} \left(r \frac{\partial^{\beta_3} p_v}{\partial r^{\beta_3}} \right) - \lambda_{mv}(p_v - p_m) - \lambda_{fv}(p_v - p_f). \tag{136}$$

We reduce this system by applying semigroup properties with respect to the order of the Weyl derivative, assuming: $0 < \alpha_i \leq 1$ and $1 < \alpha_i + \beta_i \leq 2$. Let $\omega = 1 - \omega_f - \omega_v$; $\kappa = 1 - \kappa_f - \kappa_v$, putting $p_m = p$; $p_f = f$; $p_v = u$; $r = x$; $\eta_i = \gamma_i + \beta_i$. Then, the previous system can be expressed as

$$\frac{\partial^{\alpha_1} p}{\partial t^{\alpha_1}} = \frac{\kappa}{\omega} \left(-\gamma_1 \frac{1}{x} \frac{\partial^{\eta_1 - 1} p}{\partial x^{\eta_1} - 1} + \frac{\partial^{\eta_1} p}{\partial x^{\eta_1}} \right) + \frac{\lambda_{mf}}{\omega}(f - p) + \frac{\lambda_{mv}}{\omega}(u - p), \tag{137}$$

$$\frac{\partial^{\alpha_2} f}{\partial t^{\alpha_2}} = \frac{\kappa}{\omega_f} \left(-\gamma_2 \frac{1}{x} \frac{\partial^{\eta_2 - 1} f}{\partial x^{\eta_2} - 1} + \frac{\partial^{\eta_2} f}{\partial x^{\eta_2}} \right) - \frac{\lambda_{mf}}{\omega_f}(f - p) + \frac{\lambda_{fv}}{\omega_f}(u - f), \tag{138}$$

$$\frac{\partial^{\alpha_3} u}{\partial t^{\alpha_3}} = \frac{\kappa}{\omega_v} \left(-\gamma_3 \frac{1}{x} \frac{\partial^{\eta_3 - 1} u}{\partial x^{\eta_3} - 1} + \frac{\partial^{\eta_3} u}{\partial x^{\eta_3}} \right) - \frac{\lambda_{mv}}{\omega_v}(u - p) + \frac{\lambda_{fv}}{\omega_v}(u - f). \tag{139}$$

The above approach can be solved by numerical methods as finite differences along with a predictor-corrector, such as Daftardar-Gejji works, for example in [19] and compared with previous ones, such as that presented by Camacho et al. [18, 28], the approximations are significantly improved. However, there is still work to be completed; the optimal solution method has not been found and the best way to determine the appropriate order, so far numerical methods, has been used to estimate the order that best approximates measurements.

The application of the fractional calculation can be very useful for the modeling of anomalous diffusion phenomena in which the fractal structure better reflects the real conditions of the medium, as it is the case of the reservoirs in which because of its very nature it is difficult to find a structure Euclidian.

Author details

Benito F. Martínez-Salgado[1*], Rolando Rosas-Sampayo[1], Anthony Torres-Hernández[1] and Carlos Fuentes[2]

*Address all correspondence to: masabemx@yahoo.com.mx

1 Faculty of Sciences, National Autonomous University of Mexico, Mexico City, Mexico

2 Mexican Institute of Water Technology, Jiutepec, Morelos, Mexico

References

[1] Camacho-Velázquez R, Vásquez-Cruz MA, Castrejón-Aivar R, Arana-Ortiz V. Pressure transient and decline curve behaviors in naturally fractured vuggy carbonate reservoirs. SPE Reservoir Evaluation & Engineering. 2005;**8**(02):95-112

[2] Camacho-Velázquez R, Fuentes-Cruz G, Vásquez-Cruz MA. Decline-curve analysis of fractured reservoirs with fractal geometry. SPE Reservoir Evaluation & Engineering. 2008;**11**(03):606-619

[3] Carlos Fuentes-Ruíz et al. Reservoirs as a fractal reactor: A model with triple porosity and triple permeability of the fractured media (matrix-vug-fracture). Fondo sectorial conacyt-sener-hidrocarburos s0018-2011-11, Universidad Autónoma de Querétaro

[4] Alexander S, Orbach R. Density of states on fractals: ``fractons". Le Journal de Physique Lettres. 1982;**43**(17):625-631

[5] O'Shaughnessy B, Procaccia I. Analytical solutions for diffusion on fractal objects. Physical Review Letters. 1985;**54**(5):455

[6] Metzler R, Glöckle WG, Nonnenmacher TF. Fractional model equation for anomalous diffusion. Physica A: Statistical Mechanics and its Applications. 1994;**211**(1):13-24

[7] Havlin S, Ben-Avraham D. Diffusion in disordered media. Advances in Physics. 1987;**36**(6):695-798

[8] Samko SG, Kilbas AA, Marichev OI. Fractional Integrals and Derivatives: Theory and Applications. Singapore: Gordon and Breach Science Publishers; 1993

[9] Oldham KB, Spanier J. The Fractional Calculus. Theory and Applications of Differentiation and Integration to Arbitrary Order. Vol. 111. Elsevier Science; 1974

[10] Podlubny I. Fractional Differential Equations. An Introduction to Fractional Derivatives, Fractional Differential Equations, Some Methods of Their Solution and Some of Their Applications, volume 198 of Mathematics in Science and Engineering. Academic Press; 1999

[11] Caputo M. Elasticità e dissipazione. Zanichelli Publisher; 1969

[12] Nigmatullin RR. Fractional integral and its physical interpretation. Theoretical and Mathematical Physics. 1992;**90**(3):242-251

[13] Nigmatullin RR, Méhauté AL. Is there geometrical/physical meaning of the fractional integral with complex exponent? Journal of Non-Crystalline Solids. 2005;**351**(33-36):2888-2899

[14] Méhauté AL, Nigmatullin RR, Nivanen L. Flèches du temps et géométrie fractale. Hermes. 1998. 151-176, 287-302

[15] Rutman RS. On physical interpretations of fractional integration and differentiation. Theoretical and Mathematical Physics. 1995;**105**(3):1509-1519

[16] Blackledge JM, Evans AK, Turner MJ. Fractal Geometry: Mathematical Methods, Algorithms, Applications. Great Britain: Elsevier; 2002

[17] Bear J. Dynamics of Fluids in Porous Media. New York: Dover; 1988. 119-129

[18] Baleanu D, Diethewlm K, Scalas E, Trujillo JJ. Fractional calculus: Models and numerical methods, volume 3 of series on complexity, nonlinearity and chaos. Singapore: World Scientific. 2012. 123-140

[19] Daftardar-Gejji V. Fractional Calculus: Theory and Applications. New Delhi: Narosa Publishing House; 2014

[20] Hilfer R. Applications of fractional calculus in physics. Singapore: World Scientific. 2000. 1-86

[21] Meerschaert M. Mathematical Modeling. 4th ed. Boston: Academic Press; 2013

[22] Miller KS, Ross B. An Introduction to the Fractional Calculus and Fractional Differential Equations, volume 111 of Mathematics in Science and Engineering. New York: Wiley-Interscience; 1993

[23] Klages R, Radons G, Sokolov IM. Anomalous Transport: Foundations and applications. Germany: John Wiley & Sons; 2008

[24] Meerschaert MM, Sikorskii A. Stochastic models for fractional calculus, volume 43 of De Gruyter studies in mathematics. Germany: Walter de Gruyter. 2012

[25] Ibe OC. Elements of Random Walk and Diffusion Processes. New Jersey: John Wiley & Sons, 2013

[26] Hardy HH, Beier RA. Fractals in reservoir engineering. Singapore: World Scientific; 1994

[27] Herrmann R. Fractional calculus: An introduction for physicists. Singapore: World Scientific; 2011

[28] Diethelm K, Ford NJ, Freed AD, Luchko Y. Algorithms for the fractional calculus: A selection of numerical methods. Computer Methods in Applied Mechanics and Engineering. 2005;**194**(6-8):743-773

[29] Letnikov AV. Theory of differentiation of fractional order (in Russian). Matematicheskii Sbornik 1868;**3**:1-68

[30] Peaceman DW. Fundamentals of Numerical Reservoir Simulation. New York, NY: Elsevier Scientific Publishing Co.; 1977

Section 2

Industry, Antennas, Spacecraft, Radar, Images, Measure

Chapter 3

Fractals in Antennas and Metamaterials Applications

Wojciech Jan Krzysztofik

Additional information is available at the end of the chapter

http://dx.doi.org/10.5772/intechopen.68188

Abstract

Recently, telecommunication systems have been requiring more advanced features in the design and operation. Among others a smaller size of devices, which can be integrated for multiple mobile communication systems, applied in one user's device board, such as PDA or smart phone. Moreover, the cost of mass production should be minimized as much as possible. To meet part of that request, the antennas of these devices should have small size, lower weight, operating in multiple frequency bands and/or be broadband. There are many research methods to achieve this goal, one of which is using the fractal geometries for the shape of antenna elements. In recent years, there are many fractal shapes that have been proposed for such applications, and the designed antennas have significantly improved antenna features such as smaller size, operating in multi-frequency bands, with improved power gain and efficiency. In recent years, the new approach for modern antenna the metamaterials, MTM, is adopted, and sometimes that based on the fractal geometry is adopted.

Keywords: fractal geometries, fractal dimension, IFS, fractal antenna, antenna elements, multiband, power gain, efficiency, compact size, antenna array, metamaterials, MTM

1. Introduction

Antenna structures are well known, it takes them a lot of research and development centres. Relatively rarely appearing solutions that represent a new approach in the view of the traditional methods of design are applied to modern radio communication systems [9]. So far, the antenna had simple shapes that are described in Euclidean geometry. In the past dozen years, many scientists around the world tried to make the structure of fractal geometry for applications in the field of electromagnetism, which led to the development of new innovative construction of antennas and artificial dielectric/magnetic materials, so-called metamaterials, MTM.

Structure based on fractal geometry cannot be described in the traditional Euclidean geometry. They are built with sequence of replicas based on the same shapes, which could be scaled, rotated and displaced in the space. In the research environment in the field of electromagnetism, intriguing was the answer to the question whether the properties of fractals can be useful in the design of elements of specific electromagnetic properties.

Fractals are abstract objects that, as satisfying in the strict sense of mathematical description, cannot be physically implemented, because these are infinite. However, you can make some assumptions in relation to the ideal fractal, which allow for constructing electromagnetic instruments, such as antennas, filters and microwave substrates having high surface impedance, circuit filters, electronic components of frequency-stop-band properties and others. Typically, these shapes are called pre-fractals or truncated fractals. You can apply different geometries, such as configurations of multiple triangles or other complicated constructions to build antennas that may be similar to fractal shapes and extract some of their advantages, which in theory can be obtained as a result of mathematical abstraction. Generally, the fractal antenna technology is a term used to describe these antenna engineering techniques, which are based on such mathematical concepts that provide generation of new antennas with some features that were not possible even in the mid-1980s.

To sum up the results of the work carried out so far, you can formulate the following benefits resulting from the application of the fractal technology:

1. Self-similarity is useful in the design of multi-frequency antenna, as, for example, the devices based on the Sierpinski fractal gasket or carpet, Minkowski fractal loop or patch that have been used in the design of multi-band antenna.

2. The self-filling property is useful for the design of electrically small antennas, such as Hilbert fractal, Peano fractal and Koch fractal monopoles or loops and the microstrip antennas.

3. The mass fractals and the boarder fractals are useful in obtaining high directivity and low sidelobes antenna elements and arrays.

4. Recently, the self-filling in space curves like Hilbert and Peano fractals were used to obtain high-impedance ground plane EBG, so-called metamaterials, used to design high performance, low profile, conformal antennas with enhanced radiation characteristics and improved power gain of various communication and radar applications.

Fractal geometry has many applications in life and open up new research directions in many fields such as biology and economics. In many EM devices, the self-similarity and plane-filling nature of fractal geometries are often qualitatively linked to its frequency characteristics, i.e. multi-frequency operation, or small size in low frequency bands. In addition, fractal geometries have been used in the electromagnetic radiation, and especially in the design of antennas with compact size and operating at multiband frequency. Some important characteristics of fractal geometries can be applied to design antenna. The self-similar design is used to operate the multi-band antenna. The self-space-filling structure of the fractal is used to design the small antennas. Mass fractals and boundary fractals are used to design array antennas.

http://dx.doi.org/10.5772/intechopen.68188

1.1. Brief introduction to fractal geometry

The original inspiration for the development of fractal geometry came largely from an in-depth study of the patterns of nature (see **Figure 1**).

The fractals, for example, have been successfully applied to the modelling of complex objects found in nature, such as systems of galaxies and strokes of clouds, mountain ranges, coastlines, snowflakes, trees, foliage plants and many others. By millions of years of evolution, nature optimizes architecture biological structures for effective distribution and use of energy, and, in principle, fractal shapes can be found in each of these structures.

Mandelbrot realized [10] that very often it is impossible to describe the objects that occur in nature using only Euclidean geometry, by means of straight lines, circles, cubes and the like. He proposed that fractals and fractal geometry can be used to describe real objects, such as trees, flashes of lightning, meandering rivers and coasts, to name just a few. Fractal dimensions need not be expressed using integers, so intuitively, we present them as a measure of how much space does the fractal occupies.

Fractals can be found in nature or could be generated using mathematical rules. Probably the easiest way to define a fractal is to describe it as an object, which is similar in varying degrees of zoom, and as a result, with the symmetry of the whole scale, with each a small part of the entire structure of the replacement object. Some examples of objects that have the characteristics of the self-similarity are shown in **Figure 2**.

Here are five properties, which most fractals have:

- They are made up of elements with any small scale,
- They are usually defined by a simple recursive processes,
- They are too irregular to be described using traditional Euclidian geometry,
- They have some type of self-similarity,
- They have fractal dimensions.

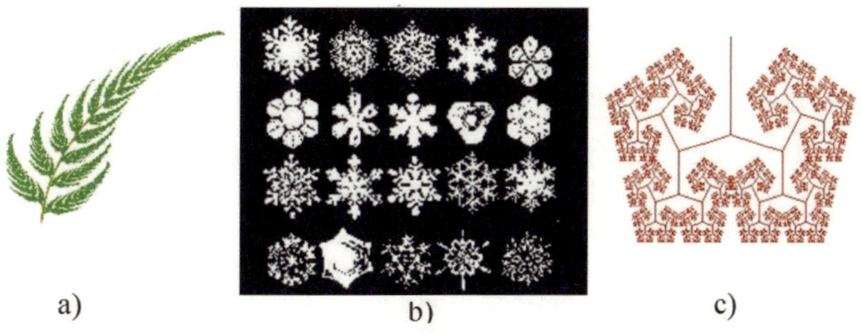

Figure 1. Fractal objects in nature: leaf of fern (a), the most popular—snowflakes (b) and human lungs formed by fractal canopies (c).

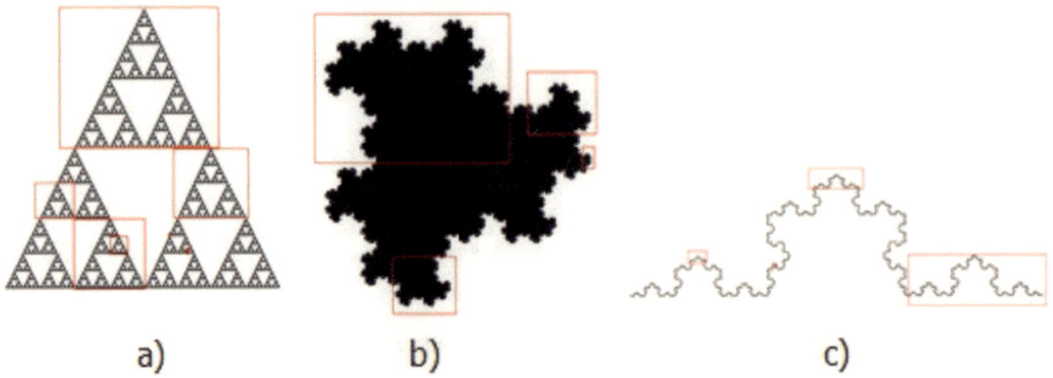

Figure 2. The self-similar components of different fractals: Sierpinski gasket (a), dragon (b) and Koch's curve (c).

1.2. Why use the antenna of fractal geometry?

Antennas are generally narrowband devices. Their properties depend on the size of the referenced to the wavelength. This means that for fixed antenna sizes, its parameters: power gain, input impedance, the radiation patterns, the side lobe level and distribution of surface currents will be continued strong changes when the operating frequency will change. Frequency dependency also means that the antenna should retain the minimum size in relation to the wavelength of operation, to work effectively. This means that for a given frequency, the antenna cannot be arbitrarily small, usually of the minimum size level of a quarter wavelengths.

These well-known rules for antenna engineers are being used for dozens of years in constructing the antennas of telecommunications systems. However, the dependence of the wavelength antenna size is still a problem in many systems where previously used antennas are not particularly useful. In this sense, the application of fractal geometry in the design of antennas and antenna arrays can help in dealing with the problems of designing antennas that meet the requirements of modern communication systems not found so far. The reason *why* the fractal design of antennas and MTM appears as an attractive way to make it is few-fold (**Figure 3**).

First, this is because you should expect each antenna similar to other, which is built with multiple copies of the same in different scales in size, to act in a similar way, for a few wavelengths. This means that the antenna should keep similar parameters of radiation for a few wavelengths.

Second, because the properties of the self-filling to an area of some fractals (fractal size) may allow small objects in the fractal shape to make better use of the small surrounding space. Fractal antenna and antenna arrays design derives from a mixture of two seemingly unconnected disciplines, namely, the theory of electromagnetism and geometry. From the first spiral and log-periodic antennas, developed in the early sixties of the 20th century by D.E. Isbell, R. DuHamelet and variants by P. Mayes [48] and from the works of Benoit Mandelbrot fractal geometry, the fractal antenna appears as a natural way to solve the problem of antenna working on multiple frequencies and antennas of reduced size.

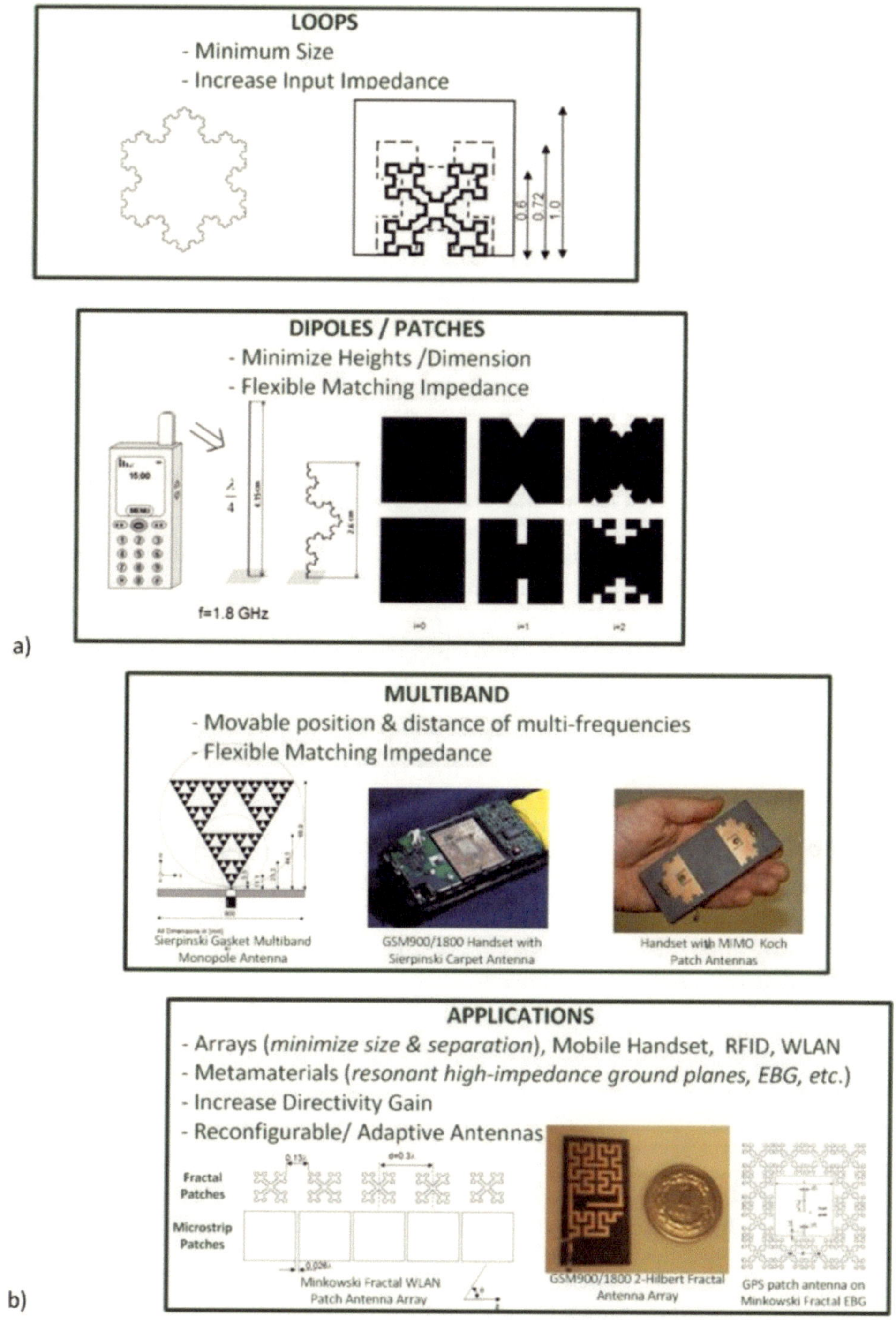

Figure 3. Fractal antennas (a) and the various fractal geometries, fall into few main categories: loops, dipoles, multiband fractal patches, antenna arrays, metamaterials (b).

1.3. How can you adopt fractal shapes in the design of antennas?

While Euclidean geometry is limited to points, lines, surfaces, and volumes, fractals include geometry, which come down between the two shapes. Therefore, a fractal can be a line, which is coming to the surface. The line can bend in such a way so as to effectively fill almost the entire surface. These properties of self-fill space lead to curves that are electrically very long but are located in a small space. This property can in effect lead to the miniaturization of antenna elements. In the previous section, it was mentioned that pre-fractals limit the complexity of fractal geometry, which is not recognized for specific applications. For antennas, it can mean that the curvatures, which are much smaller than the wavelength of the frequency range in use, can be dropped [11].

This causes the infinitely complex structure, which can be analysed only mathematically, not to be created physically. You can, however, show that the number of iterations required to achieve the benefits of miniaturization is limited to a few, before there will be the complexities difficult to distinguish in the so created structure. Not yet released many interesting works, which relate to the nascent field of the fractal electrodynamics.

Much of the pioneering work has been documented in Refs. [12, 13]. Most of them apply to mathematical foundations, as well as researches in the field of fractal antennas and/or EM wave reflection analysis of fractal surface. The area self-filling properties by the Hilbert fractal and related curves, e.g. the Peano fractal, make them attractive candidates for use in the design of antennas.

The Hilbert fractal self-filling properties were tested in Ref. [11] as an effective method to design a compact resonant antenna. The first four steps (iterations) of the construction of the Hilbert curve are shown in **Figure 4**.

The fractal structure of self-affine geometry [14] is shown in **Figure 4b**. It is constructed by scaling of the square, with a factor 3 in the horizontal direction, and a factor 2 in the vertical direction, creating six rectangles, out of which was removed in the top of the middle rectangle. This is the first iteration. The second iteration of this procedure is repeated for the remaining rectangles, and it can continue indefinitely. Such a way of generating fractal structures is defined by using the iterative function system (IFS).

1.4. The language of fractals

1.4.1. Iterated function system

A universal method to generate a variety of fractal structures is the iterative function system (IFS) introduced in Refs. [10, 15–22], that is based on the use of a series of affine transformations, w, defined by

$$w\begin{pmatrix} x \\ y \end{pmatrix} = \begin{pmatrix} a & b \\ c & d \end{pmatrix}\begin{pmatrix} x \\ y \end{pmatrix} + \begin{pmatrix} e \\ f \end{pmatrix} \tag{1}$$

where real number coefficients (a, b, c, d, e, f) are responsible for movement of fractal element in space: a, d—scaling, b, c—rotation by φ_1, φ_2 angles with respect to axis of coordinating system

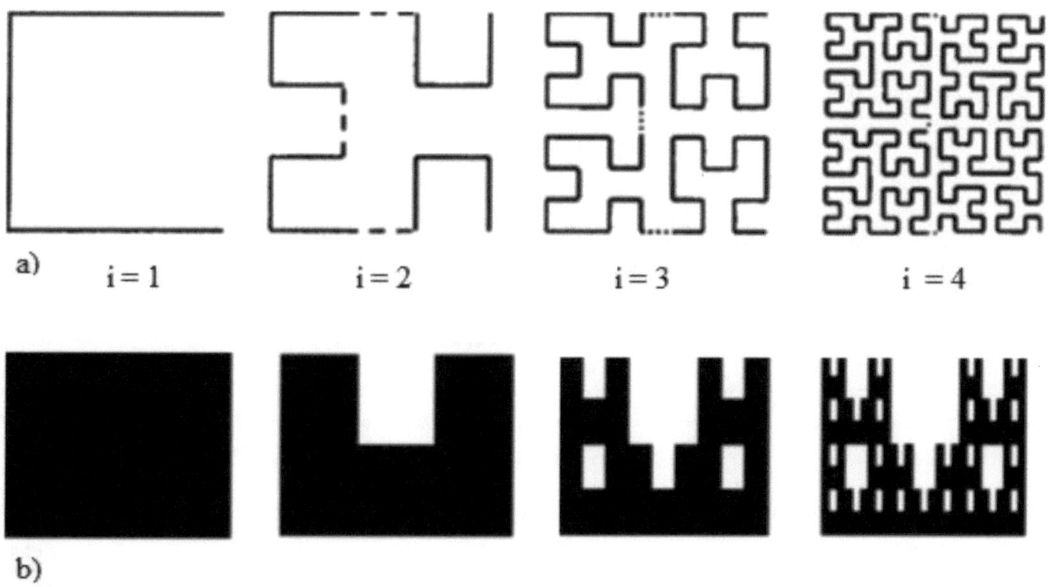

Figure 4. The four iterations of Hilbert fractal self-filing curve (a) and the self-affine fractal multiband antenna (b).

and *e*, *f*—linear translation by the vector (**e, f**), respectively (see **Figure 5**). They are expressed as follows: $a = \delta_1 \cos\varphi_1;\ d = \delta_2 \cos\varphi_2;\ b = \delta_2 \sin\varphi_2;\ c = \delta_1\ \sin\varphi_1$.

Suppose now that you should consider $w_1, w_2, \ldots, w_i$ as a system of linear affine transformations, and let *A* be the initial geometry. A new geometry you create by applying a series of transformations to the initial geometry and collecting the results of $w_1\ (A), w_2\ (A), \ldots, w_i\ (A)$, which can be represented as follows:

$$W(A) = \cup_{i=1}^{I} w_i(A) \tag{2}$$

where *W* is known as the Hutchinson operator [15, 22].

The structure of the fractal can be obtained by using the operator *W* repeatedly, starting from the initial through geometry. For example, when the initial geometry represents A_0, the next generation it gains as a result of

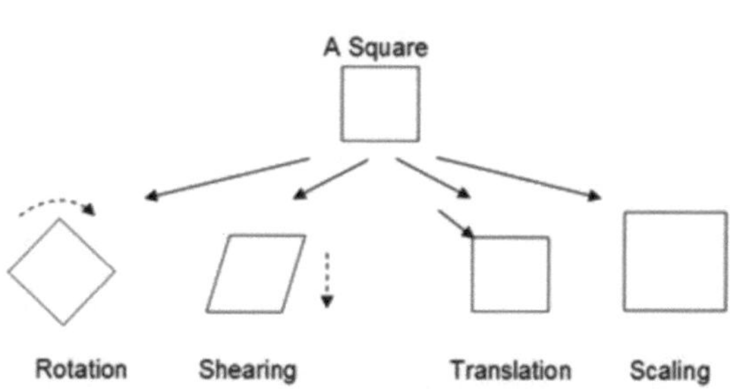

Figure 5. The affine transforms.

$$A_1 = W(A_0);\ A_2 = W(A_1);\ \ldots;\ A_{k+1} = W(A_k) \tag{3}$$

Iterative system IFS generates sequences that converge to the final image A_∞, and it is that the

$$W(A_\infty) = A_\infty \tag{4}$$

The result obtained is called an attractor of the IFS and represents a "fixed point" operator W. For the Koch fractal curve (**Figure 2c**), the affine transformation matrix has the form

$$w_q\begin{pmatrix} x \\ y \end{pmatrix} = \begin{pmatrix} \delta_{q1}\cos\theta_{q1} & -\delta_{q2}\sin\theta_{q2} \\ \delta_{q1}\sin\theta_{q1} & \delta_{q2}\cos\theta_{q2} \end{pmatrix}\begin{pmatrix} x \\ y \end{pmatrix} + \begin{pmatrix} t_{q1} \\ t_{q2} \end{pmatrix} \tag{5}$$

where the scaling factor is expressed as $\delta_q = (2 + 2\cos\theta_q)^{-1}$, θ_{qi} is the inclination angle of the second subsection with respect to the first and t_{qi} is an element displacement on the plane.

Figure 6 illustrates the iterated function system procedure for generating the well-known Koch fractal curve.

In this case, the initial set, A_0, is the line interval of unit length, i.e., $A_0 = \{x : x \in [0, 1]\}$; $\theta_q = 60^0$, and $\delta_q = \frac{1}{3}$.

The results of the four linear transformations are combined together to form the first iteration of the Koch's curve, described by A_1. The second iteration, A_2, of the Koch's curve can be obtained by applying to the A_1 same four affine transformations.

The Koch's curves of higher order are generated by repeating the iterative process, until you reach the desired resolution. The first four iterations of the Koch curve are shown in **Figure 6c**. **Table 1** shows the collection of the basic fractal structures, with particular importance for the construction of antennas for the low-frequency bands, multi-frequency operated as well as for the construction of metamaterials.

1.4.2. Self-affine sets

Self-affine sets form an important class of sets, which include self-similar sets as a particular case. An affine transformation $\boldsymbol{S} : \mathfrak{R}^{\boldsymbol{n}} \to \mathfrak{R}^{\boldsymbol{n}}$ is as follow [15]:

$$w(x, y) = T(x, y) + t \tag{6}$$

where T is the affine transformation on $\mathfrak{R}^n$, which can be represented in (n x n)-matrix notation, and **t** is the vector in space $\mathfrak{R}^n$.

As you can see that the affine transformation is a combination of translation, rotation, dilation and reflection (**Figure 4b**). In particular, w could convert the sphere to an ellipsoid, squares to parallelograms, etc. Unlike the similarity, affine transformations are accomplished with different coefficients in different directions.

If $w_1, \ldots, w_m$ are self-affine construction on $\mathfrak{R}^n$, so it is a unique, unchanging set of F for the w_i is referred to as the self-affine set of equations. In **Figure 5**, w_1, w_2 and w_3 are defined as

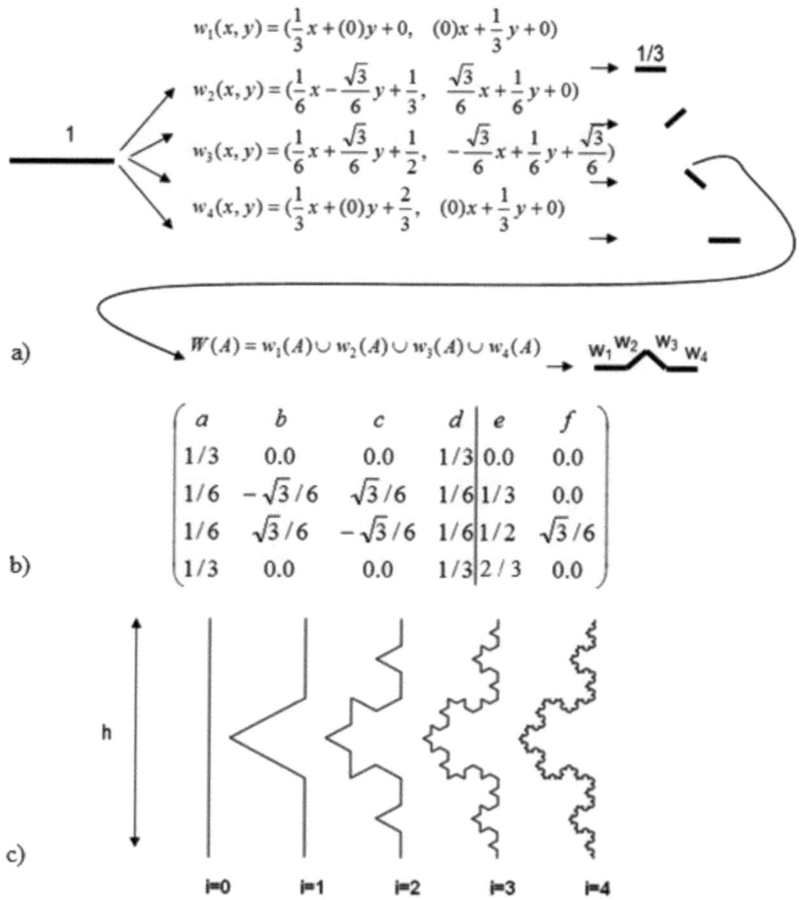

Figure 6. The IFS (a), the affine transformation matrix (b) and the first four iteration steps of construction (c) of the Koch fractal curve.

transformations that map the section into three equal sections, obviously. It is natural, therefore, to search for the description for size self-affine transformations, which will serve as the generalization of the formula for systems and geometries with similar affine transformations.

We hope that the dimension depends on the formula of affine transformation set in a relatively simple way, easily presented to the matrix and/or the vectors that represent transformation. Unfortunately, the situation is more complicated. If the affine transform change continuously, the dimensions of the set of self-affine do not need to change continuously. Such intermittent behaviour becomes more difficult for the more complex affine transformation; therefore, it may be difficult to obtain a general expression in the form of affine sets.

1.4.3. The fractal dimension

In general, we can imagine objects that have a zero dimension, 0D (points), 1D (lines), 2D (plane) and 3D (solid). We feel comfortable with the objects of such dimensions. We create a 3D

Fractal type	Similarity dimension	Iterated function system	Sketch of iterated structure
Koch curve	1.26186	$w_1(x,y) = \left[\frac{1}{3}x;\ \frac{1}{3}y\right]$ $w_2(x,y) = \left[\frac{1}{6}x - \frac{\sqrt{3}}{6}y + \frac{1}{3};\ \frac{\sqrt{3}}{6}x + \frac{1}{6}y\right]$ $w_3(x,y) = \left[\frac{1}{6}x + \frac{\sqrt{3}}{6}y + \frac{1}{2};\ -\frac{\sqrt{3}}{6}x + \frac{1}{6}y + \frac{\sqrt{3}}{6}\right]$ $w_4(x,y) = \left[\frac{1}{3}x + \frac{2}{3};\ \frac{1}{3}y\right]$	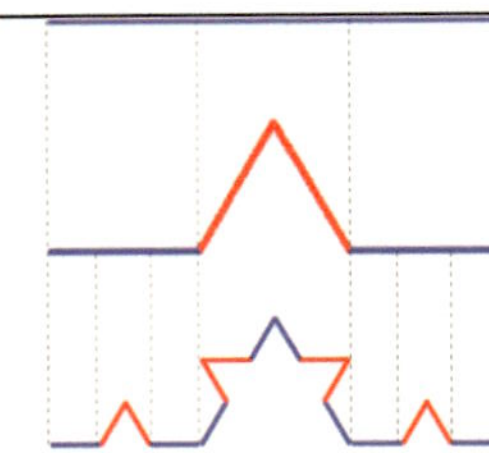
Koch snowflake	2	$w_1(x,y) = \left[\frac{1}{2}x - \frac{\sqrt{3}}{6}y;\ \frac{\sqrt{3}}{6}x + \frac{1}{2}y\right]$ $w_2(x,y) = \left[\frac{1}{3}x + \frac{1}{\sqrt{3}};\ \frac{1}{3}y + \frac{1}{3}\right]$ $w_3(x,y) = \left[\frac{1}{3}x;\ \frac{1}{3}y + \frac{2}{3}\right];\ w_4(x,y) = \left[\frac{1}{3}x - \frac{1}{\sqrt{3}};\ \frac{1}{3}y + \frac{1}{3}\right]$ $w_5(x,y) = \left[\frac{1}{3}x - \frac{1}{\sqrt{3}};\ \frac{1}{3}y - \frac{1}{3}\right]$ $w_6(x,y) = \left[\frac{1}{3}x;\ \frac{1}{3}y - \frac{2}{3}\right]$ $w_7(x,y) = \left[\frac{1}{3}x + \frac{1}{\sqrt{3}};\ \frac{1}{3}y - \frac{1}{3}\right]$	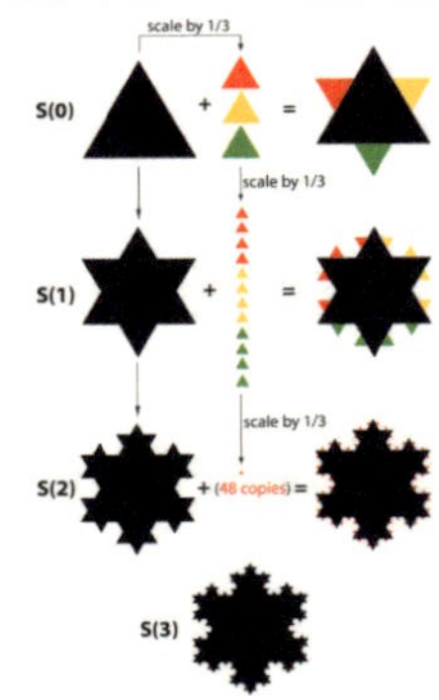
Minkowski curve	1.5	$w_1(x,y) = \left[\frac{1}{4}x;\ \frac{1}{4}y\right];\ w_2(x,y) = \left[-\frac{1}{4}y + \frac{1}{4};\ \frac{1}{4}x\right]$ $w_3(x,y) = \left[\frac{1}{4}x + \frac{1}{4};\ \frac{1}{4}y + \frac{1}{4}\right]$ $w_4(x,y) = \left[-\frac{1}{4}y + \frac{1}{2};\ \frac{1}{4}x + \frac{1}{4}\right]$ $w_5(x,y) = \left[-\frac{1}{4}y + \frac{1}{2};\ \frac{1}{4}x\right];\ w_6(x,y) = \left[\frac{1}{4}x + \frac{1}{2};\ \frac{1}{4}y - \frac{1}{4}\right]$ $w_7(x,y) = \left[-\frac{1}{4}y + \frac{3}{4};\ \frac{1}{4}x - \frac{1}{4}\right];\ w_8(x,y) = \left[\frac{1}{4}x + \frac{3}{4};\ \frac{1}{4}y\right]$	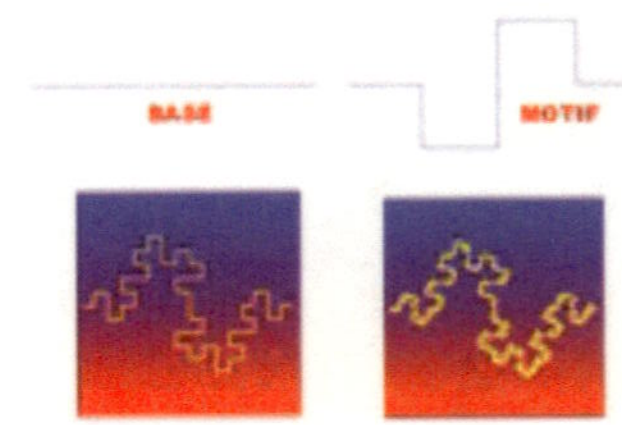
Hilbert curve	1.2619	$w_1(x,y) = \left[\frac{1}{2}y - \frac{1}{2};\ -\frac{1}{2}x - \frac{1}{2}\right]$ $w_2(x,y) = \left[\frac{1}{2}x - \frac{1}{2};\ \frac{1}{2}y + \frac{1}{2}\right]$ $w_3(x,y) = \left[\frac{1}{2}x + \frac{1}{2};\ \frac{1}{2}y + \frac{1}{2}\right];\ w_4(x,y) = \left[-\frac{1}{2}y - \frac{1}{2};\ \frac{1}{2}x - \frac{1}{2}\right]$	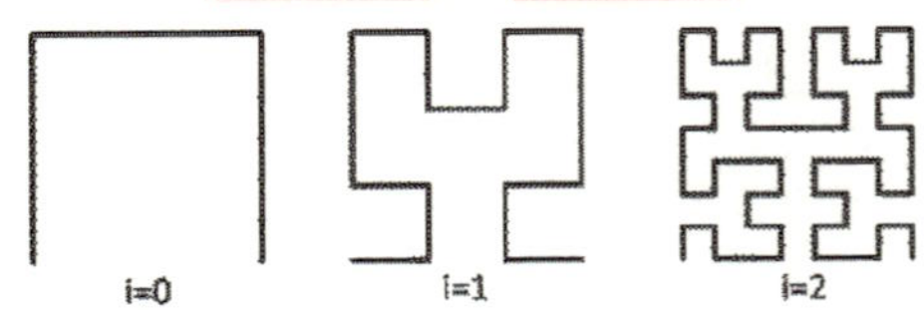

Fractal type	Similarity dimension	Iterated function system	Sketch of iterated structure
Sierpinski gasket	1.58496	$w_1(x,y) = \left[\frac{1}{2}x; \frac{1}{2}y\right]$; $w_2(x,y) = \left[\frac{1}{2}x + \frac{1}{2}; \frac{1}{2}y\right]$ $w_3(x,y) = \left[\frac{1}{2}x + \frac{1}{4}; \frac{1}{2}y + \frac{\sqrt{3}}{4}\right]$	
Sierpinski carpet	1.89279	$w_1(x,y) = \left[\frac{1}{3}x; \frac{1}{3}y\right]$; $w_2(x,y) = \left[\frac{1}{3}x; \frac{1}{3}y + \frac{1}{3}\right]$ $w_3(x,y) = \left[\frac{1}{3}x; \frac{1}{3}y + \frac{2}{3}\right]$; $w_4(x,y) = \left[\frac{1}{3}x + \frac{1}{3}; \frac{1}{3}y\right]$ $w_5(x,y) = \left[\frac{1}{3}x + \frac{1}{3}; \frac{1}{3}y + \frac{2}{3}\right]$ $w_6(x,y) = \left[\frac{1}{3}x + \frac{2}{3}; \frac{1}{3}y\right]$; $w_7(x,y) = \left[\frac{1}{3}x + \frac{2}{3}; \frac{1}{3}y + \frac{1}{3}\right]$ $w_8(x,y) = \left[\frac{1}{3}x + \frac{2}{3}; \frac{1}{3}y + \frac{2}{3}\right]$	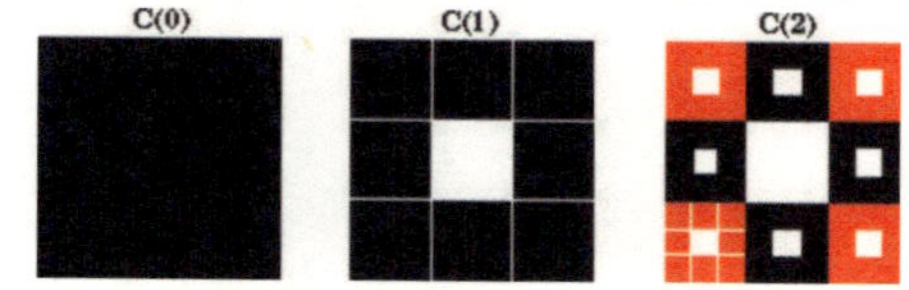
Sierpinski pentagon	1.67228	$w_1(x,y) = [0.382x; 0.382y]$ $w_2(x,y) = [0.382x + 0.618; 0.382y]$ $w_3(x,y) = [0.382x + 0.809; 0.382y + 0.588]$ $w_4(x,y) = [0.382x + 0.309; 0.382y + 0.951]$ $w_5(x,y) = [0.382y - 0.191; 0.382y + 0.588]$	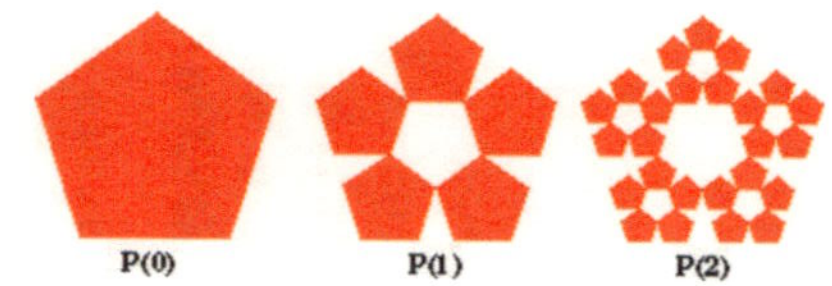
Fractal Tree	1.5849	$w_1(x,y) = \left[\frac{1}{3}x; \frac{1}{3}y + 2\right]$; $w_2(x,y) = \left[\frac{1}{3}x - \frac{4}{3}; \frac{1}{3}y + 2\right]$ $w_3(x,y) = \left[\frac{1}{3}x + \frac{4}{3}; \frac{1}{3}y + 2\right]$	
Cantor Set	0.63092	$w_1(x,y) = \left[\frac{1}{3}x; \frac{1}{3}y - W\right]$ $w_2(x,y) = \left[\frac{1}{3}x + \frac{2}{3}; \frac{1}{3}y - W\right]$	

Table 1. Collection of the fractal structures useful in construction of antenna and metamaterials.

image of our world by combining 2D images from each of our eyes. You can also imagine multidimensional objects, that is, 4D, 5D, 6D. And, what about objects which have non-integer dimensions, such as 2.12 D, 3.79 D or 36.91232 … (D)? Classical methods of geometry and calculus are not suitable for studying fractals, and that is why we need alternative techniques.

There are many definitions of fractal dimensions [22], many of them are evaluated in Ref. [15] that include: similarity dimension, *DS*; the division dimension, *DD*; the Hausdorff dimension, *DH*; the boxing counting dimension, *DB*; the correlation dimension, *DC*; the information dimension, *DI*; the point wise dimension, *DP*; the averaged point wise dimension, *DA*; and the Lyapunov dimension, *DL*. The last seven dimensions listed are particularly useful in identifying the fractal structures in the form of strange attractors, related to the chaotic dynamics.

So, the main tool for the description of fractal geometry is a dimension, in many forms. In a large simplification, the dimension determines how much space has been filled. This is a measure of how many are exposed irregularities when we look at in a very small scale. The dimension contains a lot of information about the geometric properties of the fractal structure.

In the classification of fractals, one of the most important is the Hausdorff dimension [22]. In fact, Mandelbrot suggested that fractal can be defined as an object that has a Hausdorff dimension, exceeding its topological dimension. A complete mathematical description of Hausdorff dimension is beyond the scope of this text [15]. In addition, the Hausdorff dimension is not particularly useful for an engineer or a scientist wanting to evaluate the fractal object, because it is virtually impossible to designate the actual data.

The Hausdorff dimension, for example, and the box counting dimension can be defined for any sets and you can show that they are equal to the similarity dimension. We will focus on the use of the similarity dimension, *DS*, to characterize the properties of regular fractals. The dimension *DS* is a key parameter for describing the structure of the self-similar fractal and is defined by the segmenting of the volume covering the fractal, on cubes of side δ. We assume that at least one of the cubes will contain the described fractal

$$N(\delta) \sim \delta^{-D} \tag{7}$$

The concept of dimension is closely associated with scaling. Consider lines, surfaces, and solids, divided by self-similar shorter segment, smaller surfaces and little volumes, with a side δ. For simplicity, assume that the length *L*, the area *A* and the volume *V* are equal to unity. First, consider the line, divided into *N* smaller segments, each with a length of δ. In this case δ is the factor scale that means $\delta/L = \delta$ with line of a unity length of *N* equal segments, scaled by $\delta = 1/N$.

Now consider the unit area. If we divide it into *N* segments, each with an area δ^2, because $A = N\delta^2 = 1$, i.e. per unit area consists of *N* identical scaled by $\delta = 1/N$ ^ (1/2). Using similar reasoning, the unit for volume $V = n\,\delta$ ^ 3 = 1, which is the unit of volume consists of *N* identical cubes scaled *by* $\delta = 1/N$ ^ (1/3) self-similar parts. Comparing the above expressions, we see that the exponent δ in each case is a measure of the similarity of the object. We can generally describe it as

$$N\,\delta^{DS} = 1 \tag{8}$$

In logarithmic measure, it leads to an expression,

$$DS = \frac{\log(N)}{\log(\frac{1}{\delta})} \tag{9}$$

Note that here the letter "*S*" denotes the similarity dimension. The above expression has been derived using familiar objects, which have the same integer Euclidean, topological and similarity dimensions, i.e. a straight line, planar surface and solid object, where $DE = DS = DT$. However, Eq. (9) may also be used to produce dimension estimates of fractal objects where DS is non-integer. The DS dimensions of all representative fractals are introduced in **Table 1**.

2. Useful fractal geometry in antenna engineering

After reviewing the properties of fractal geometry, you might want to explain what benefits you can get with such geometry used in the construction of antennas. Fractals are abstract objects, which cannot be implemented physically. You can apply some simplification in approach to an ideal of fractals, which will allow you to use them in the construction of antennas. Such simplified geometries are called pre-fractals or truncated fractals.

Configurations, consisting of many triangles or multilevel meandered structures, that are close to the fractal shapes and have some advantages, which in theory can be obtained only as a result of mathematical abstraction, can be used as antennas. Generally, the term "fractal antenna technology" is used to describe these antenna engineering techniques, which are based on mathematical concepts, enabling the creation of a new generation of antennas, which, even in the mid-1980s, were often regarded as impossible to achieve.

As a result of the works carried out so far, you can point to the benefits of using the fractal antenna technology: (1) self-similarity is useful in the design of the antenna working on multiple frequencies, for example, the Sierpinski triangle, fractal tree and the Cantor set; (2) self-filling reduces the overall dimension and is useful for the design of small antennas, such as monopoles/loops/microstrip patches of Hilbert, Peano, Minkowski, Koch fractal shapes; (3) mass fractals and boundary fractals are used in the construction of antenna of high directivity and low sidelobes level.

Probably the earliest publication in which the term fractal radiating element, fractal antenna for the determination of elements made of crooked wires appeared was released in May 1994 [23].

2.1. Fractal dipole antenna

An interesting study of the space-filling properties of fractal antennas is to investigate fractal dipole antennas. Three types of Koch fractals were compared as dipoles. They are depicted in **Figure 6c** and **Table 1**, for the first stages of growth. The Koch fractal dipole is shown in **Figure 7**.

The output structure for construction of fractal Koch [24] and fractal tree [25] was dipole antenna of height h. As previously calculated (**Table 1**), Koch fractal has a fractal (similarity) size of $DS = 1.2619$. Fractal tree is similar to a real tree, in the sense that at the top each branch is divided into more branches. In the tree, flat (two-dimensional) each of the three branches is

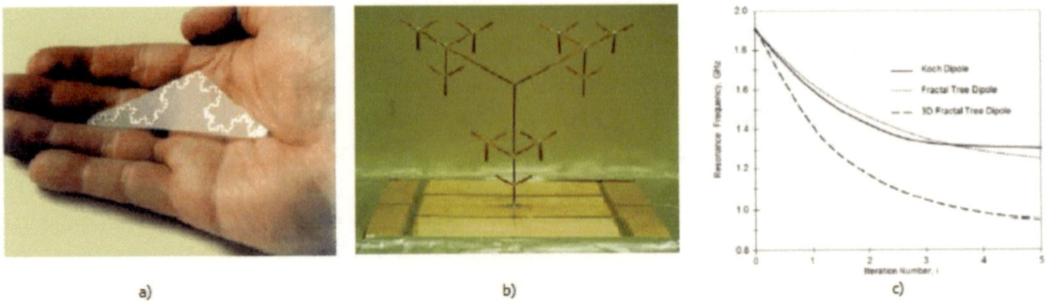

Figure 7. Koch fractal dipole printed on microwave laminate (a), wire fractal tree (b) and resonant frequencies of Koch, two-dimensional and three-dimensional trees fractals versus iteration (c).

divided into two sections. The length of each branch remains the same, in this way, going from the base of the tree to the top of the branches have the same length as the initiator. Fractal dimension (the similarity) is equal to DS = 1.395 for flat tree and 1.5849 for the three-dimensional tree.

Decrease in the resonance frequency has the same effect as the miniaturization of antenna at a constant frequency. The same can be said about the flat antenna structures. The use of non-planar structures, 3-D, increases this effect.

As you can see from the research, the benefits of the application of fractal for miniaturization antennas are already achieved by the first few iterations. By observing the resonance frequency of the antenna, it is also very interesting to look at the quality factor Q of these antennas [15, 26]. It can be observed if due to fact that fractal wire antenna fills the space more effectively than linear dipole, it has lower Q-factor. Quality factor is reduced for greater number of iterations, what can be expected. Any increase in the generation of iteration brings the geometry of fractal away from the linear one-dimensional dipole and closer to the ideal fractal.

As mentioned many times before, an important feature of much fractal geometry is the self-filling. The size of the antenna is a critical parameter, because the electromagnetic properties depend on the antenna dimensions related to the wavelength (λ). In many applications, the amount of space occupied by an antenna is a factor limitations, therefore, may not be comparable to the wavelength, but less, that are called the small antennas. It is said that the antenna is electrically small, when its largest dimension is smaller than the diameter of the sphere of which the radius r is of specified dependency $r = \lambda/2\pi$ [15]. Wheeler and Chu were the first who showed the fundamental limitations of such antennas. Hilbert, Koch, Peano fractals are especially useful in the design of small antennas. The purpose of this section is to present characteristics of miniature microstrip (printed) antennas in the shape of a Hilbert curves and Koch patches.

2.2. Fractal patch antennas

Fractals can be used for miniaturization of surface antenna, as well as linear elements. Applies the same concepts in order to increase the electrical dimensions of flat radiating element as wire were [15–21, 26, 27]. Flat antenna can be seen, as a microstrip line with extended transverse dimensions. Therefore, if the electric current is forced to flow along the meandered fractal path instead of a simple Euclidean path, the area required for the placement of resonant transmission line can be greatly reduced. This technique has been applied to the design of, so-called microstrip antennas of different shapes of patches of fractal geometry, which are fabricated (printed) by means of photo etching technology or in the fast laser prototyping processes.

One of the fractal structures was discovered in 1916 by Polish mathematician Wacław Sierpiński. So called, Sierpinski gasket applied to the construction of the antenna has the properties of multi-frequency operations, due to the fact that it is made up of similar triangles, in each of the roof in the heights scaled with the factor δ. This design was first proposed by Puente et al. [28] (**Figure 8**).

From the point of view of antenna engineering, interpretation shown in **Figure 8a** is such that the dark areas of the triangle represent the conductive metallic surfaces, white areas of the triangle represent surfaces where metal has been removed. With few exceptions (notably log-periodic aerials), typically we use a single antenna for our system (frequency) as shown in **Figure 8b**. So, in this case, we have three antennas in one compact configuration. The geometry of fractal antenna in the form of a Sierpinski gasket is completely determined by four parameters: the height h of the triangle, the countersink angle α, the number of iterations I, and by the

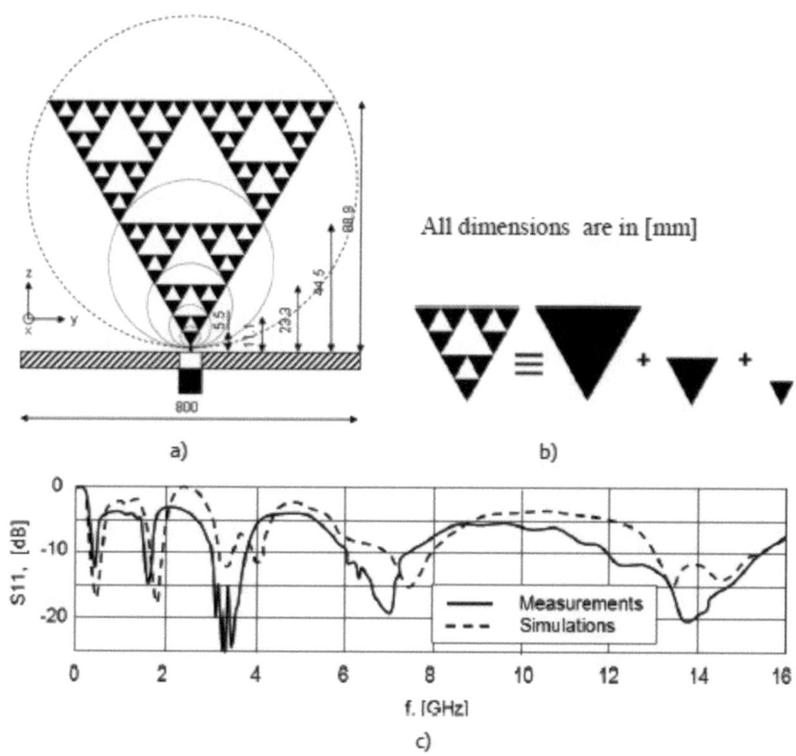

Figure 8. Sierpinski fractal gasket triangle of fifth-iteration as monopole antenna (a), second-iteration Sierpinski triangle as equivalent antenna to an array of three triangles (b) and input reflection coefficient of antenna (a) versus frequency (c).

scaling factor ($\delta = h_i/h_{i+1}$, where h_i is the height of triangle). As it was described in Ref. [20], the Sierpinski monopole behaves like the antenna of logarithmic-periodic geometry [16, 19, 29] (**Figure 8c**), with the each next resonant frequencies separated by the distant relative but reversed to self-similarity scale factor $\delta = 2$. The antenna has similar parameters for the next resonant frequency, with a moderate bandwidth $BW = 21\%$ [15–17, 28, 29, 48].

$$f_{ri} = 0.3\cos\left(\frac{\alpha}{2}\right)\sqrt{\frac{2.5}{\varepsilon_{reff}}\frac{c}{h}}\delta^{i} \tag{10}$$

where c is the speed of light in vacuum, h is the height of the largest gasket, δ is the log period scaling factor, i is a natural number of iterations, and $\varepsilon_{reff}\, r$ is the effective permittivity of the antenna substrate.

Figure 9 shows the new flat IFA antenna (inverted-F-antenna) made in the form of the Hilbert fractal curve for use in handset of mobile cellular phone GSM 900/1800 system. Such geometry of the antenna effectively increases the length of the path for the electrical current comparably to a simple antenna IFA, occupying the same surface. Such solution allows significant miniaturization of the antenna. To support the two frequency bands of GSM, two antenna elements are designed in the shape of a fractal Hilbert.

Antenna, consisting of two elements in the form of a Hilbert pre-fractal, as shown in **Figure 9**, is designed for GSM 900/1800 mobile system handset [9]. The smaller element, near the power point, works on a higher frequency of 1.8 GHz, while the bigger one at $f = 900$ MHz. Each element has been built in the form of a Hilbert curve of second iteration $i = 2$. The antenna was fabricated on one-side metalized microwave laminate of type DUROID-5880 ($h_s = 0.125$ mm, $\varepsilon_{rr} = 2.2$, $tan\delta = 0.009$) and placed on the foam, Rochacell 51

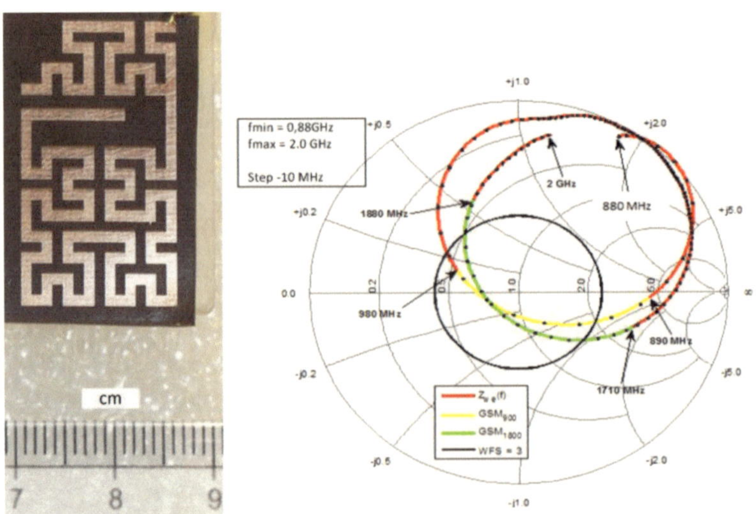

Figure 9. Photo of miniaturized Hilbert fractal PIFA antenna (a) and input impedance plot on Smith Chart (b).

IG/A (h_f = 9 mm, ε_{rr} = 1.071, $tan\delta$ = 0.0031). This design is about 50% (~4.3 cm^3) of the volume occupied by the conventional microstrip planar inverted-F, PIFA antenna.

Another microstrip patch antenna with edges in the shape of a Koch pre-fractal [30] is shown in **Figure 10**. It was designed for the handset terminal, smartphone, cooperating with several mobile communication systems: GSM1800, UMTS and HiperLAN2. Application of PIFA antenna in conjunction with the fractal geometry reduces antenna size by 62% compared to the conventional PIFA.

Two identical PIFA antennas, in the form of Koch pre-fractal (as former one), are placed above the screen of finite sizes (100 × 45 mm^2—the typical size of a smartphone) have been applied for MIMO (Multiple Input Multiple Output) system.

The Sierpinski fractal carpet (**Figure 10b**) was used by the Spanish company FRACTUS, to build the build-in antenna of a mobile cellular system GSM 900/1800 handset.

a)

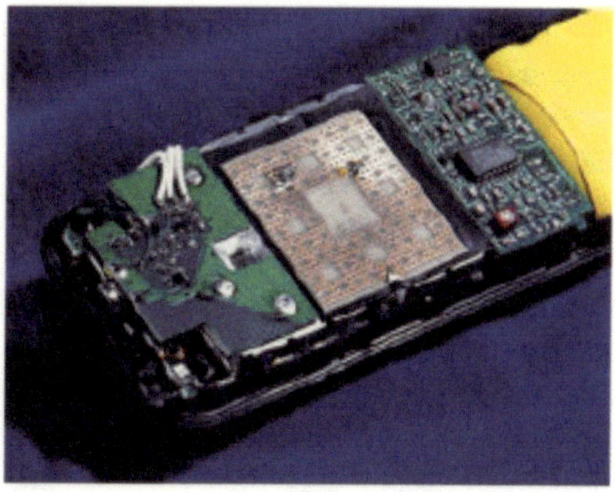

b)

Figure 10. Photos of the microstrip patches of Koch fractal double-PIFA (a) and Sierpinski carpet (b) antennas.

3. Metamaterials in electromagnetics applications

3.1. Introduction

Basic research and applications of electromagnetism have undergone different phases of development, keeping pace with the general socio-technology growth [3]. Currently, there is a huge demand for small antennas for mobile communication systems [31]. In the microwave frequency range, It is tried to solve the problem by the use of artificial dielectrics and or magnetic materials, so-called metamaterials, MTMs. They have been used in place of traditional one as surrounding environment, e.g. the substrates or ground planes to build antennas, filters and other microwave devices. The MTMs are also used, as energy absorbers of electromagnetic waves in order to eliminate the effects of surface currents, which can cause unwanted EM-coupling between nearly situated elements or produce the reflected signals (echoes) in radar applications.

We distinguish two main categories of metamaterials: (1) SNGs are such one, which are characterized by single-negative ($-\varepsilon$ or $-\mu$) and DNG the double-negative ($-\varepsilon$ and $-\mu$) material parameters, built with the cells of small sizes (~λ/10); (2) EBGs are so-called electromagnetic bandgap materials, and FSSs are the frequency selective materials, where the periodic structure is built out of cells of size comparable with the wavelength (typical λ/2).

The MTMs completely revolutionized the construction technologies of small microwave components such as antennas, filters, and others that provide high directivity (antenna) attenuation (filters) in a wide range of frequencies that are comparable to those they had previously used in microwave devices of much larger sizes.

3.2. Electromagnetic environment of negative material parameters

Electromagnetic materials in nature play a major role in operation of the instruments operating on the basis of the phenomena of electricity and magnetism, such as capacitors, inductors, resonant circuits and, in general, the physics of propagation of EM waves. The discovery of artificially made materials with unusual properties such as negative electrical permittivity and/or permeability, causing the occurrence of negative refractive index, negative propagation of EM waves and the inverse Doppler effect [3, 33–37] changed the way of thinking of scientists and engineers.

It is well known that the reaction of the system for the presence of the electromagnetic field depends to a large extent on the properties of the materials used and the parameters of the environment in which they occur. Therefore, the behaviour of the material (environment), in the presence of the electromagnetic field, depends on the macroscopic material parameters: permittivity ε, permeability μ and conductivity σ. **Figure 11** shows the classification of materials on the basis of the parameters ε and μ.

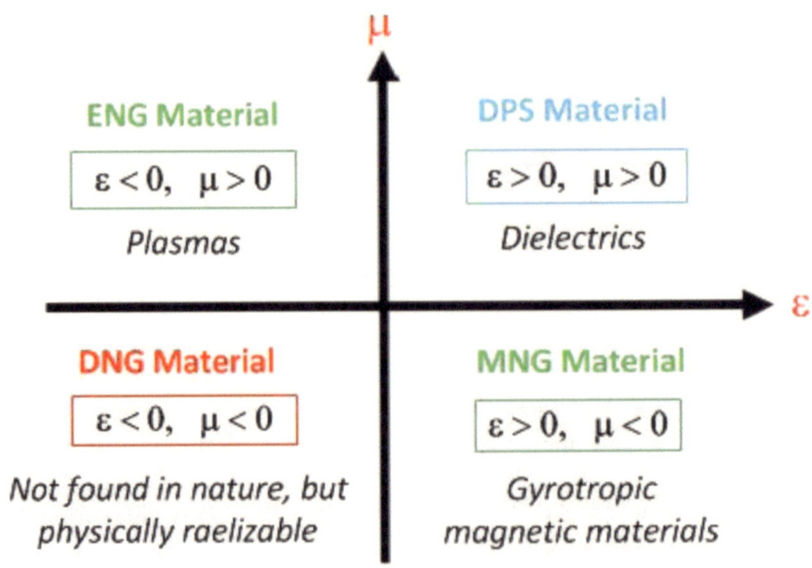

Figure 11. Material classifications.

Concept of metamaterial (MTM) technology appeared in the 1990s, with the work of about macroscopic composites of synthetic cells, periodic over the matter [2]. However, the initial attempts to use "artificial" materials were initiated in the end of nineteenth century in 1898 by Jagadis Chunder Bose carried out by his experiment with coiled structures [38]. Lindman was next in the artificial media explorers, who researched the artificial chiral media in 1914 [39].

In the last 20 years since the end period of twentieth century, an increased interest in MTM technologies has been observed. This involves researches on super lens, applications in telecommunications, including antennas and transmission lines [40], and nowadays in the direction of improving the parameters of microwave devices, which will then satisfy the electromagnetic compatibility of communication systems [41]. As mentioned previously, MTM is the composite of macroscopic cells of periodically or non-periodically distributed all around structure, whose parameters depend on the cells architecture and the chemical composition [42]. MTMs are formed through the introduction of integration (e.g. molecules) in the structure of the material of the dielectric substrates. These inclusions change the electromagnetic properties of dielectric material of interest. In this work, different MTM patterns as fractal cells, and CLL cells, are applied. Almost all bulk MTMs used at the present state of art are based on cell structures that consist of a dense array of thin wires DWA, a ring resonator—RR (Ring Resonator) or CLL (Capacitively Loaded Loop) in SRR (Split Ring resonator) or CSRR (Complementary Split Ring Resonator) configuration topologies. They are oriented according to the wave direction.

V. G. Veselago suggested a novel aspect at this new kind of materials [3]. Theoretical foundation of the occurrence of the "*electromagnetic materials that exhibit both negative values for ε and μ*" was published in 1968 [32]. It describes some of the properties of these materials, such as (1) a negative refractive index, (2) phase reverse, (3) left-handed triad, (4) reversed the effect of the Vavilov-Cerenkov, (5) inverted Doppler effect, (6) anisotropy frequency dispersion and so on. Using them, it means the MTM, to build electromagnetic instruments makes the lens completely flat, and the waves on the border of media with such materials are reflected in the same direction where they come from. Nowadays a number of names for the MTM with negative permittivity and negative permeability have been used such as (1) LH—left-handed media; (2) NRI—a negative refraction index media; (3) the BWM backward wave media; and (4) DNG MTM double negative-MTM.

3.3. Dipoles and loops as the building cells of metamaterial

The structures proposed in Ref. [32] do not exist in nature; therefore, try to create them artificially in the laboratory. The MTMs are created in the form of small cells as inclusions or homogeneities embedded in the host medium. They could also be attached or embedded on the surface of the host substrate. One of the most widely used structures MTM are arrays of densely spaced elementary electrical dipoles, in the form of thin wires [43] (**Figure 12**).

If the operating frequency is lower than the cut-off frequency of an array ("plasma frequency"), the equivalent effective permittivity (1) is negative (ENG). When a lattice constant a is much smaller than a wavelength ($a \ll \lambda$), the wire array can be thought as a continuous plasma like material described by an equivalent macroscopic ε—relative permittivity [4]:

$$\varepsilon_{reff,z} = \varepsilon'_{reff,z} - j\varepsilon''_{reff,z} = 1 - \frac{f_p^2}{f^2 - j\gamma f} \tag{11}$$

where, $\varepsilon_{reff,z}$ denotes the effective relative permittivity in the z-direction, f and f_p represent the signal frequency and the cut-off-frequency of the array, respectively, whereas γ is a factor that represents losses.

Eq. (11) applies only if the propagation takes place in transversal (x-y) plane.

The plasma frequency generally depends on the geometry of the system (a lattice constant, wire radius). The relative ε in the transversal directions (x and y) is always positive and, in a case of thin wires, is approximately equal to that of vacuum ($\varepsilon_{ref\ x,y} \cong 1$). Permittivity is positive above the plasma frequency (EPS). However, if the frequency is only slightly higher than the plasma frequency, the structure will support the propagation of waves with a wavelength much longer than the length of the wave in free space, because of the relative permittivity that takes very small positive values ($\varepsilon_r \ll 1$). In this region, the structure behaves like a low-loss of epsilon-near-zero, ENZ MTM.

Such MTMs give rise to the phenomena of ultra-reflection and spatial filtering, which can be used to increase the directivity of simple antennas. Currently, the MTMs are implemented in the form of an array of metallization properly oriented in space, which is therefore internally homogeneous and microscopic. On the other hand, the size of the metallization and the distance between them are very small compared to the wavelength (electrically small). Relying on the theory of effective media, it is reasonable, therefore, to seek the bulk properties, in this case, the bulk permittivity and permeability that characterize the macroscopic behaviour of the medium.

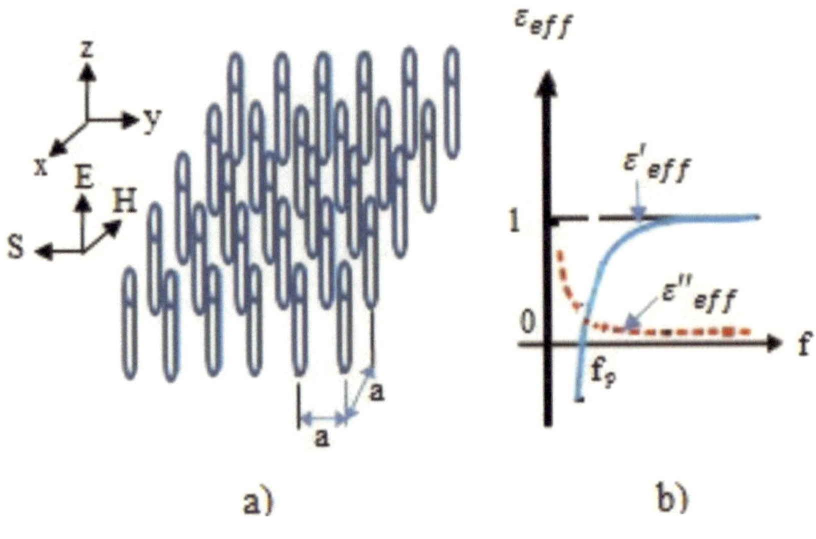

Figure 12. An array of dipoles in the form of thin-wires-based MTM (a) and effective permittivity of such MTM (b).

Since the first theoretical introduction in Ref. [6], an array of SRR inclusions (**Figure 13a**) has been widely used for the synthesis of MNG metamaterials [4].

A SSR can be thought of as a small, CLL-loaded antenna [4]. If the antenna is working just above the resonance frequency, local distributed magnetic field will be almost in the opposite phase relative to the incident field. Therefore, the resulting local magnetic field will be smaller than the incident field. It leads to a negative magnetic polarity and negative effective permeability of the resulting metamaterial. It has been shown that the effective permeability of such metamaterials can be specified on the basis of an analysis of loop antenna system used in the cells that make up the MTM [4].

$$\mu_{reff} = \mu'_{reff} - j\mu''_{reff} = 1 - \frac{f_{mp}^2 - f_0^2}{f^2 - f_0^2 - j\gamma f} \tag{12}$$

where f is the frequency of the signal, f_{mp} denotes the frequency in lossless case when $\mu_{eff} = 0$ ("magnetic plasma frequency"), f_0 is the frequency at which μ_{eff} diverges (the resonant frequency of SRR) and γ represents the losses.

Eq. (12) describes the simplified SRR model that does not take into account the minor electrical polarization [44] and some small bi-anisotropic effects [45], which obviously influences the effective permeability. The SRR is treated as a purely magnetic particle.

Figure 13b shows the dependence of the effective permeability μ_{eff} as a function of frequency. The frequencies of the f_{mp} and f_0 depend on architecture of SSR distribution, as well as its geometric parameters (internal and external radiuses and the width of the slits) [44]).

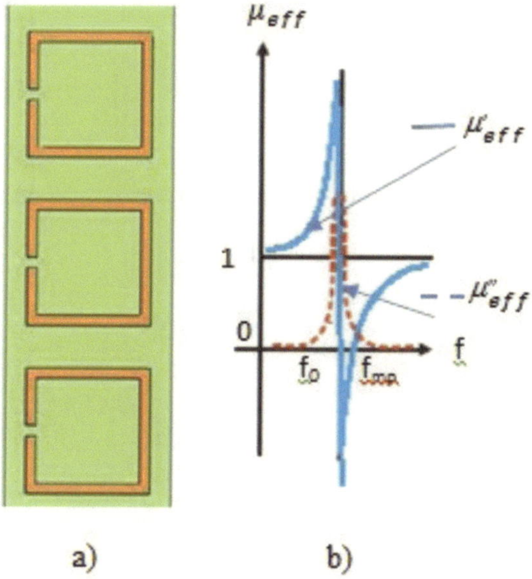

Figure 13. An array of SRRs (a) and effective permeability of array (b).

A general SRR-array-based MNG MTM can be described by a 2 × 2 uniaxial permeability tensor:

$$\overline{\mu} = \mu_0 \begin{bmatrix} \mu_{tr} & 0 \\ 0 & \mu_{lr} \end{bmatrix} = \mu_0(\overline{\mu}_r' - j\overline{\mu}_r'') = \mu_0\overline{\mu}_r \quad (13)$$

where μ_0 is the absolute permeability and μ_{tr} and μ_{lr} are the relative permeability in the transversal (x) and longitudinal (y) directions, respectively.

Current implementations of MTM rely on "infinite" rods and split-ring resonators (SRRs), as shown in **Figure 14**, to achieve a negative permittivity and a negative permeability, respectively.

The MTM, built from the rings (loop antenna) and rods (small electric dipole) from microscopic view, is seen as a homogeneous effective medium, characterized by the material parameters ε and μ. This approach is possible because the sizes of the rings and rods as well as the distances between them are very small compared to the wavelength.

The shapes of the rings, their effective radii, width of metallization and many other factors directly translate into MTM properties: the position of the resonance frequency and plasma frequency that control directly into frequency bandwidth in which appear the negative values of $-\varepsilon$ and/or $-\mu$ [44]. Therefore, the design and optimization of the geometry of the rings have still been a current area of research.

For lossy DNG MTM material, negative permeability and permittivity can be determined by using the Drude model [44] as follows:

$$\varepsilon(\omega) = \varepsilon_0 \left[1 - \frac{\omega_{pe}^2}{\omega(\omega - j\Gamma_e)}\right] \quad (14)$$

$$\mu(\omega) = \mu_0 \left[1 - \frac{\omega_{pm}^2}{\omega(\omega - j\Gamma_m)}\right] \quad (15)$$

where ω_{pe}, ω_{pm} and Γ_e, Γ_m mean plasma frequency and the damping frequency, respectively.

Figure 14. Synthesis of DNG MTM.

Analysis based on the Drude model can be implemented in the FDTD method by introduction of appropriate density of electric current to the equations that describe their temporal behaviour. For modelling of E (M) NG MTM, the dispersion permittivity or permeability is replaced by the frequency-independent constants.

Many research groups in the world have been studying various aspects of this class of MTM, where a number of ideas and suggestions for future applications of these materials have been proposed. Initially, it was to enter the isotropic material parameters, namely scalar permeability (ε) and scalar permeability (μ). The isotropic approximation, however, is valid for single polarization of excitation of MTM containing rings in the two dimensions, such as in the original experiment. Later versions of the MTM were built by using rings in only one direction only, which violates the principle of isotropy and creates a MTM internally anisotropic. The relative parameters of the material, therefore, must be described in the form of tensors, characterizing the model of biaxial media.

3.4. Negative refractive index

The phenomenon of EM-waves refraction with a negative refraction coefficient has been studied taking into account the EM wave scattering incident at any angle to the boundary surface of the DPS-DNG materials, as shown in **Figure 15**.

Enforcing the electromagnetic boundary conditions at the interface, one obtains the law of reflection and Snell's Law from phase matching:

$$\theta_{ref1} = \theta_{ref2} \tag{16}$$

$$\theta_{trans} = sgn(n_2) \sin^{-1}\left(\frac{n_1}{|n_2|} \sin \theta_{inc}\right) \tag{17}$$

It should be noted that if the refractive index of medium is negative, it means the angle of refraction, in accordance with of the Snell's law, should be "negative". This suggests that the refraction is abnormal, and the angle of refraction is on the same side of the normal to the interface as the incident wave angle is (see **Figure 16**).

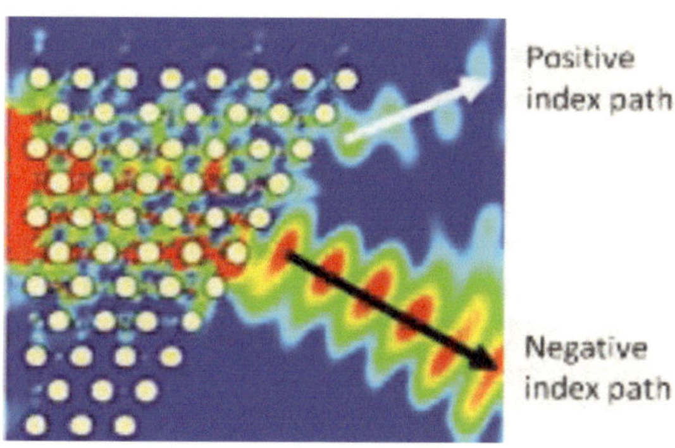

Figure 15. Refraction properties of the resonant MTM.

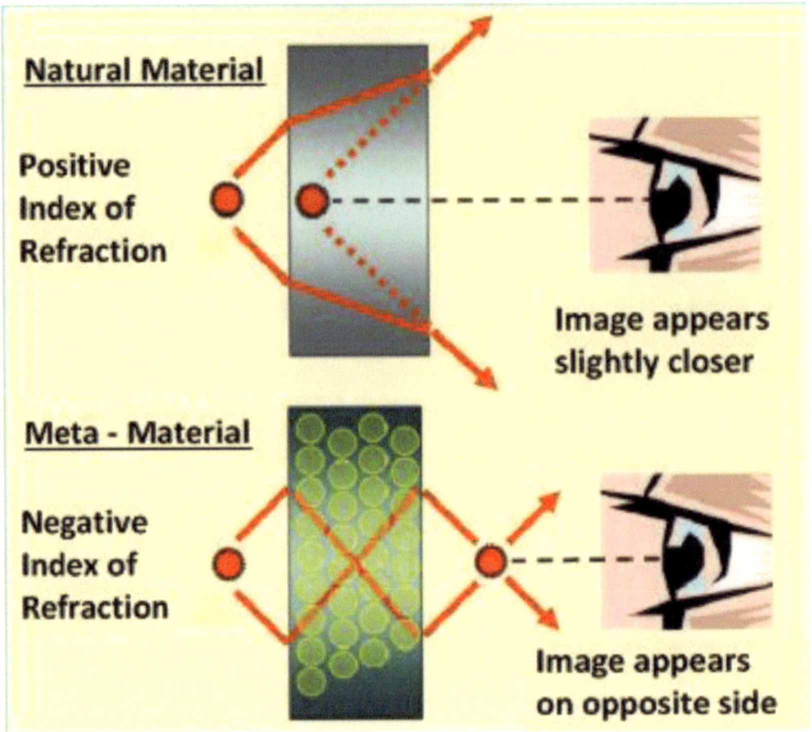

Figure 16. Imaging with DPS and DNG materials.

3.5. Metamaterials of high impedances surface

MTMs with high-impedance surface HIS are also known as artificial magnetic conductors that drew the attention of investigators in the past few years [5]. These surfaces have a reflectivity coefficient $\Gamma \cong +1$ when plane wave illuminate it, in contrast to the typical $\Gamma \cong -1$ in the case of the surfaces of perfectly electrically conductive (PEC). These structures can meet the interesting application in the design of the antenna [46] and the thin absorbing screens.

For example, a dipole antenna located above such HIS MTM has an image current with the same phase as the current on the host dipole, such phenomena increase the level of EM-waves radiation as well as the efficiency of dipole antenna. Several different types of HIS as the modified ground planes have been studied by a number of research groups [46, 47].

Because the magnetic conductive surfaces do not occur naturally, it is necessary to create them artificially to meet the surface conduction properties in selected frequency ranges. This can be achieved by utilizing the resonant properties of inclusions on the non-conducting substrate host layer in parallel with the ground plane. Near resonance frequency on inclusions of strong surface currents is induced and together with the conductive ground plane, this structure behave as an equivalent magnetic conductor PMC also known as HIS, for a frequency range corresponding to the vicinity of a resonance of the surface.

Periodical MTM can be divided into two groups: (1) the three-dimensional volumetric structure and (2) the two-dimensional surfaces. The latter, have such advantages as low profile, low weight and low-cost production. Therefore, such a solution is proposed in mobile wireless communication systems [4].

The artificial surfaces have been testing for many years and so far have developed such as frequency-selective surfaces (FSSs) [9]; artificial soft and hard surfaces LIS and HIS, respectively [10], and micro-machined substrates [4]. Recently, the planar EBG surfaces are proposed [4], which have some specific properties relative to incident waves, respectively when

i. an incident wave is a surface wave ($k_x^2 + k_y^2 \geq k_0^2$, where the k_x and k_y are wavenumbers in horizontal direction, while the k_0 is the wavenumber in a free space), analyses of such structure shows if it has the property of bandgap (EBG) material, it means that a surface wave cannot propagate along it for any incident angle and polarization, and

ii. an incident wave is plane wave ($k_x^2 + k_y^2 < k_0^2$), the reflection coefficient of a material is equal $\Gamma = +1$ for a specific frequency, which resembles to the PMC that does not exist in nature. The simplest example of a textured electromagnetic surface is a metal plate, in which milling out the grooves (corrugations) with a depth of $d = \lambda/4$ [32, 38], is shown in **Figure 17a**.

This structure depending on the polarization and direction of propagation of the EM-wave is described as a soft or hard surface [4]. The action of such surface, relative to EM-fields we can understand treating the corrugations as transmission quarter-wavelength lines, short-circuited at the bottom of each groove, which is transformed in the open-circuit at the top of the surface. This provides a high impedance (hard surface) for electric fields polarized perpendicular to the grooves and low impedance (loft surface) for parallel-polarized electric fields. Soft LIS and hard HIS surfaces are used in various applications, such as modification of radiation patterns of aperture antennas (open-end waveguide, horn, parabolic dish, slot, microstrip, etc.), the elimination of surface currents between the radiating elements of flat antenna arrays

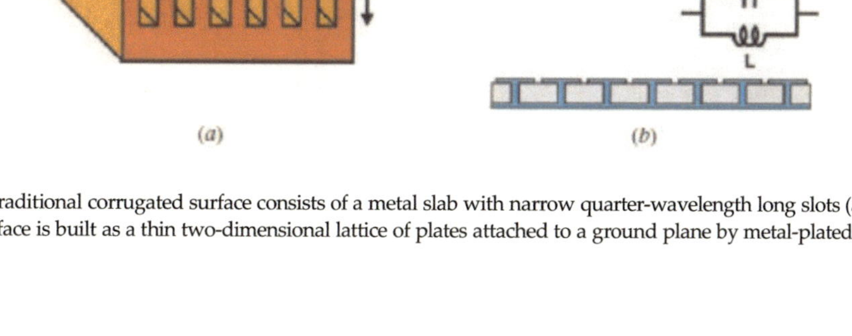

Figure 17. A traditional corrugated surface consists of a metal slab with narrow quarter-wavelength long slots (a) and a high-impedance surface is built as a thin two-dimensional lattice of plates attached to a ground plane by metal-plated vias (b).

(minimization of mutual couplings) or modelling of parameters of the edges on which the diffraction of EM-waves occur.

More compact metal constructions, with a thickness much less than $\lambda/4$, that change the boundary conditions for the EM-wave, were proposed in Ref. [4]. They are built of metal in the shape of mushrooms (flat metal square over short-pin), about the size of a much smaller wavelength, as shown in **Figure 17b**. Each of these "mushrooms" can be modelled as an LC-resonant circuit. The structure has reduced thickness that is achieved by means of capacitive load. These materials provide high-impedance boundary conditions for both polarization and all directions of propagation of waves, and act as hard surfaces. EM waves reflecting from such surface do not change the phase, $\varphi = 0$, as opposed to reflection from the perfect conductor, for which $\varphi = \pi$.

It is sometimes known as artificial magnetic conductor, because tangential component of magnetic field H is equal to zero on the surface, such as the E-field tangential component on the surface of a perfect conductor (metal). In addition to the unusual behaviour of phase of the reflected EM-wave, these materials have a bandgap of the surface wave in which they prevent the occurrence of TM and/or TE-surface waves. Hence, it can be treated as a kind of bandgap structure for electromagnetic waves (EBG) or photonic crystal for surface waves [4]. Although the surface waves cannot propagate in such structures, leaky TE-modes of EM-wave can propagate within the bandgap, which can be useful in some applications.

One of the possibilities to form inclusions that are resonant but have an electrically small footprint at their resonant frequency is the use of the space-filling curves of fractal geometry.

4. Metamaterials based on fractal geometries

At the beginning of the twenty-first century appeared the new MTM configurations, based on fractal geometry. This allows for a significant reduction in the size of individual cells, which are built of artificial materials, which, in turn, translates into higher homogeneity of the material and the reductions in the profile of artificial ground [1].

Many shapes have been proposed for artificial magnetic materials [6]. **Figure 18** shows some of these topologies.

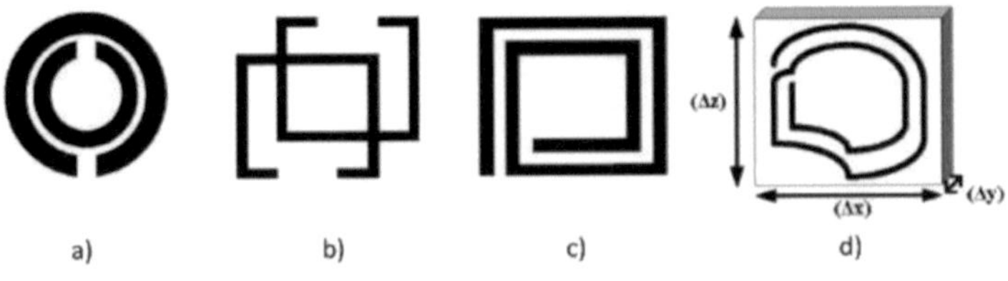

a) b) c) d)

Figure 18. Some particles used in magnetic MTM: SRR (a), meta solenoid (b), split spiral (c), and ring with arbitrary shape (d).

New structures, despite of the other shapes, provide unchanged mechanism of its operation. When the resonant-cells are immersed in the media, where there an external magnetic field H_{ext} exists, the changes in magnetic flux density, covered by these inclusions in the shape of the rings, induces in them the electromotive force (*emf*). Voltages of *emf* cause the current flow on the ring surface, which, in turn, induces a dipole magnetic moment. In turn, the dipole magnetic moment is the source of the effective permeability. We will try to determine the magnetic moments of any shape cell-inclusions (see **Figure 18d**). The electromagnetic properties of the proposed structures were examined in detail and compared to a property of spiral-shaped cells.

4.1. MTM based on Hilbert fractal curves

In this chapter, we introduce combination of the square spiral loop configuration with fractal Hilbert curves as inclusions for artificial magnetic material to further increase the miniaturization potential (**Figure 19**).

It is shown that by using third-order fractal Hilbert curves, inclusions as small as $\lambda/55$ (**Figure 19c**) (λ the wavelength in the host dielectric at the resonance frequency) can be realized.

To analytically predict the behaviour of Hilbert curve inclusions, the equivalent circuit model parameters are provided in Ref. [6]. The unit cell shown in **Figure 19c** has dimension (as in **Figure 18d**) has dimension Δx, Δy and Δz, in the x, y and z directions, respectively. The resultant permeability is described by following expression [6]:

$$\mu_{rf} = 1 - \frac{S}{\Delta z \Delta x} \frac{j\omega L_{eff}}{R_{eff} - \frac{1}{\omega C_{eff}} + j\omega L_{eff}} \tag{18}$$

where ω is the external field frequency, S is the surface enclosed by inclusion, and R_{eff}, L_{eff} and C_{eff} are equivalent circuit model components, derived in [6] for the proposed Hilbert curve inclusions.

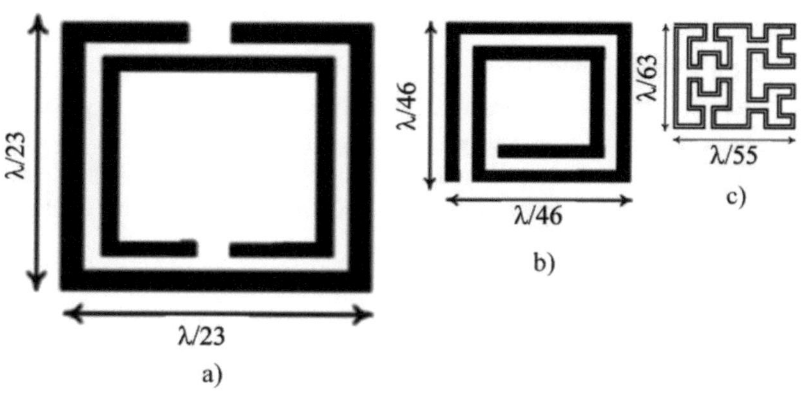

Figure 19. A unit cell of inclusion with SRR (a), square spiral (b) and third order Hilbert fractal (c), respectively.

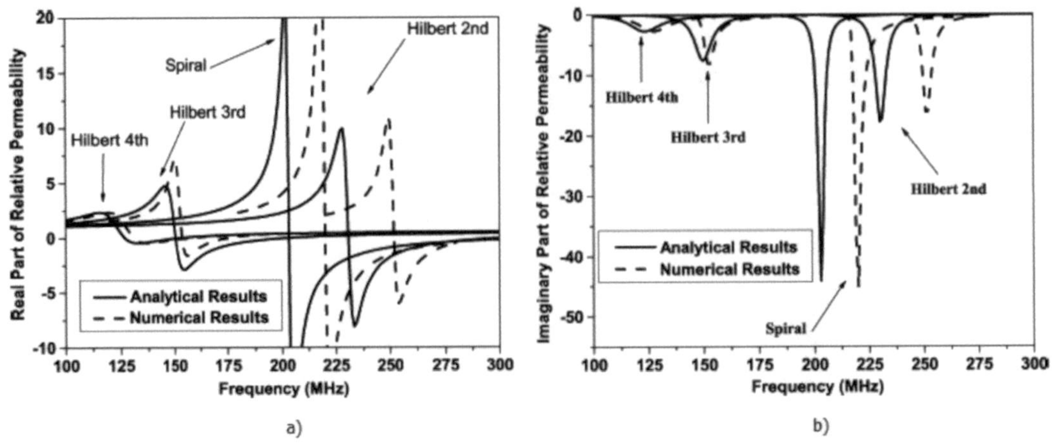

Figure 20. Real (a) and imaginary (b) part of permeability of MTM with Hilbert fractal curve inclusion.

The relative permeability of different order Hilbert fractal curve unit cells was derived in Ref. [6] and shown in **Figure 20**.

Figure 20 shows of the permeability plots versus frequency of single cell of MTM structure in the form of the Hilbert fractal curve (**Figure 19c**). They are quite similar to the characteristics of the MTM cell in the form of a square spiral (**Figure 19b**), which was used before fractals [6]. It should be noted that the external dimensions of fractal cells ($\lambda/55$) are much smaller than the dimensions of the split spiral cell ($\lambda/46$), having the same copper strip width w and separation s between them ($s = w = 0.127$ mm), as well as the host dielectric substrate ($\varepsilon_r = 3.38$, $\tan\delta = 0.0027$). For higher iterations of the Hilbert fractal, the differences become greater.

In addition, as shown in **Figure 20**, when the number of iterations of the Hilbert fractal is increased, the permeability plot becomes smoother. Therefore, the use of fractal structures causes decrease in changes of the permeability depending on the frequency, which, in turn, reduces the coefficient of dispersion in an artificial medium. For frequencies below the resonance, the imaginary part of the permeability takes larger values for fractal Hilbert inclusions than for rectangular spiral cells. Moreover, the Hilbert cells with higher iterations have higher values for the imaginary component μ of the permeability.

Therefore, at frequencies below resonance, the Hilbert inclusions bring greater losses compared to the spiral cells. It has been shown that using four-order fractal Hilbert, we can build a very small MTM cells, with dimensions of the order of 0.014λ in the dielectric substrate. This size corresponds to 63% of the size of the spiral cell and 32% of the size of the SRR cell (see **Figure 19**).

4.2. MTM composed of fractal tree sphere

Properties of an artificial dielectric MTM depend primarily on the type of inclusions, as well as their distribution in the host environment. Some types of cell geometry considered in the past to build for artificial dielectric materials include small dipoles and loops. In this chapter, we will consider the possibility of using for this purpose the fractal tree, which could approximate the sphere [7]. It will be a sort of a combination of aspects of both the geometry of the dipole and the sphere.

MTMs will be considered here that are composed of a new type of molecule called fractal tree sphere. A fractal sphere is collection of symmetric self-similar fractal tree dipoles arranged to form a sphere-like structure as illustrated in **Figure 21**.

The effective dielectric constant of artificial dielectric MTM is evaluated from effective electric and magnetic polarizabilities, respectively:

$$\boldsymbol{p} = \varepsilon(\bar{\bar{a}}_{ee}\boldsymbol{E} + \bar{\bar{a}}_{em}\eta\boldsymbol{H}) \tag{19}$$

$$\boldsymbol{m} = \frac{\bar{\bar{a}}_{me}}{\eta}\boldsymbol{E} + \bar{\bar{a}}_{mm}\boldsymbol{H} \tag{20}$$

The effective dielectric constant of fractal sphere MTM is evaluated as follows:

$$\varepsilon_{eff} = \varepsilon\frac{(1 + 0.66\ Na_{ee})\ (1 - 0.33\ Na_{mm}) - 0.22N^2a_{em}^2}{(1 - 0.33\ Na_{ee})\ (1 - 0.33\ Na_{mm}) * 0.11N^2a_{em}^2} \tag{21}$$

where N is the inclusion volume density.

The real and imaginary parts of the effective constant for the three iterations of the fractal sphere are shown in **Figure 22**.

The plots of an effective dielectric constant for an artificial dielectric MTM consisting of conventional dipole inclusions are inserted in **Figure 22a, b** for comparison purposes. In this case, maximum number $N_{\max}$ of inclusions that can be placed in a cubic meter volume without intersection was considered. It could be seen that each, successive iteration of the fractal tree sphere molecules produces a characteristic downward shift in the corresponding peak value of the ε_{eff}. Also, it is seen that a medium composed of third iteration fractal tree spheres behaves as a dual-band artificial dielectric MTM. It has been shown that fractal

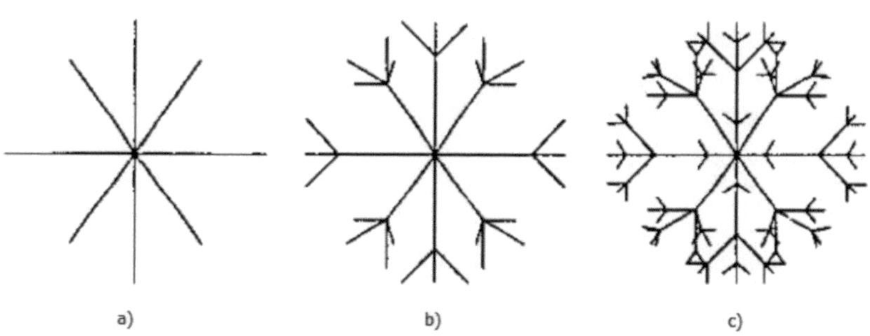

a) b) c)

Figure 21. Fractal trees of first (a), second (b), and third iteration (c), approximating the spheres, as viewed along any axes.

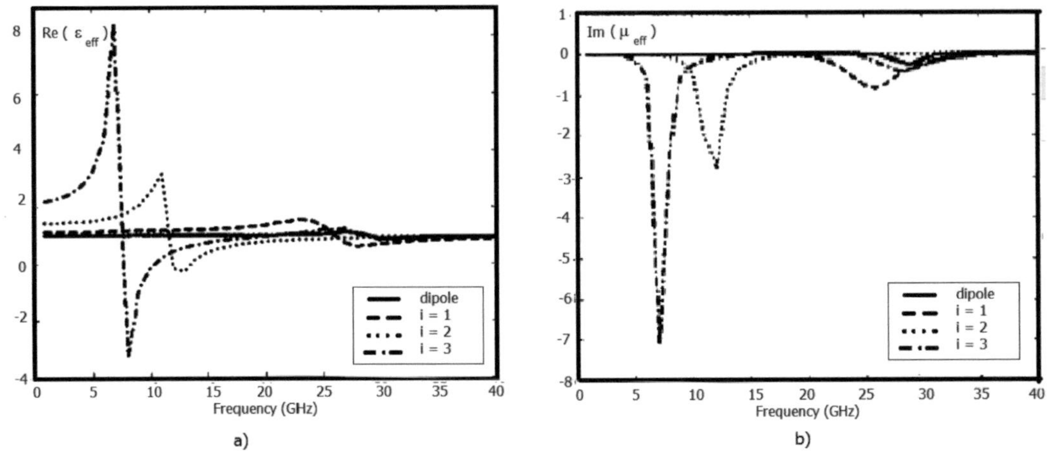

Figure 22. Real (a) and imaginary (b) part of effective dielectric constant of the fractal tree sphere.

spheres have a lower resonant frequency than those solid spheres of the same physical size. These properties have been exploited to develop a new class of low-frequency multiband MTM.

4.3. Fractal self-space-filling curve high-impedance ground planes

Because the magnetic conductive surfaces do not occur naturally, as mentioned earlier, it is necessary to artificially create such a surface that meet the requirements in the specified frequency range [4]. This can be achieved using the resonant properties of inclusions in non-conducting layer of the substrate in parallel with a conductive plane host substrate. Near resonance of the inclusions, the strong currents are induced on the surface of the material, which causes the material to behave as an equivalent magnetic conductor. One of the possible inclusions, which possess the resonant properties and are electrically small ($\ll\lambda$) is the use of self-filling fractal curves.

The self-filling of the surfaces curves are, in general, the continuous maps of the area by an interval [0, 1] on standardised square [0, 1] × [0, 1]. In the year 1890, Peano first proposed the self-filling curve, now called the Peano curve [12]. In 1891, David Hilbert introduced its version of the same—filling curve [4]. The curves are iterative, both fill the entire surface, when the number of iterations (so called curve order) tends to infinity; this means that the curves pass through each points on the surface of the square. While both curves pass through each point on the surface for $i = \infty$, it realizes this in a quite different way, as you can see for the first three iterations of the Peano curves and Hilbert, shown in **Figure 23**.

These curves offer certain attractive properties; that is, a structure of this shape can be made from an electrically long metallic wire compacted within a very small footprint area. As the iteration order of the curve increases, a space-filling curve may maintain its footprint size while its length increases. Moreover, these space-filling geometries can be planar structures, thus allowing for ease of fabrication.

The total lengths of these Peano, D_P, and Hilbert, D_H, curves as a function of the iteration order number i, are given by Eqs. (22a) and (22b), respectively.

$$D_P = L\frac{3^{2i} - 1}{3^i - 1} \tag{22a}$$

$$D_H = L\frac{2^{2i} - 1}{2^i - 1} \tag{22b}$$

where L is a linear side length of the curve.

As you can see, the Peano curve has a higher compression ratio (e.g. greater total length) than Hilbert curve for a fixed order. The space-filling curve elements can resonate at frequencies where the curve side dimension can be considered as electrically very small. The higher order curve has a lower resonant frequency; thus, it takes up less surface area in relation to the resonant wavelength. The cost of the implementation of the compact resonant is seen in narrower bandwidth.

This effect has also been observed in the design of electrically small antenna based on Peano or Hilbert curves [4]. The element in the form of Peano curve with the same surface and order of iteration as an element of a Hilbert curve has a much lower resonance frequency, at the expense of a much narrower bandwidth, due to the higher compression ratio of Peano curve.

To construct a high impedance surface, HIS, the Peano curve elements can be placed in a planar, 2D array as shown in **Figure 24**.

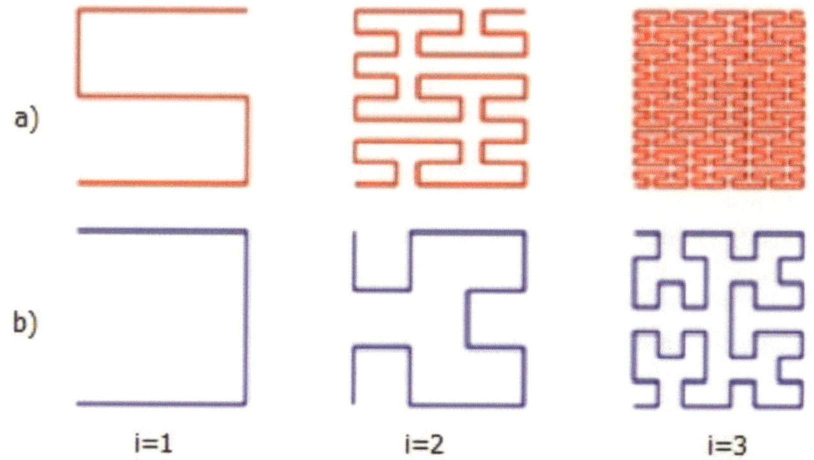

Figure 23. Peano (a) and Hilbert (b) fractal space-filling curves for first three iterations.

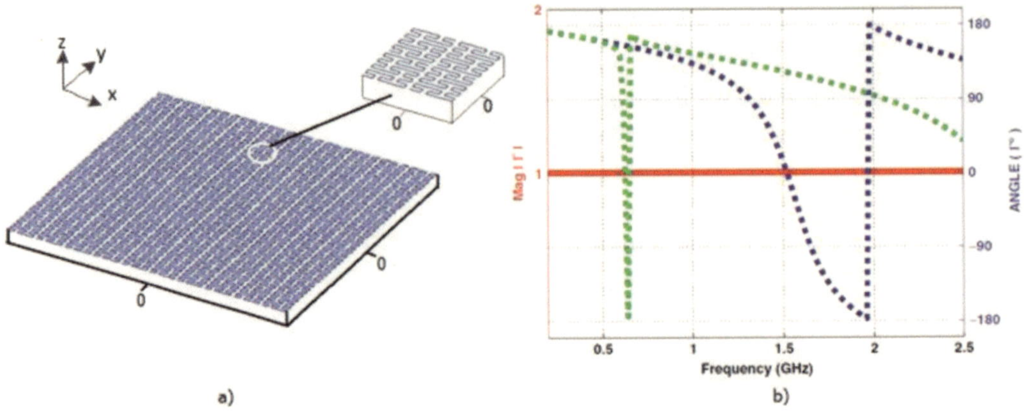

Figure 24. The HIS build-up of second order Peano fractal above conducting ground plane (a) and reflection coefficient for normally incident wave with polarizations in x- and y-planes (b).

In this case, each element was designed as a thin metallic strip with a strip width of 0.5 mm. The footprint dimensions remain identical (30 × −30 mm). The Peano array was placed a distance (15 mm) above a conducting ground plane of infinite extent. The supporting dielectric substrate is considered air here, although any other dielectric can be considered.

A separation distance was of 3.75 mm between each Peano curve inclusion within the array. This distance was chosen to be equal to the length of a single section of the curve itself. **Figure 24b** shows the magnitude and phase of the reflection coefficient Γ versus frequency. For the frequency of f_{HIS} = 1.53 GHz, in the case of the incident EM-wave polarized in the x-direction, where phase is $\varphi = 0$ over the entire surface of HIS, above the ground plane the reflection coefficient is $\Gamma = +1$; that is, it satisfies the conditions of a magnetic conductor MTM. Beyond the resonance across the entire surface of HIS, the reflection coefficient is $\Gamma = -1$ and, therefore, behaves as an electric conductor. Note that the cells of MTM in the form of a fractal Hilbert curve are of external sizes appropriately as 0.063 λ_{HIS} and 0.153 λ_{HIS} and thickness of substrates around 0.031 λ_{HIS} and 0.076 λ_{HIS}, respectively, at f_{HIS} (**Figure 24b**). That is, these cells are small electrically for both polarities [4].

MTM cells in the form of self-filling Peano and Hilbert curves can offer a lot of interesting features in electromagnetic applications. One of the most interesting features is the ability to work in resonant conditions where electrical dimensions of area filled by the curve are very small. This means that the self-filling curves can be excellent candidates for building surface and bulk media, where electrically small inclusions are desirable, for the design of specific electromagnetic properties of MTM. One of the most important uses of HIS surface is to use them to improve the performance of low-profile antennas.

4.4. EBG structures based on fractal cells

Recently, a novel electromagnetic bandgap (EBG) structure design based on fractal, the Sierpinski gaskets are proposed. They are arranged by repetition of 60° each to introduce the hexagonal unit cells [8]. By changing the gaps G_2 between two adjacent Sierpinski triangles inside EBG unit cell, it can separately introduce two EBG structures that have broadband and dual band-gap. Single cell geometry exhibits symmetry in six-planes, which makes it polarisation angle-independent. The proposed EBG structures, photo-etched on the FR4 microwave laminate ($\varepsilon_r = 4.4$, $h = 1.6$ mm, $tan\delta = 0.02$) are shown in **Figure 25**.

The Sierpinski gasket is a well-known fractal. Dark parts in this figure represent the metallic periodic structure, which is etched on a dielectric substrate. Unit cell dimensions are $W = 10$ mm, $W_1 = W - 2(G_1 + G_2)/\sqrt{3}$, $W_2 = 0.5\ W_1$, $W_3 = 0.25\ W_1$, $W_4 = W_1/8$, $G_1 = 1$ mm, $G_2 = 0.5$mm, and the metallization thickness is 18 μm. The Sierpinski gasket triangles are arrangement to

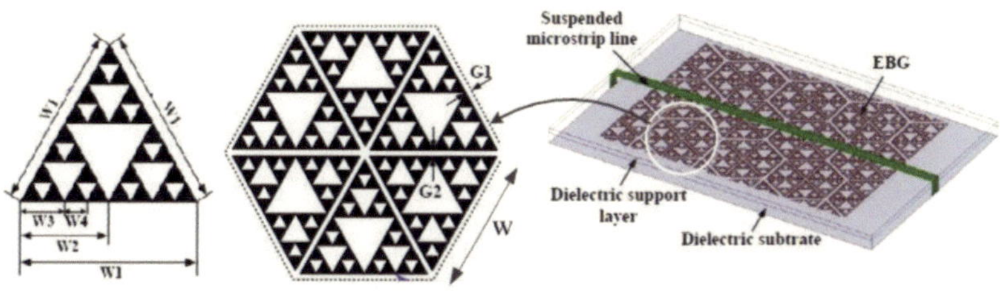

Figure 25. Sierpinski gasket fractal triangle (a), two-dimensional sketch of unit hexagonal cell (b) and three-dimensional of EBG array (c).

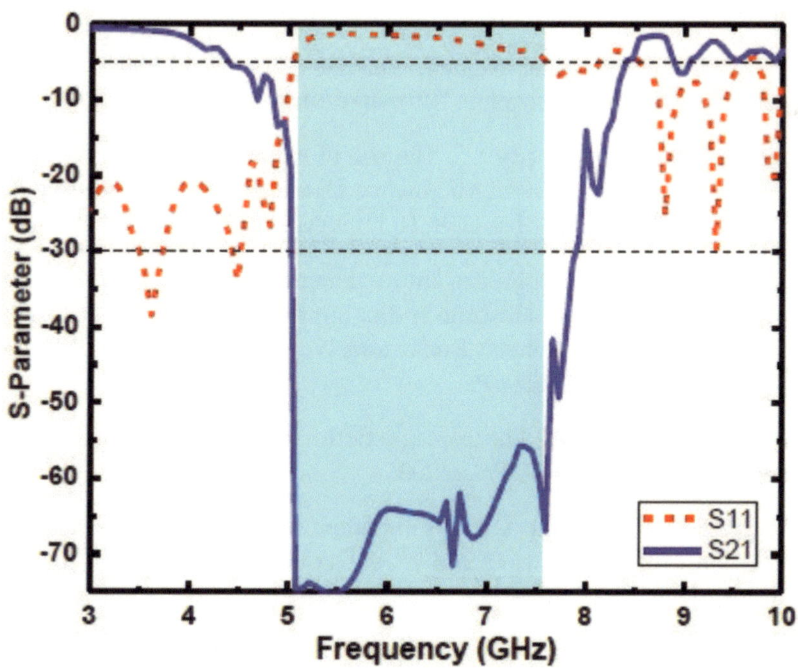

Figure 26. Operation bandwidth of HIS MTM based on an array of hexagonal cells of Sierpinski fractal triangles.

repeat each 60°. The resonance frequencies and the EBG operation bandwidth can be modified by changing the width and the gap G_1 between the hexagonal lattices. Moreover, by changing the gap G_2 between Sierpinski triangles units, it can introduce two EBG structures separately. The first one, which is called broadband EBG structure (BEBG), has a broader bandgap with the value of *G2* up to 0 mm. The simulation results of the EBG structures based on Sierpinski are shown in **Figure 26**.

The EBG structure has a bandgap, in the frequency range from 5.07 to 7.58 GHz, for $G_2 = 0.5$ mm. A wider bandwidth of operation frequency, with smaller EBG cell-size was obtained due to significantly reduced capacity C of an equivalent parallel resonant circuit, as well as limitation of the relative permittivity.

Author details

Wojciech Jan Krzysztofik

Address all correspondence to: wojciech.krzysztofik@pwr.edu.pl

Faculty of Electronics, Wroclaw University of Science & Technology, Wroclaw, Poland

References

[1] Youefi L, Ramahi OM. Artificial magnetic materials using fractal hilbert curves. IEEE Transactions on Antennas and Propagation. 2010;**58**(8):2614-2622. DOI: 10.1109/TAP.2010.2050438

[2] de Araujo HX, Barbin SE, Kretly LC. The use of metamaterial technology to improve a GTEM chamber. In: International Microwave Optoelectronics Conference (IMOC), 2013 SBMO/IEEE MTT-S. 2013. pp. 1-5. DOI: 10.1109/IMOC.2013.6646506

[3] Krzysztofik WJ. Antenna properties improvement by means of modern technology - metamaterials as a modified substrate and/or superstrate. In: MIKON'2014 20th International Conference on Microwaves, Radar and Wireless Communications. 2014. pp. 1-4. DOI: 10.1109/MIKON.2014.6899975

[4] Engheta N. Ziolkowski RW. Electromagnetic Metamaterials: Physics and Engineering Explorations. USA: Wiley-IEEE Press; 2006

[5] McVay J, Engheta N, Hoorfar A. High impedance metamaterial surfaces using Hilbert-curve inclusions. IEEE Microwave and Wireless Components Letters. 2004;**14**(3):130-132. DOI: 10.1109/LMWC.2003.822571

[6] Yousefi L, Ramahi OM. New artificial magnetic materials based on fractal Hilbert curves. International Workshop on Antenna Technology: Small and Smart Antennas Metamaterials and Applications, 2007. IWAT '07. 2007. pp.237-240. DOI: 10.1109/IWAT.2007.370119

[7] Facchinen MJ, Werner DH. Metamaterials composed of fractal sphere molecules. In: IEEE Press, editor. IEEE Antennas and Propagation Society International Symposium, IEEE APS-2003; 2003. pp. 102-105. DOI: 10.1109/APS.2003.1217411

[8] Phuong HNB, Chien DN, Tuan TM. Novel design of electromagnetic bandgap using fractal geometry. International Journal of Antennas and Propagation. 2013;**2013**:1-8/ 162396. DOI: 10.1155/2013/162396

[9] Krzysztofik WJ. Fractal geometry in electromagnetics applications - from antenna to metamaterials. Microwave Review. 2013;**19**(2):3-14. ISSN: 14505835.

[10] Mandelbrot BB. The Fractal Geometry of Nature. New York: W.H. Freeman; 1983.

[11] Zaad MZ, Ali M. A miniature hilbert PIFA for dual-band mobile wireless applications. IEEE Antennas and Wireless Propagation Letters. 2005;**4**:59-62

[12] Best SR, Morrow JD. The effectiveness of space-filling fractal geometry in lowering resonant frequency. IEEE Antennas and Wireless Propagation Letters. 2002;**1**:112-115

[13] Kritikos HN, Jaggard DL, editors. Recent Advantages in Electromagnetic Theory. New York: Springer-Verlag; 1990

[14] Sachendra SN, Jain M. A self-affine multiband antenna. IEEE Antennas and Wireless Propagation Letters. 2007;**6**:110-112

[15] Krzysztofik WJ. Printed multiband fractal antennas. In: Sanchez-Hernandez DA, editors. Multiband Integrated Antennas for 4G Terminals. 1st ed. Norwood, MA, USA: Artech House, Inc; 2008. pp. 95-150.

[16] Krzysztofik WJ. Modified sierpinski fractal monopole for ISM handset applications. IEEE Transaction on Antenna and Propagation. 2009;**57**(3)

[17] Krzysztofik WJ. Fractal monopole antenna for dual-ISM-bands. In: European Microwave Conference (EuMC-2006); UK, Manchester; 2006. pp. 1461-1464

[18] Krzysztofik WJ. Terminal antennas of mobile communication systems – some methods of computational analysis. Wroclaw, Poland: Wroclaw University of Science and Technology Press; 2011. p. 239

[19] Krzysztofik WJ, Baranski M. Fractal structures in antenna technology. Scientific Books of Gdansk University of Technology, Series: Radio Communication, Radiobroadcasting & Television. 2007; Gdansk, Poland **(1);** 2007. pp. 381-384

[20] Krzysztofik WJ. Fractal antenna for WLAN/bluetooth multiple-bands applications. In: EUCAP 2009, 3rd European Conference on Antennas and Propagation; 2009; Berlin, Germany; 2009. pp. 2407-2410

[21] Krzysztofik WJ, Komaryczko M. Fractal antennas in terminals of mobile communication systems. In: KKRRiT-2004 National Conference on Radio Communication, Radiobroadcasting & Television; 2004; Warsaw, Poland; 2004. pp. 436-439

[22] Falconer KJ. Mathematical Foundations and Applications. 2nd ed. Chinchester, New York: John Wiley & Sons; 2003.

[23] Werner DH. Fractal radiators. In: Proceedings of the 4th Annual 1994 IEEE Mohawk Valley Section Dual-Use Technologies & Applications Conference; 23–26 May 1994; New York: SUNY Institute of Technology at Utica/Rome; 1994. pp. 478-482

[24] Puente C, Baliarda CP, Romeu J, Cardama A. The koch monopole: A small fractal antenna. IEEE Transactions Antennas and Propagation. 2000;**AP-48**(11):1773-1781

[25] Werner DH, Bretones AR, Long BR. Radiation characteristics of thin-wire ternary fractal trees. Electronics Letters. 1999;**35**(8):09-10

[26] Rahmat-Samii Y, Gianvittorio JP. Fractal antennas: A novel antenna miniaturization technique and applications. IEEE Antenna's and Propagation Magazine. 2002;**44**(1):20-36

[27] Werner DH, Gangul S. An overview of fractal antenna engineering research. IEEE Antennas and Propagation Magazine. 2003;**45**(1):38-57

[28] Puente C, Romeu J, Pous R, Garcia X, Benitez F. Fractal multiband antenna based on the sierpinski gasket. IEE Electronics Letters. 1996;**32**(1):1-2

[29] Rahim MKA, Jaafar AS, Aziz MZAA. Sierpinski gasket monopole antenna design. In: Asia-Pacific Conference on Applied Electromagnetic; 20–21 December 2005; Malaysia; 2005. pp. 49-52

[30] Guterman J, Mortira AA, Peixeiro C. Microstrip fractal antennas for multistandard terminals. IEEE Antennas and Wireless Propagation Letters. 2004;**3**:351-354

[31] Mittra R. A critical Look at the Performance Enhancement of Small Antennas using Metamaterials. International Workshop on Antenna Technology: Small and Smart Antennas Metamaterials and Applications, 2007, IWAT'07, pp.7-10, DOI: 10.1109/IWAT.2007.370069

[32] Veselago V G. The electrodynamics of substances with simultaneously negative values of J and μ. Soviet Physics Uspekhi. January–February 1968;**10**:509-514

[33] Cui T.J. , ed., Metamaterials. Springer Science+Business Media B.V. Netherlands, DOI: 0.1007/978-1-4419-0573-4 _13, 2010

[34] Falcone F., et al. Babinet principle applied to the design of metasurfaces and metamaterials. Physical Reviews Letters. 2004;**93**:197401

[35] Abdalla MA, Hu, Z. Comapct and tunable metamaterial antenna for multi-band wireless communication applications. In: IEEE International Symposium on Antennas and Propagation (APSURSI); July 2011; pp. 1054-1057

[36] Zhang H, Ziolkowski RW, Xin H. A compact metamaterial-inspired mmW CPW-Fed antenna. In: IEEE International Workshop on Antenna Technology (iWAT); 2009; pp. 1-4

[37] Mei ZL, et al. A half Maxwell fish-eye lens antenna based on gradient-index metamaterial. IEEE Transactions on Antennas and Propagation. January 2012;**60**(1):398-401

[38] Bose JC. On the rotation of plane of polarization of electric waves by a twisted structure. Proceedings of Royal Society. 1898;**3**:146-152

[39] Lindell IV, Sihvola AH, Kurkijarvi J. Karl F. Lindman: The last Hertzian, and a Harbinger of electromagnetic chirality. IEEE Antennas and Propagation. Magazine. 1992;**34**(3):24-30

[40] Cui TJ, Smith DV, Liu R. Metamaterials: Theory, Design and Applications. New York, US: Springer; 2010.

[41] Araujo HX, Kretly LC. The effect of metamaterial patterning to improve the septum GTEM chamber performance. In: 29th Progress in Electromagnetics Research Symposium; March 2011; Marrakesh. 2011

[42] Cui T.J. Electromagnetic metamaterials: Recent advances on the theory, experiments, and applications. 2009 International Conference on Microwave Technology and Computational Electromagnetics (ICMTCE 2009), DOI: 10.1049/cp.2009.1248, pp. 12-13, 2009

[43] Hrabar S, et al. Application of wire-based metamaterials for antenna miniaturization. In: 3rd European Conference on Antennas and Propagation (EuCAP) 2009; pp. 620-623

[44] Shelby RA,Smith DR, Schultz S. Experimental verification of a negative index of refraction. Science. April 2001;**292**(5514):77-79

[45] Pendry JB. Negative refraction makes a perfect lens. Physical Review Letters. October 2000;**85**(18):3966-3969

[46] Sievenpiper D, Zhang L, Broas RFJ, Alexopolous NG, Yablonovitch E. High-impedance electromagnetic surfaces with a forbidden frequency band. IEEE Transactions on Microwave Theory and Techniques. 1999;**47**:2059-2074

[47] Zhang Y, Hagen J, Wiesbeck W. Patch array as artificial magnetic conductors for antenna gain improvement. Microwave and Optical Technology Letters. 2002;**35**(3):172-175

[48] Isbell D. E., "Log Periodic Dipole Arrays", IRE Trans., Vol. AP-8, No. 3, May 1960, pp. 260-267.

Chapter 4

ASCCC Fractal and Its Application in Antenna Miniaturization

Zeinab Eskandari, Asghar Keshtkar,
Javad Ahmadi-Shokouh and Leila Ghanbari

Additional information is available at the end of the chapter

http://dx.doi.org/10.5772/intechopen.68431

Abstract

In this chapter, ASCCC fractal is defined. The name "ASCCC" is based on the process that the fractal is built. It is made by adding and subtracting circles to the circumference of a circle. Then the necessary formulas to build up the first and higher orders of ASCCC fractal are derived. By calculating the perimeter of each order, it is shown that the ASCCC fractal has a great capability in antenna miniaturization. Based on first-order ASCCC fractal, a systematic approach is designed to miniaturize an antipodal dipole at any arbitrary frequency. Then the proposed method is applied at band LTE13 (746–787 MHz), which is controversy for mobile antenna, because it causes the size of a common antenna to become very large for a handheld mobile. It is illustrated that not only the ASCCC fractal is successful in miniaturization of dipole antenna, but also it is very good at improving the antenna's efficiency in comparison with its counterparts like Koch dipole/monopole.

Keywords: fractal antenna, antenna miniaturization, antenna's efficiency, antipodal dipole antenna, mobile antenna

1. Introduction

Nowadays, there is demand for antennas which fit in small space while have good radiation performance. Therefore, miniaturization techniques are inevitable in antenna design. Most of miniaturization techniques are based on slot loading, lumped loading, material loading, meandering, using fractal shapes or meta-materials. Generally, these techniques cause radiation efficiency and bandwidth to reduce. The antenna performance can be improved if the available volume within the Chu's sphere is used effectively. Fractal, meander and volumetric antennas are based on this method [1]. However, volumetric antennas are not suitable for planar structures. The meander antennas [2] and some fractal antennas such as Hilbert [1] and Koch dipole/monopole [3, 4] have some sections of cancelling current from adjacent conductors that cause the efficiency not to improve significantly. Furthermore, the resonance frequency cannot be found analytically because the physical length is not equivalent with electrical length [1, 2].

In this chapter, a novel fractal named adding and subtracting circles to the circumference of a circle (ASCCC) is defined and the required formulas are derived to build it. The ASCCC fractal is made by adding and subtracting an even number of circles on circumference of a circle, in brief named as adding and subtracting circles to the circumference of a circle (ASCCC). Then, a procedure is shown to miniaturize an antipodal dipole based on first order of ASCCC fractal at any arbitrary frequency. A formula is extracted to determine the resonance frequency of the ASCCC dipole with excellent precision. The proposed procedure is used to design a mobile antenna at challenging band of LTE13 (746–787 MHz). Because of low frequency nature of LTE13, the in-building penetration and area coverage are very good [6]. On the other hand, the size of antenna becomes so large at LTE13 that it is not suitable for a handheld mobile [7]. Therefore, some miniaturization techniques should be applied to the design. One of the great advantages of ASCCC dipole antenna is using the Chu's sphere so effectively that the antenna's efficiency improves considerably in addition to antenna miniaturization. Actually, the currents in adjacent teeth of ASCCC fractal dipole do not weaken the effect of each other, so very good efficiency is obtained. This advantage also makes the physical length to be approximately equal with electrical length.

The design is simulated by full-wave software (Ansoft-HFSS version 15). The results of simulation and measurement are in very good agreement. The efficiency of the proposed dipole antenna is higher than the existing works at LTE13 for handheld mobile antenna [8–25]. Also, the design obtains 40% size reduction compared with a common dipole. Furthermore, the ASCCC design has advantages of being planar and vialess [5].

In Section 2, the ASCCC fractal is explained. Then in Section 3, a procedure is shown to use ASCCC fractal in arms of an antipodal dipole. Theoretically, how to design an ASCCC dipole antenna for a special band is discussed. Next, a mobile antenna is designed, simulated and measured at LTE13. Finally, the conclusion is presented in Section 4.

2. ASCCC fractal

ASCCC fractal is based on adding and subtracting an even number of circles alternately on circumference of an initial circle. In brief, it is named as adding and subtracting circles to the circumference of a circle (ASCCC). It should be noted that radius of secondary circles (R_2) must be smaller than the radius of initial circles (R_1). To make ASCCC fractal clear, firstly consider a circle with radius R_1 as shown in **Figure 1(a)**. Then, arbitrary even numbers of circles (n_1) with radius R_2 ($R_2 < R_1$) are placed to the circumference of **Figure 1(a)** such that their centres are on the initial circle and two adjacent circles have two common points that one of them is on the circumference of initial circle and the other one is inside of it. These conditions lead n_1 to be at least 4 (for $n_1 = 2$, two adjacent circles have two common points but both of them are on the circumference of initial circle and $R_2 > R_1$) and secondary circles cover whole circumference of

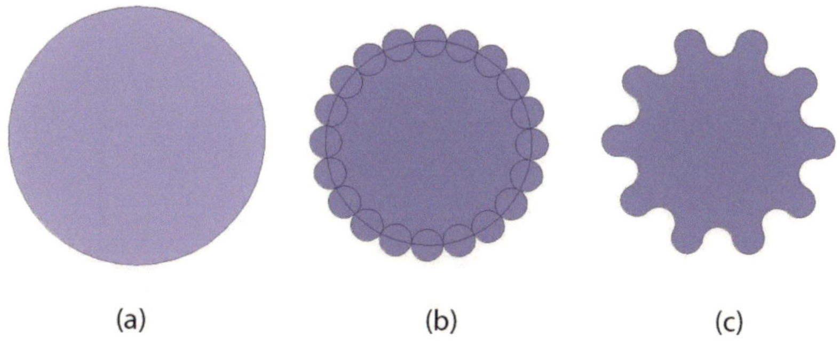

Figure 1. ASCCC fractal with $R_1 = 21$ mm and $n_1 = 20$ (a) initial circle, (b) 20 secondary circles with $R_2 = 3.295$ mm are placed on the circumference of initial circle and (c) secondary circles are united and subtracted alternately [5].

the initial circle. For example, **Figure 1(b)** shows 20 secondary circles that have been placed on an initial circle with $R_1 = 21$ m. **Figure 1(c)** illustrates how secondary circles are added and subtracted alternately. The radius of the secondary circles (R_2) is calculated as follows. Firstly, it is supposed that each secondary circle occupies 2θ angle on the initial circle as in **Figure 2(a)**. Two radiuses of R_1 and one radius of R_2 can make an isosceles triangle with a θ vertex angle as in **Figure 2(b)**. Then, its two leg length and base length are equal to R_1 and R_2, respectively. R_2 is determined by a trigonometric relationship as in Eq. (1) with respect to **Figure 2(c)**. It should be noted that the value of θ is known because 2θ is related to n_1 as in Eq. (2). If Eq. (2) is substituted in Eq. (1) and some simplifications are done, the R_2 can be written as in Eq. (3).

$$\sin\frac{\theta}{2} = \frac{R_2}{2R_1} \rightarrow R_2 = 2R_1 \sin\frac{\theta}{2} \tag{1}$$

$$2\theta = \frac{2\pi}{n_1} \rightarrow \theta = \frac{\pi}{n_1} \tag{2}$$

$$R_2 = 2R_1 \sin(\pi/2n_1) \tag{3}$$

Zero, first and second orders of ASCCC fractal for $R_1 = 21$ mm, $n_1 = 12$ and $n_2 = 10$ are shown in **Figure 3(a)–(c)**. **Figure 4(a)–(c)** illustrates the stages of producing **Figure 3(a)–(c)**.

Perimeter of the first-order ASCCC (P_1) is equal to perimeter of $n_1/2$ full circle with radius R_2. So, it is determined by Eq. (4). To understand clearly Eq. (4), firstly consider two adjacent circles shown in **Figure 1(b)**. One of them is supposed to be added (united) and another one is subtracted from the initial circle. Therefore, the effect of these two adjacent circles on the perimeter of **Figure 1(c)** is equal to the circumference of one full circle with radius R_2. Since the total number of secondary circles is n_1, the total perimeter of **Figure 1(c)** is equal to $(n_1/2)(2\pi R_2)$.

$$P_1 = (n_1/2)(2\pi R_2) \tag{4}$$

For calculating the perimeter of the second-order ASCCC fractal (P_2), as it is made of $(n_1/2)$ full-circle that each has $n_2/2$ full-circles with radius R_3; therefore, its perimeter is equal to the perimeter of a total number of $(n_1/2)(n_2/2)$ full-circle with radius R_3 as in Eq. (5).

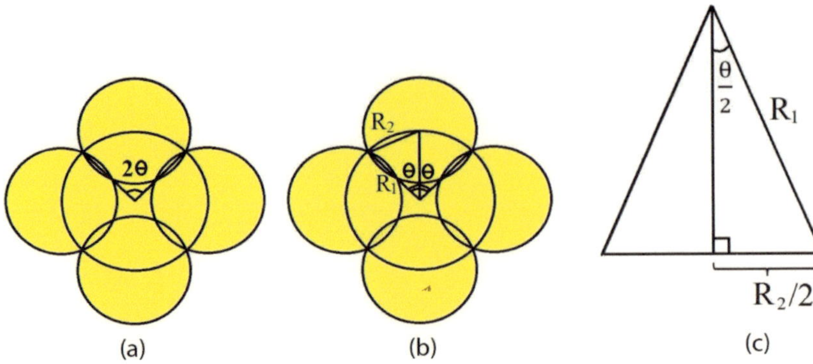

Figure 2. (a) A secondary circle occupies 2θ angle on the initial circle, (b) the radiuses of initial circle and secondary circle make an isosceles triangle with a θ vertex angle and (c) R_1 and R_2 in the isosceles triangle.

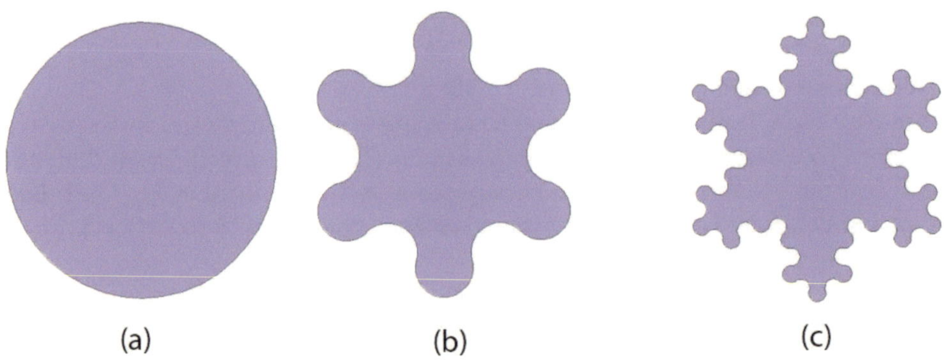

Figure 3. ASCCC fractal for $R_1 = 21$ mm, $n_1 = 12$ and $n_2 = 10$ (a) initial circle with $R_1 = 21$ mm, (b) first-order ASCCC with $n_1 = 12$ and (c) second-order ASCCC with $n_2 = 10$ [5].

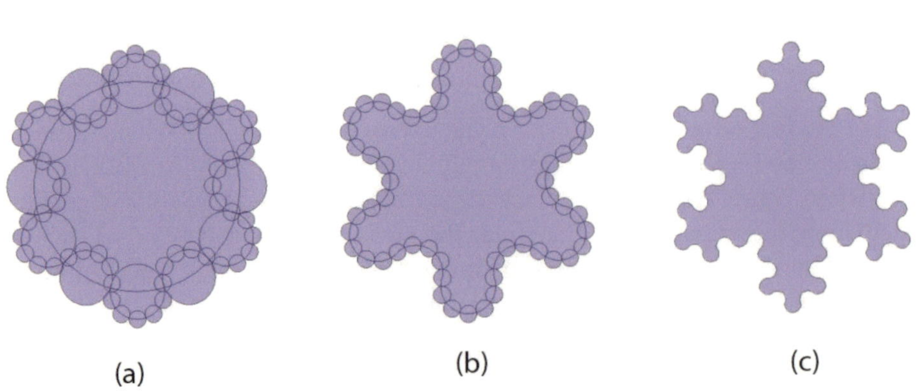

Figure 4. An illustration for building of second-order ASCCC fractal (a) for simplicity, third-order circles are placed only on inner (outer) edge of secondary circles which are supposed to be subtracted (added). (b) Secondary circles are added and subtracted alternatively and (c) third circles are added and subtracted alternatively [5].

$$P_2 = (n_1/2)(n_2/2)(2\pi R_3) = (n_1/2)(n_2/2)\left(2\pi \times 4R_1 \sin(\pi/2n_1)\sin(\pi/2n_2)\right) \tag{5}$$

Eqs. (6) and (7) show the ratio of P_1 and P_2 to the perimeter of initial circle (P_0), respectively. If n_1 and n_2 have large values, sine function could be approximated by its argument. Then, Eqs. (6) and (7) can be written as ($\pi/2$) and $(\pi/2)^2$, respectively. Therefore, the perimeter of ASCCC fractal can be multiplied by ($\pi/2$) in each order.

$$\frac{P_1}{P_0} = \frac{2n_1\pi R_1 \sin(\pi/2n_1)}{2\pi R_1} = n_1 \sin(\pi/2n_1) \tag{6}$$

$$\frac{P_2}{P_0} = \frac{2n_1 n_2 \pi R_1 \sin(\pi/2n_1)\sin(\pi/2n_2)}{2\pi R_1} = n_1 n_2 \sin(\pi/2n_1)\sin(\pi/2n_2) \tag{7}$$

Now, it is time to compare P_1 (the perimeter of the first-order ASCCC with initial (R_1) and secondary (R_2) radiuses) to the circumference C_1 of a common circle with radius $R_1 + R_2$ that occupies the same space on a board. Eq. (8) shows the ratio of P_1 to C_1. Eq. (9) presents the solutions of Eq. (8) for different n_1. When the argument of sine is much smaller than unity, the sine can be approximated to its argument. Therefore, approximation $\sin(\pi/2n_1) \approx \pi/2n_1$ is used for $n_1\ _1{\geq}\ 10$. As it is seen, the ratio [Eq. (9)] is greater than one for $n_1 \geq 6$, so P_1 is larger than C_1. Therefore, if a way is found to force the current to travel the perimeter P_1, antenna miniaturization is obtained for $n_1 \geq 6$ [5].

$$\frac{P_1}{C_1} = \frac{2n_1\pi R_1 \sin(\pi/2n_1)}{2\pi(R_1 + R_2)} = \frac{n_1 \sin(\pi/2n_1)}{1 + 2\sin(\pi/2n_1)} \tag{8}$$

$$\frac{P_1}{C_1} = \begin{cases} 0.867 & n_1 = 4 \\ 1.023 & n_1 = 6 \\ 1.122 & n_1 = 8 \\ 1.191 \leq \dfrac{\pi/2}{1 + (\pi/n_1)} \leq \dfrac{\pi}{2} & n_1 \geq 10 \end{cases} \tag{9}$$

3. An application of ASCCC fractal in antenna miniaturization

In this section, it is shown that an antipodal dipole antenna is miniaturized by applying the first-order ASCCC fractal to arms of the dipole antenna. The procedure could be applied to any arbitrary frequency [5].

3.1. The proposed design

In this section, it is shown that an antipodal dipole antenna is miniaturized by applying the first-order ASCCC fractal to arms of the dipole antenna. **Figure 5(a)–(d)** presents the utilized method. In the first step, two first-order ASCCC fractals with the same n_1 but different R_1 are designed (**Figure 5(a)**). To distinguish R_1 of fractals, R_{1i} is chosen for inner fractal and R_{1o} for outer fractal. In the next step, the inner fractal is subtracted from the outer fractal as shown in **Figure 5(b)**. Then, the shape is split into two equal halves as shown in **Figure 5(c)**. Finally, each

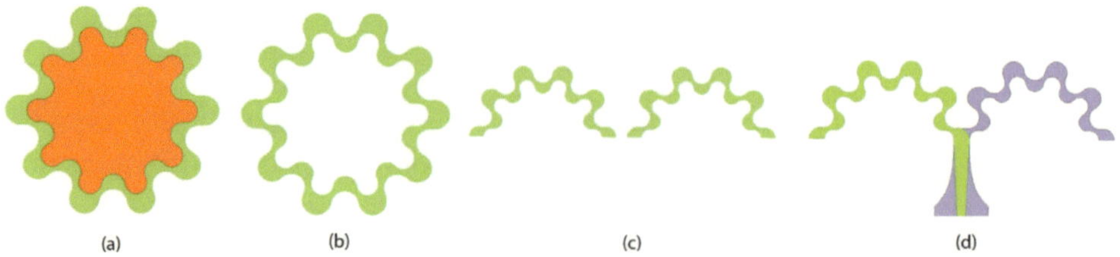

Figure 5. A method to use ASCCC fractal in arms of an antipodal dipole antenna (a) two first-order ASCCC fractal with same n_1 but different R_1 are designed, (b) inner fractal is subtracted from the outer one, (c) the shape is split into two equal halves and (d) each half is employed as an arm in an antipodal dipole antenna [5].

half is employed as an arm in a balanced antipodal dipole as illustrated in **Figure 5(d)**. The first resonance of the proposed dipole is calculated by Eq. (10). In Eq. (10), c is the speed of light and λ is the resonance wavelength.

$$f = c/\lambda \tag{10}$$

In a common dipole antenna, the length in which current travels along the two arms is equal to $\lambda/2$ of the first resonance wavelength. In calculation of $\lambda/2$, it should be noted that current tends to travel the shortest path. In **Figure 5(d)**, the current is confined to area between inner and outer fractals. The perimeter of inner fractal is shorter than the perimeter of outer one. Therefore, the inner fractal perimeter is more likely to be tracked. To be sure that current does not find any shorter path than the inner perimeter, $R_{1\text{o}}$ should be chosen as close to $R_{1\text{i}}$ as possible. As a result, the current travelling length is approximately equal to the inner fractal perimeter that is determined by using Eq. (4). Then, the resonance frequency could be written as in Eq. (11). In Eq. (11), $P_{1\text{i}}$ is the perimeter of inner fractal in **Figure 5(d)**.

$$f = \frac{c}{\lambda} = \frac{c}{2P_{1i}} = \frac{c}{4n_1\pi R_{1i}\sin\left(\pi/2n_1\right)} \tag{11}$$

To design a balanced feedline, the method described in Refs. [26] and [27] is used. The line parameters are given in **Figure 6**. The exponential part of line is made by Eqs. (12) and (13). W_{cps} is equal to $(R_{1\text{o}} + R_{2\text{o}}) - (R_{1\text{i}} + R_{2\text{i}})$. L_{exp}, L_{m}, W_{gnd} and p are arbitrary parameters that are chosen with respect to a good S_{11} result.

$$y = \pm\left[A \times \exp\left(px\right) + \left(\frac{W_{cps}}{2} - A\right)\right] \tag{12}$$

$$A = \frac{W_{gnd} - W_{cps}}{2\left(\exp\left(p \times L_{\exp}\right) - 1\right)} \tag{13}$$

3.2. Simulation and measurement results

The method described in Section 3.1 is used to design a handset mobile antenna at the LTE13 band (746–787 MHz). The antenna is printed on an FR4 substrate with $\varepsilon_r = 4.4$ and tan

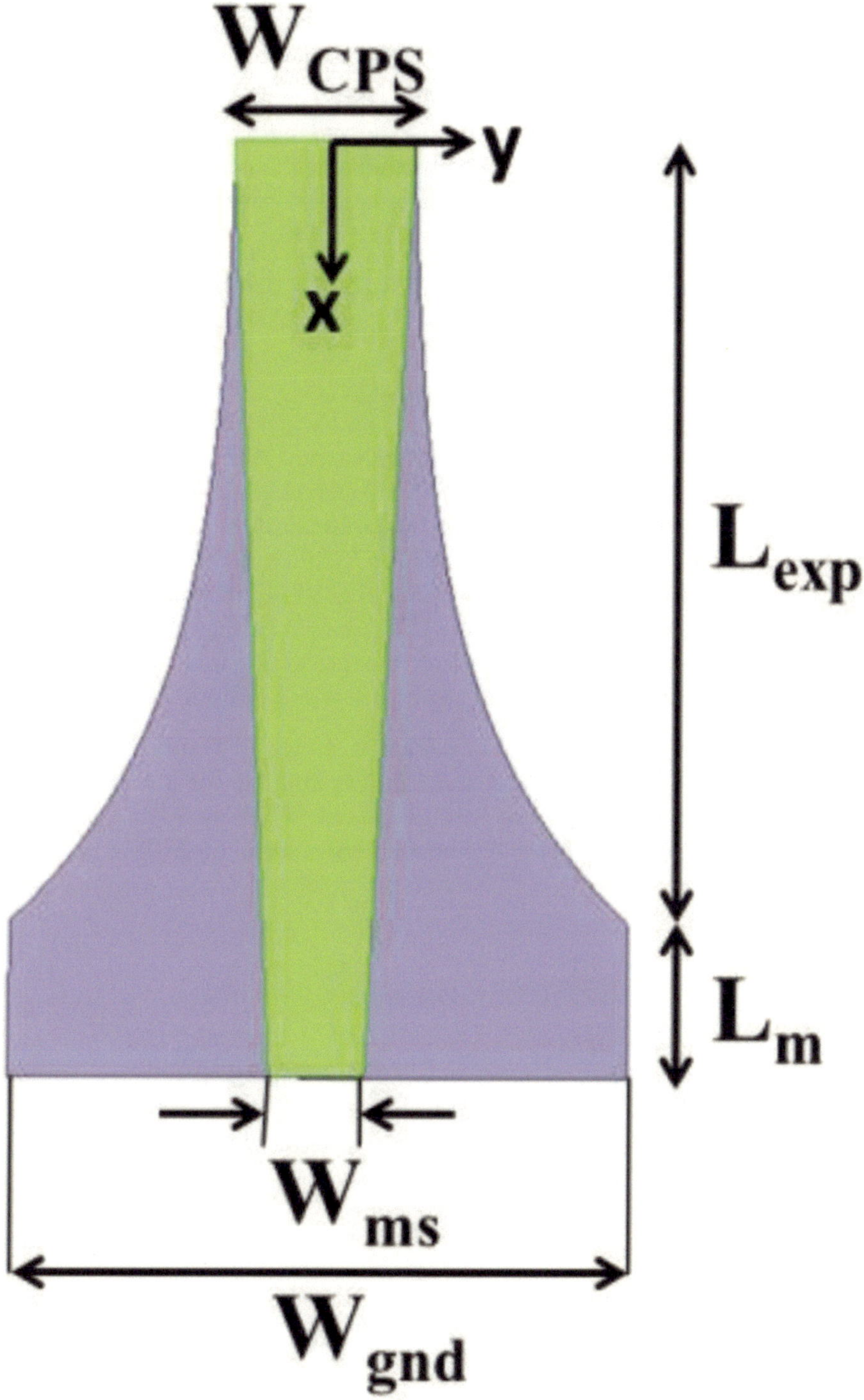

Figure 6. Geometry and parameters of the balanced feedline.

$\delta = 0.02$. Firstly, an initial resonance frequency within band LTE13 should be picked out to determine $\lambda/2$ by Eq. (10). As the length $\lambda/2$ is approximately equal to the perimeter of inner ASCCC, as shown in **Figure 5(a)**, so R_{1i} is determined by Eq. (4). To stay in safe side, a frequency of 750 MHz is picked out for the initial design because a good S_{11} in lower frequencies needs longer length in arms while preparing longer length is harder to obtain. Please note that the value of n_1 is arbitrary. The larger n_1 results in the smaller R_{2i}, so a more compact design is obtained. In the simulations, feedline parameters have been chosen as: $L_m = 5$ mm, $L_{exp} = 25$ mm, $W_{gnd} = 20$ mm, $W_{ms} = 3.04$ mm and $p = 150$.

$R_{1i} = 20.28$ mm is found for $n_1 = 20$ at 750 MHz. To determine a value for R_{1o}, some simulations are done for different R_{1o} radiuses ($R_{1o} = 23$, 24.5 and 26 mm). The simulated S_{11} results are shown in **Figure 7**. As seen, a smaller R_{1o} makes a better confinement of current to the inner fractal perimeter; therefore, the resonance frequency is closer to the initial design. On the other hand, a bigger R_{1o} results in better S_{11} and wider bandwidth because of a larger radiating area. However, $R_{1o} = 23$ mm cannot be chosen. Although resonance frequency is closer to the initial design, the whole band of LTE13 cannot be covered. The problem could be tackled as follows. If a bigger R_{1i} is chosen, the resonance frequency is lowered, so more freedom is prepared to pick out larger values for R_{1o} that could somehow compensate for lowering of frequency while enough bandwidth and good S_{11} are obtained at the LTE13 band. By a little try and error, it is found the whole LTE13 band could be covered with good S_{11} for $R_{1i} = 20.5$ mm and $R_{1o} = 25.5$ mm. As it is seen, the selected R_{1i} is so close to the initial design ($R_{1i} = 20.28$ mm) and the proposed formulas prepare very good primary guess.

A fabricated prototype of the proposed antenna is shown in **Figure 8**. The overall size of the printed antenna is $62 \times 115 \times 1.6$ mm^3 that is suitable for a handheld mobile. The simulated and measured results of S_{11} are presented in **Figure 9**. It is seen that there is very good agreement between them. The small resonance at 1.045 GHz is due to the type of feedline.

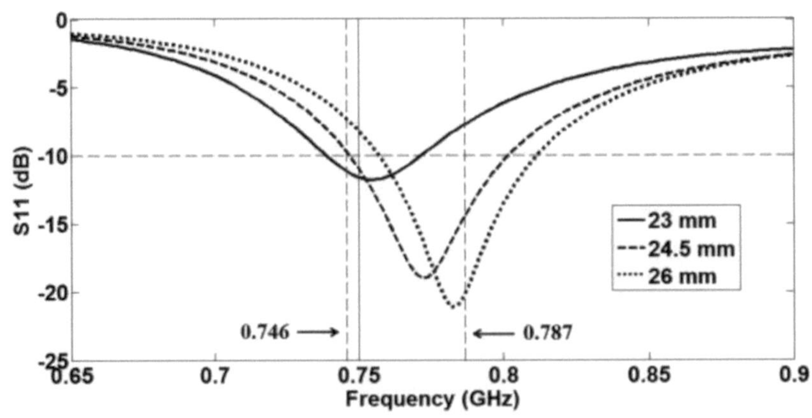

Figure 7. S_{11} parameter for different R_{1o} ($R_{1i} = 20.28$ mm, $n_1 = 20$) [5].

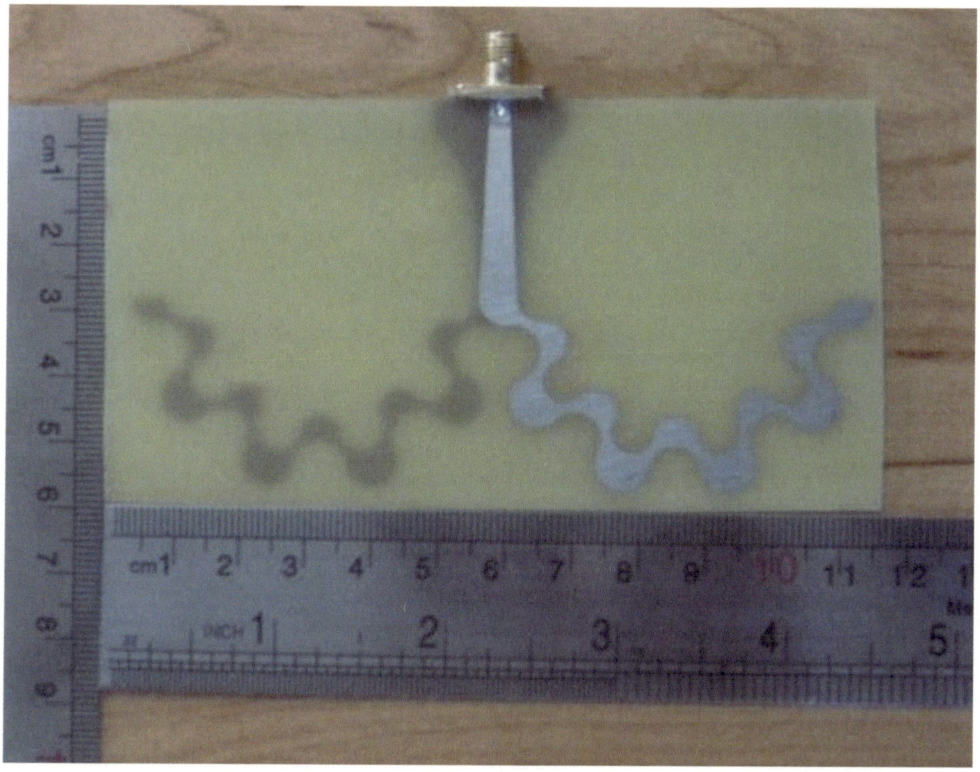

Figure 8. Fabricated prototype of the proposed design [5].

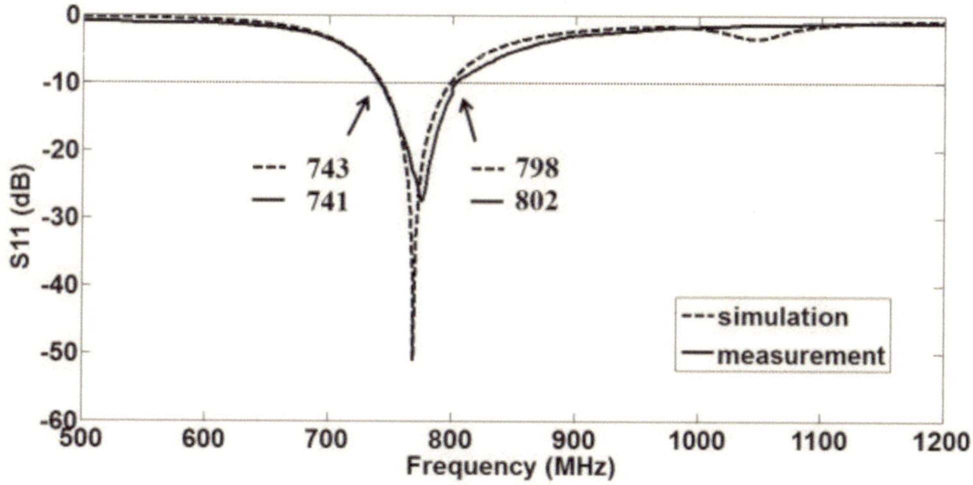

Figure 9. Simulation and measurement results of S_{11} against frequency [5].

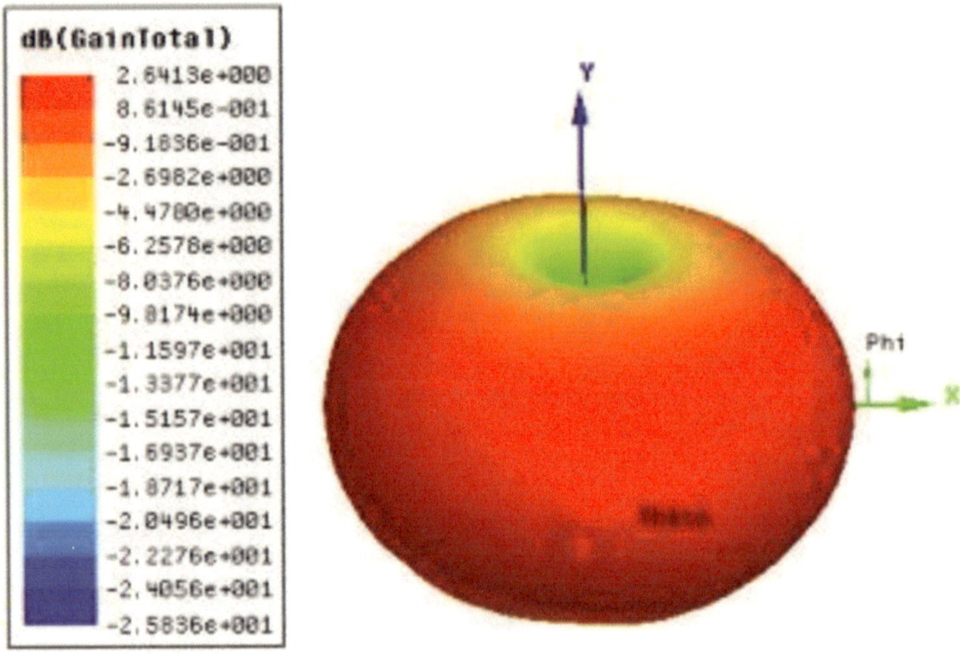

Figure 10. Simulated 3D radiation patterns at 769 MHz [5].

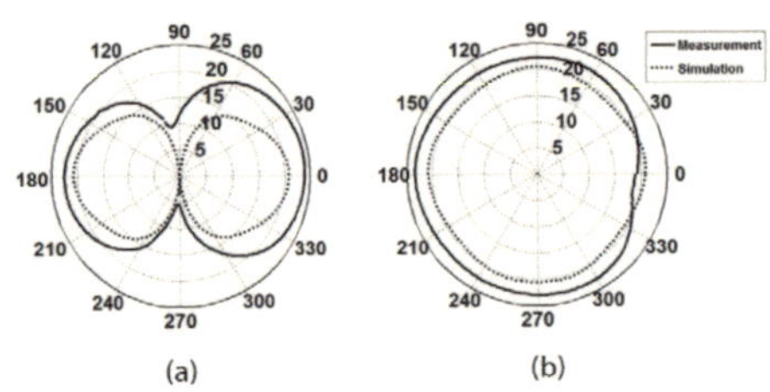

Figure 11. Simulated and measured radiation patterns at 769 MHz (a) E-plane and (b) H-plane [5].

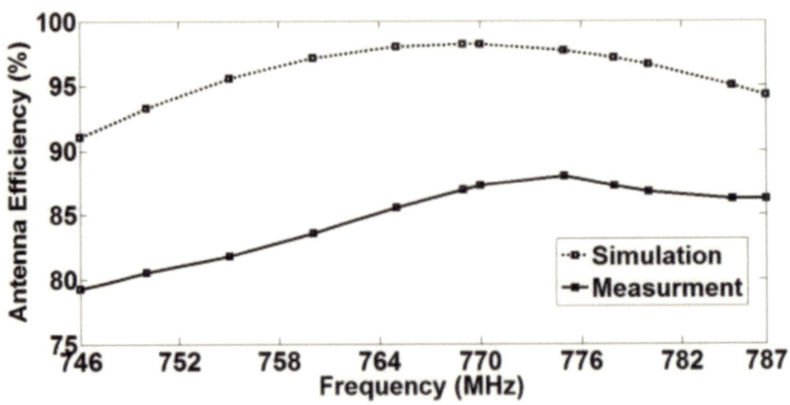

Figure 12. Simulated and measured radiation efficiency for LTE13 [5].

Refs.	[8]	[9]	[10]	[11]	[12]	[13]	[14]	[15]	[16]
Antenna efficiency (%)	45 at 762 MHz	40–60	40–60*	40–51.24	30–55*	34 at 746 MHz	45–55	56–84	52–72
Refs.	[17]	[18]	[19]	[20]	[21]	[22]	[23]	[24]	[25]
Antenna efficiency	50–65	Average 62.85	50–70	53–76	52–75*	40–53*	53–76	45.24–48.43	10–40*

Note: *means the radiation efficiency has been reported in the reference.

Table 1. Antenna efficiency at LTE 13 in different papers.

Figures 10 and **11** present the 3-D and polar radiation patterns of the proposed antenna at 769 MHz, respectively. As they show, the antenna has a dipolar radiation pattern. **Figure 12** shows the efficiency of antenna. The measured efficiency is obtained by the improved Wheeler-cap method [28]. Antenna efficiency varies from 79.28 to 88.01%. As it is seen, the antenna has very high efficiency at LTE13, on the contrary of the other designs for this band that are listed in **Table 1** [8–25].

Finally, the antenna exhibits 40% size reduction in comparison with a common dipole. This is evidence that the proposed procedure is a good technique in antenna miniaturization [5].

4. Conclusion

In this chapter, ASCCC fractal is defined and its driving formulas are extracted. It is shown that ASCCC fractal has a great potential in antenna miniaturization and improving efficiency. A miniaturization method was designed for a dipole antenna at any arbitrary frequency. Then, the method applied to the dipole antenna at band LTE13 which is very challenging for reduction in size of mobile antennas. The total size of antenna is $62 \times 115 \times 1.6\ \text{mm}^3$, which is appropriate for handheld mobiles. The efficiency of antenna is greater than 79% with S_{11} better than –10 dB. The amount of efficiency is considerably higher than existing works.

Author details

Zeinab Eskandari[1]*, Asghar Keshtkar[1], Javad Ahmadi-Shokouh[2] and Leila Ghanbari[1]

*Address all correspondence to: zeinab.eskandari@hotmail.com

1 Imam Khomeini International University, Qazvin, Iran

2 University of Sistan and Baluchestan, Zahedan, Iran

References

[1] Volakis J, Chen CC, Fujimoto K. Small Antennas: Miniaturization Techniques & Applications. 1st ed. New York: McGraw-Hill; 2010. p. 428

[2] Hansen RC, Collin RE. Small Antenna Handbook. 1st ed. Hoboken, New Jersey: John Wiley & Sons, Inc.; 2011. p. 346

[3] Best SR. On the performance properties of the koch fractal and other bent wire monopoles. IEEE Transactions on Antennas and Propagation. 2003;**51**(6):1292–1300. DOI: 10.1109/TAP.2003.812257

[4] Vinoy KJ, Abraham JK, Varadan VK. On the relationship between fractal dimension and the performance of multi-resonant dipole antennas using Koch curves. IEEE Transactions on Antennas and Propagation. 2003;**51**(9):2296–2303. DOI: 10.1109/TAP.2003.816352

[5] Eskandari Z, Keshtkar A, Ahmadi-Shokoh J, Ghanbari L. A novel fractal for improving efficiency and its application in LTE mobile antennas. Microwave and Optical Technology Letters. 2015;**57**(10):2429–2434. DOI: 10.1002/mop.29346

[6] Sharawi MS, Numan AB, Khan MU, Aloi DN. A dual-element dual-band MIMO antenna system with enhanced isolation for mobile terminals. IEEE Antennas and Wireless Propagation Letters. 2012;**11**:1006–1009. DOI: 10.1109/LAWP.2012.2214433

[7] Zhang Z. Antenna Design for Mobile Devices. 1st ed. Singapore: John Wiley & Sons (Asia) Pte Ltd; 2011. p. 280

[8] Meshram MK, Animeh RK, Pimpale AT, Nikolova N K. A novel quad-band diversity antenna for LTE and Wi-Fi applications with high isolation. IEEE Transactions on Antennas and Propagation. 2012;**60**(9):4360–4371. DOI: 10.1109/TAP.2012.2207044

[9] Yu Y, Kim G, Ji J, Seong W. A compact hybrid internal MIMO antenna for LTE application. In: 2010 Proceedings of the Fourth European Conference on Antennas and Propagation (EuCAP); 12–16 April 2010; Barcelona, Spain: IEEE; 2010.

[10] Lopez N, Lee C-J, Gummalla A, Achour M. Compact metamaterial antenna array for long term evolution (LTE) handset application. In: IEEE International Workshop on Antenna Technology(iWAT) 2009; 2–4 March 2009; Santa Monica, CA, USA: IEEE; 2009. DOI: 10.1109/IWAT.2009.4906933

[11] Bae H, Harackiewicz FJ, Park MJ, Kim T, Kim N, Kim D. Compact mobile handset MIMO antenna for LTE700 applications. Microwave and Optical Technology Letters. 2010;**52**(11):2419–2422. DOI: 10.1002/mop.25507

[12] Zhao X, Choi J. Design of a MIMO antenna with low ECC for a 4G mobile terminal. Microwave and Optical Technology Letters. 2014;**56**(4):965–970. DOI: 10.1002/mop.28196

[13] Cheon Y, Lee J, Lee J. Quad-band monopole antenna including LTE 700 MHz with magneto-dielectric material. IEEE Antennas and Wireless Propagation Letters. 2012;**11**:137–140. DOI: 10.1109/LAWP.2012.2184517

[14] Ban YL, Chen JH, Yang S, Li JLW, Wu YJ. Low-profile printed octa-band LTE/WWAN mobile phone antenna using embedded parallel resonant structure. IEEE Transactions on Antennas and Propagation. 2013;**61**(7):3889–3894. DOI: 10.1109/TAP.2013.2258879

[15] Wong KL. 4G/Multiband handheld device ground antennas. In: Microwave Conference Proceedings (APMC); 5–8 November 2013; Asia-Pacific: IEEE; 2013. DOI: 10.1109/APMC.2013.6695215

[16] Wong KL, Chang YW. Internal WWAN/LTE handset antenna integrated with USB connector. Microwave and Optical Technology Letters. 2012;**54**(5):1154–1159. DOI: 10.1002/mop.26788

[17] Jeon S, Kim H. Mobile terminal antenna using a planar inverted-e feed structure for enhanced impedance bandwidth. Microwave and Optical Technology Letters. 2012;**54**(9):2133–2139. DOI: 10.1002/mop.27035

[18] Jo Y, Park K, Lee J, Kim HH, Kim H. Mobile handset antenna with parallel resonance feed structures for wide impedance bandwidth. In: Proceedings of the International Symposium on Antennas & Propagation (ISAP); 23–25 October 2013; Nanjing, China: IEEE; 2013.

[19] Lee J, Liu Y, Kim HH, Kim H. PIFA with dual-resonance feed structure for enhancement of impedance bandwidth. Electronics Letters. 2013;**49**(15):921–922. DOI: 10.1049/el.2013.1432

[20] Chu FH, Wong KL. Planar printed strip monopole with a closely-coupled parasitic shorted strip for eight-band LTE/GSM/UMTS mobile phone. IEEE Transactions on Antennas and Propagation. 2010;**58**(10):3426–3431. DOI: 10.1109/TAP.2010.2055807

[21] Wong KL, Chen WY, Wu CY, Li WY. Small-size internal eight-band LTE/WWAN mobile phone antenna with internal distributed LC matching circuit. Microwave and Optical Technology Letters. 2010;**52**(10):2244–2250. DOI: 10.1002/mop.25431

[22] Ban YL, Liu CL, Chen Z, Li JLW, Kang K. Small-size multiresonant octaband antenna for LTE/WWAN smartphone applications. IEEE Antennas and Wireless Propagation Letters. 2014;**13**:619–622. DOI: 10.1109/LAWP.2014.2313353

[23] Lu JH, Guo JL. Small-size octaband monopole antenna in an LTE/WWAN mobile phone. IEEE Antennas and Wireless Propagation Letters. 2014;**13**:548–551. DOI: 10.1109/LAWP.2014.2311797

[24] Luo J, Gong SX, Duan P, Mou C, Long M. Small-size wideband monopole antenna with CRLH-TL for LTE mobile phone. Progress in Electromagnetics Research C. 2014;**50**:171–179.

[25] Takemura N. Tunable inverted-L antenna with split-ring resonator structure for mobile phones. IEEE Transactions on Antennas and Propagation. 2013;**61**(4):1891–1897. DOI: 10.1109/TAP.2012.2232894

[26] Antoniades MA, Eleftheriades GV. Multiband compact printed dipole antennas using NRI-TL metamaterial loading. IEEE Transactions on Antennas and Propagation. 2012;**60** (12):5613–5626. DOI: 10.1109/TAP.2012.2211324

[27] Bourqui J, Okoniewski M, Fear EC. Balanced antipodal vivaldi antenna with dielectric director for near-field microwave imaging. IEEE Transactions on Antennas and Propagation. 2010;**58**(7):2318–2326. DOI: 10.1109/TAP.2010.2048844

[28] Geissler M, Litschke O, Heberling D, Waldow P, Wolff I. An improved method for measuring the radiation efficiency of mobile devices. In: Antennas and Propagation Society International Symposium. IEEE; 22-27 June 2003; Columbus, Ohio, USA: IEEE; 2003. DOI: 10.1109/APS.2003.1220380

Application of Fractal Analysis While Designing of Family of Spacecraft for Needs of Space Industry

Andrew V. Sedelnikov and Ksenia I. Potienko

Additional information is available at the end of the chapter

http://dx.doi.org/10.5772/67922

Abstract

This paper deals with methods of fractal analysis, which allow creation of a line of spacecraft from different classes as addition to classical methods for needs of space industry. It is shown that the fractal analysis that widens opportunities of classical methods can become a base for solution of the modern problems of space technologies. Besides, the results of fractal analysis in this chapter are practically valued recommendations for designing of the line, algorithms of control over equipment to measure microaccelerations and of majority control over measuring data. It clearly demonstrates potential and applicability of methods of fractal analysis in practice.

Keywords: fractal analysis, fractal property, spacecraft, microaccelerations, majority control

1. Introduction

Space technologies have given a lot of new materials and medicines with unique features. For instance, semiconducting germanium for purposes of radio industry was firstly gotten under space conditions. Its purity achieved 99.9999%. Ultrastrong and thermostable monocrystals for development of blades of a turbine for aviation and rocket engines were also firstly grown in space laboratories. The metal foam is of great interest for engineering industry. It has a weak structure and combines high durability with lightness.

Such unique features can be achieved for materials, which are developed out of the field of terrestrial gravity. The absence of the field allows thorough mixing of melted metal with air for creation of foam metal with weak structure or exclusion of admixtures from the melt to significant reduction of crystallization centers to get large monocrystals, etc. Such technological processes, which require almost absolute absence of force action for their successful realization, are called as gravity-sensitive processes.

The best resources and significant funds were wasted for realization of ambitious space projects. For instance, in USA, it was developed and realized grand project of space laboratory "Skylab." The laboratory was placed into near-earth orbit by legendary carrier rocket "Saturn-5" on 14 May 1973 for realization of more than 300 series of different experiments connected with cross-cutting technologies [1]. There were realized growth of monocrystals, experiments on directed crystallization and others among gravity-sensitive processes. The project "Skylab" was unprecedented in the scales of funds—its budget is considered to be 2.5 bn of dollars.

The large numbers of gravity-sensitive processes in Russia were conducted during realization of large-scale project of orbital complex "MIR." This platform was prepared better for technological and biomedicine experiments than space laboratory "Skylab." "Skylab" was the one big laboratory but "MIR" represented the complex of separated orbital modules with detail specialization of experiments conducted in each module. During realization of the project "MIR," the experience of "Skylab" was taken into account as well as finished by that moment Russian projects of manned orbital stations of the "Salut" series. For providing the research part of the project, the unique equipment was created such as the ovens "Krater-VM" and "Gallar," the complex to control conducting of experiments "Onyx," glass oven "Optyzon" experimental devices "Katalyzator" and "Titus," gauges "Alice" and others.

Evidently that high pace of development of scientific-technical progress today and also significant changing of earth technologies are the results of large-scale space projects. It seemed like space technologies would make a real revolution, but revolution was postponed due to small knowledge of the new conditions which determined realization of technological processes.

Investigations showed that it is wrong to believe that there is absolute weightless into inner environment of space laboratory. An influence of outer (gravitational field, aerodynamic drag, light pressure, etc.) and inner (running of operating elements, proper oscillations of big elastic elements, running of system of orientation, and control of spacecraft's movement) disturbing factors causes a microacceleration field in inner environment.

It influences the results of gravity-sensitive processes negatively. Today, the row of such processes, which demand microacceleration level no higher than 10.7 g for their successful realization, has been already developed. However, modern developmental stage of space technic has not yet allowed realization of the processes on a board of space laboratory. It is necessary to investigate the structure and nature of microaccelerations, to find out their new features to provide the necessary level for successful realization of gravity-sensitive processes. It is necessary to understand that requirements for microacceleration level will become tougher to the extent of development of space technologies.

Investigations show that microacceleration field as the test subject has several features, which significantly complicate its investigation. One of the features consists in overloads of microaccelerometers during launching of spacecraft/these overloads can achieve 10 g. It is approximately in 7–8 orders bigger than microacceleration values [2]. Such situation significantly complicates getting the reliable experimental data using onboard gauges. The following situation [3] was registered when testing of Canadian vibration isolator MGIM (microgravity isolation mount) during 2 years in orbital complex "MIR" and also during the flight of space shuttle STS-85 by gauges. In one series of experiments which were considered to be successfully realized, the oscillation amplitude inside was protected by the MGIM area that was significantly lower than the outside area. However, in another series of experiments which were marked as unsuccessful, the oscillation amplitude inside the protected area was higher than the outside area. The values of amplitudes (in successful and unsuccessful series) differed more than in 20 times. Later, during measurements of microaccelerations on the board of specialized technological spacecraft "FOTON-11" by French equipment BETA, the situation was similar [4]. Microaccelerometers, installed at different axes of bound coordinate system, showed data that differed more than 20 times. Nevertheless, there were not any obvious causes for it. Sazonov et al. [4] suppose that the software of BETA equipment was applied incorrectly. However, failure of some microaccelerometers is quite possible too. It is also remarkable that the equipment was not applied more on the spacecraft from "FOTON" series. Thus, the reliable way to control adequacy of microaccelerometers on the operational stage of spacecraft, after launching overloads, is required.

The second important feature of microacceleration field is impossibility to measure microaccelerations directly like temperature or angular velocity. That is why the measuring results are indirect estimations based on different methods. The gauges created according to the methods dive only qualitative-similar data. It significantly complicates the investigation of microacceleration field too. For instance, tested on orbital complex "MIR," complex of gauges which included "Alice-2" (MEMS-technologies) and "DACON" (convection sensor) showed that "DACON" is insensitive to high-frequency microaccelerations. Although, "Alice-2" managed to register it [5]. The changing of working body of modification of convection sensor "DACON-M" allowed changing of characteristics of the device [6]. The equipment "GRAVITON" based on measurement of magnetic field of the Earth and also row of other gauges were developed [7]. All mentioned facts show that experimental data are the results of mathematical modeling but not of the pure experiment. Thus, the role of mathematical modeling for investigation of microaccelerations is very important and stimulates application of new mathematical methods, particularly, of fractal analysis for investigation of microacceleration field.

2. Main part

There are a lot of different classifications of microaccelerations. For instance, microaccelerations are divided into quasi-static (low-frequency) and vibrational (high-frequency) according to its frequency. According to the nature of the source, they are divided into inner and outer. There is offered the classification according to the method of their control [8] because it is important to provide and control designed microacceleration level while creating new space technic.

Microaccelerations can be divided into three groups in accordance with the classification feature:

- Metastable component of microacceleration is caused by the influence of permanent factors of cosmic space such as gravitational and electromagnetic fields, aerodynamic forces and momentums, light pressure, etc.

- Random component of microaccelerations is caused by multifocal uptake and oscillations of gravitational and electromagnetic fields, changes in solar activity and also by influence of random factors of space such as micrometeorites, multifocal uptake of the atmosphere, etc. Extraordinary situations, connected with failure of equipment, are also in the list.
- Constructive component of microaccelerations is caused by inner disturbances such as orbit correction of spacecraft, running of orientation engines, different operational elements, equipment for passive and active orientation, vital activity of crew, etc.

The constructive component is of the greatest interest among the three components because it is connected with construction of the spacecraft and the way of orbital control from the early stages of designing and development.

In a way, the constructive component of microacceleration field of the inner environment of spacecraft is formed while creating the constructional layout of space laboratory and developing the system of its orbital orientation in the early stages of designing.

Constructive component, which often mainly contribute to microacceleration field of inner environment, should be controlled for achieving of favorable microacceleration level.

Investigations of constructive component of microacceleration field show [9] that it has the feature which can be compared with fractal quality of scaling. For instance, the scaling of cosine part of Weierstrass-Mandelbrot function (WMF):

$$ReW(t) = C(t) = \sum_{n=-\infty}^{\infty} \frac{1 - cosb^n t}{b^{(2-D)n}} \quad (1)$$

where D is fractal dimension and b is scaling parameter of WMF by the replacement of t to b^4t, and $C(t)$ to $b^{4(2\text{-}D)}C(t)$ does not change the form of function (**Figure 1**).

According to the scale invariance of differential equations which describe oscillations of big elastic elements of spacecraft time-dependence of microaccelerations has the similar feature. However, not graph of function but the spacecraft itself is scaled: its sizes and inertia-mass parameters (**Figure 2**).

The results are described in detail below. For investigation of the fractal quality of constructive component, the simple scheme of spacecraft was chosen with the purpose of qualitative demonstration of the quality. The spacecraft represents the central absolutely rigid body that is rigidly attached to its elastic elements (**Figure 3**).

The elastic elements are considered to be the Euler-Bernoulli beams to simplify the investigation process. The estimation of microaccelerations using such models gives a bit overevaluated results [10].

The vibrations of elastic element are caused by inner and outer disturbing factors on spacecraft (**Figure 3**). Just the vibrations make the most important contribution to constructive component of microacceleration field. The impulse firing of engines of orientation and control of orbital motion of spacecraft [11], permanent running of operational equipment [2, 12] or the other factors, including the outer, can be considered to be the sources of vibrations.

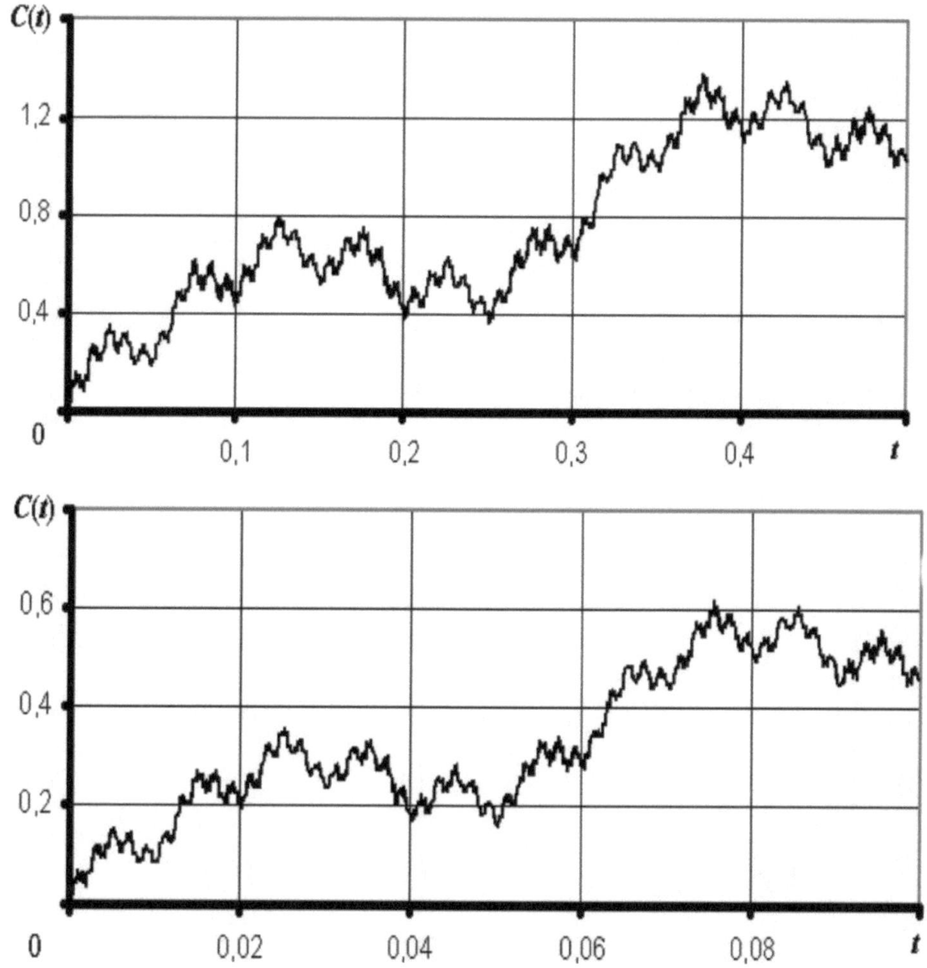

Figure 1. The self-affine property of the cosine part of Weierstrass-Mandelbrot function.

Proper frequency when cantilever attaching is performed by the following formula:

$$\omega_i = \frac{\eta_i^2}{l^2}\sqrt{\frac{EI}{\mu}}, \quad (2)$$

where η_i is i-equation root, $cos\eta_i ch\eta_i + 1 = 0$; i is number of oscillation mode; l is length; EI is flexural rigidity; and μ is linear mass of elastic element.

Linear mass and rigidity are constant for the designed material. Roots η_i are also constant. Thus, frequency is the function of beam's length. So, double decrease of the length will lead to fourfold increase of frequency or fourfold decrease of period of vibrations. At that, microacceleration's amplitude will change. Beam with less length and mass will create less microaccelerations when central body of space laboratory will be permanent. Tangential inertia force influences each point of a beam:

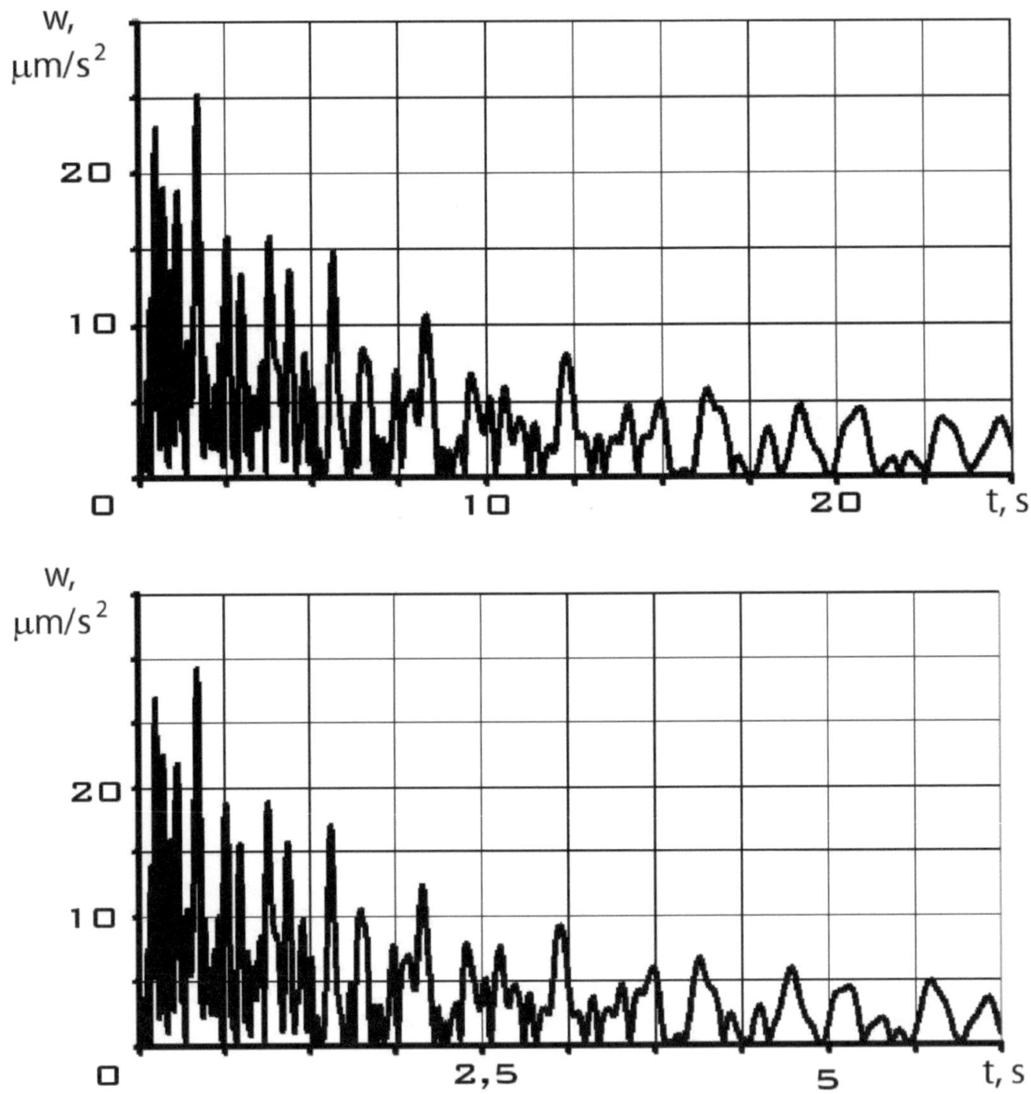

Figure 2. The feature of constructive component of microacceleration field when scaling of inertia-mass parameters and sizes of spacecraft.

$$d\Phi_\tau = \mu \frac{M}{I}(x+s)dx, \tag{3}$$

where M is moment of orientation engine, s is projection of radius vector of fixing point of elastic element relatively to center of mass of space laboratory on x-axes (**Figure 3**) and $I = I_0 + \left(\frac{1}{3}\mu\iota^3 + \mu\iota s^2\right)$—inertia moment of space laboratory. At that, beam is considered to be homogeneous. I_0 is inertia moment of central body. Then, the general inertia force of elastic element is:

http://dx.doi.org/10.5772/67922

$$\Phi_\tau(l) = \mu \frac{M}{2I}(x+s)^2 = \frac{\mu}{2} \frac{M(l+s)^2}{I_0 + \left(\frac{1}{3}\mu\iota^3 + \mu\iota s^2\right)}. \tag{4}$$

When length of elastic element decreases in two times, this formula has the following appearance:

$$\Phi_\tau\left(\frac{l}{2}\right) = \frac{\mu}{2} \frac{M\left(\frac{l}{2}+s\right)^2}{I_0 + \left(\frac{1}{24}\mu\iota^3 + \mu\frac{\iota}{2}s^2\right)}. \tag{5}$$

Equations (4) and (5) give more than double decrease of the real tangential inertia force of spacecraft "NIKA-T" [13] after substitution of basic parameters of the spacecraft. At that, energy losses are not taken into account there. It means the reduction of amplitude of reaction force of embedding. The reaction force to the spacecraft's body and causes microaccelerations by its moment around the spacecraft's center of mass.

Losses of energy in the attaching lug of elastic element are not taken into account. It means reduction of the amplitude of reaction force of embedding. The reaction force transmits to the body pf spacecraft and causes microaccelerations, because it creates momentum around center of mass of spacecraft. Mass and size of spacecraft should be reduced so as angular accelerations from tangential inertia forces are similar to get initial amplitude of microaccelerations. It is considered to be a condition of constancy of the form of constructive component caused by vibrations of elastic element (**Figure 2**). Thus:

$$\varepsilon(l) = \frac{M[\Phi_\tau(l)]}{I} = \frac{\Phi_\tau(l)s}{I} = \frac{\mu}{2} \frac{M(l+s)^2 s}{\left(I_0 + \frac{1}{3}\mu\iota^3 + \mu\iota s^2\right)^2}, \tag{6}$$

$$\varepsilon\left(\frac{l}{2}\right) = \frac{M\left[\Phi_\tau\left(\frac{l}{2}\right)\right]}{I_1} = \frac{\Phi_\tau(l)s_1}{I_1} = \frac{\mu}{2} \frac{M\left(\frac{l}{2}+s_1\right)^2 s_1}{\left(I_{10} + \frac{1}{24}\mu\iota^3 + \mu\frac{\iota}{2}s_1^2\right)^2}. \tag{7}$$

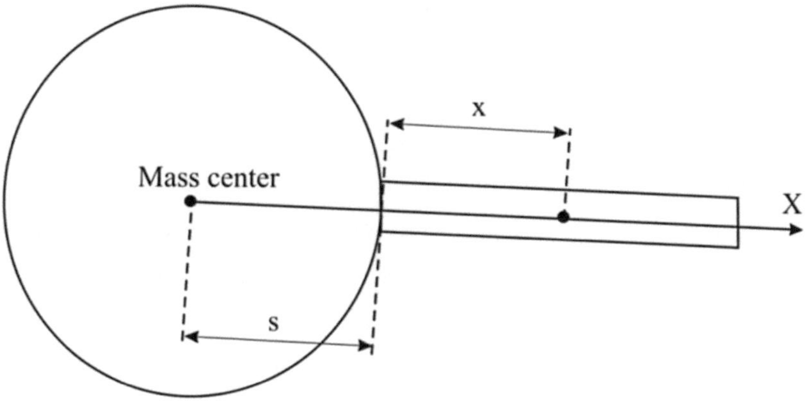

Figure 3. Circuit of space laboratory.

Inertia moment of central body is e according to the formula: $I_{i0} = \frac{m_i R_i^2}{4}$, where m_i and R_i—mass and radius of central body for i-iteration accordingly.

The initial spacecraft determined by scheme on **Figure 3** is considered to be the zero iteration ($i = 0$). The spacecraft with elastic element length of which is less than two times the initial length $\left(l_1 = (0.5)^1 l_0\right)$ and inertia-mass parameters and sizes are matched so as angular accelerations in Eqs. (6) and (7) are equal ($\varepsilon(l) = \varepsilon\left(\frac{l}{2}\right)$) is considered to be the first iteration. The spacecraft with length of elastic element which is more than two times the initial length ($l_{-1} = (0.5)^{-1} l_0$) and inertia-mass parameters and sizes are matched so as angular accelerations in Eqs. (6) and (7) are equal $\left(\varepsilon(l) = \varepsilon\left(\frac{l}{2}\right)\right)$ is considered to be the minus first iteration. Thus, there is view that the following change is lower than the elastic element while iterations:

$$l_i = \left(\frac{1}{2}\right)^i l_0. \tag{8}$$

Besides, the values of i can be positive (reduction of spacecraft's sizes relatively to initial) as well as negative (increase of spacecraft's sizes relatively to initial).

It is necessary to question: Would inertia-mass parameters of spacecraft change as synchronously so length of elastic element does? Is formulation of changing low of these parameters in the form similar to Eq. (8) possible? If the answers to the questions are positive, there are quality which is similar to the self-affine property of the cosine part Weierstrass-Mandelbrot function Eq. (1). Oppositely, if parameters will change disproportionally and angular accelerations would not be equal after each iteration, the analogy with fractal properties would be inapplicable. So, it is impossible to apply fractal functions for study and modeling of microacceleration field.

To search answers for these questions, the generalized parameter that characterizes inertia-mass features of spacecraft was formed [14]:

$$z = \frac{\sum_{i=1}^{N} m_i}{m_0 + \sum_{i=1}^{N} m_i}, \tag{9}$$

where m_0 is mass of central body; m_i is mass of i-elastic element and N, is number of elastic elements of spacecraft.

The generalized parameter in Eq. (9) has the following form for investigated situation:

$$z = \frac{m_1}{m_0 + m_1}. \tag{10}$$

In fact, generalized parameter z characterizes mass percent of elastic elements to the total mass of spacecraft. The equality of angular accelerations is the condition of constancy of appearance of time-dependence of microaccelerations which correspond to **Figure 2**:

$$\varepsilon(l_{-1}) = \varepsilon(l_0) = \ldots = \varepsilon(l_i). \tag{11}$$

The value l obeys the law in Eq. (8). It is necessary to provide equality of Eq. (11) without changing of geometrical parameters of central body of spacecraft (radius and coordinates of fixing points of elastic element). In fact, it means multiple changing of elastic element's length and mass of central body to provide in Eq. (11) independently from sizes of spacecraft's body. Results of these actions for basic parameters of spacecraft "NIKA-T" ($l_0 = 2$ m, $\mu = 10$ kg/m, $M = 1$ H m, $m_0 = 6000$ kg) are shown in **Figure 4**. The following equation plays the role of coefficient, which determines multiplicity of changing of generalized parameter z:

$$kz_i = \frac{z_{i-1}}{z_i}. \tag{12}$$

According to **Figure 4**, kz_i coefficients significantly differ for different iterations. However, it is absolutely predictable, because changing of mass was conducted while permanent size of spacecraft's central body. Only length of elastic element was changed accordingly to the law in Eq. (8). It is hard to imagine significant changing of mass of spacecraft without changing of radius of central body. Thus, search of answers to be mentioned above questions after conducted in which the analysis will be leaded to the following. Is it possible to realize such changing law of central body's radius to achieve the following:

- values of kz_i do not significantly differ for different iterations;
- values of analogical coefficient for radius changing $kR_i = \frac{R_{i-1}}{R_i}$ do not significantly differ for different iterations;
- changing of mass and radius are connected to each other, it means that they do not contradict to each other (reduction of mass leads to reduction of radius and vice versa).

According to the investigations, conducted for spacecraft, shown in **Figure 3**, such law exists. The sample dispersion of values kz_i was analyzed for different values $kR_i = Const$, using the

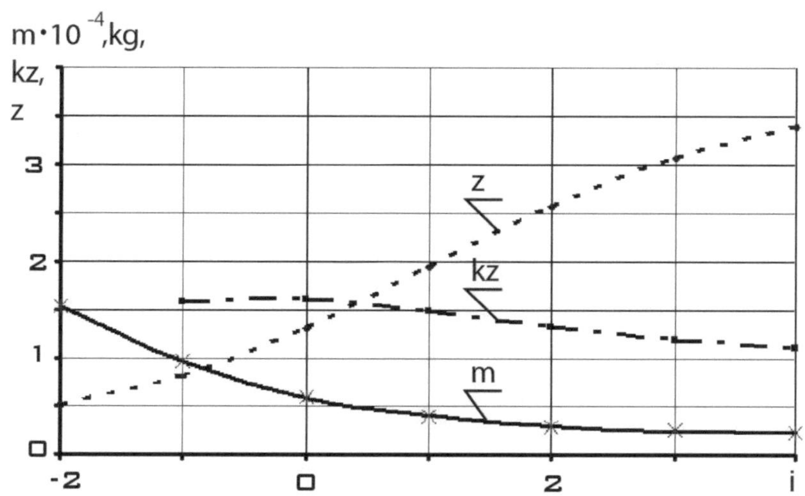

Figure 4. Changing dynamic of spacecraft's mass and generalized parameter if Eq. (11) is prominent for iterations $-2 \le i \le 4$.

same quantitative data, which correspond to basic parameters of spacecraft "NIKA-T." The results of quantitative analysis are shown in **Figure 5**.

Minimum of the dispersion is achieved closely to the value $kR = 0.4$. Thus, changing law of radius of spacecraft's central body, which is analogical to Eq. (8), has the following form:

$$R_i = (0.4)^i R_0. \tag{13}$$

At that, generalized parameter z changes accordingly to approximated law:

$$z_i = (1.873)^i z_0. \tag{14}$$

Then, changing dynamic of the parameters with taking into account radius's changing, which is analogical to **Figure 4**, has the form shown in **Figure 6**.

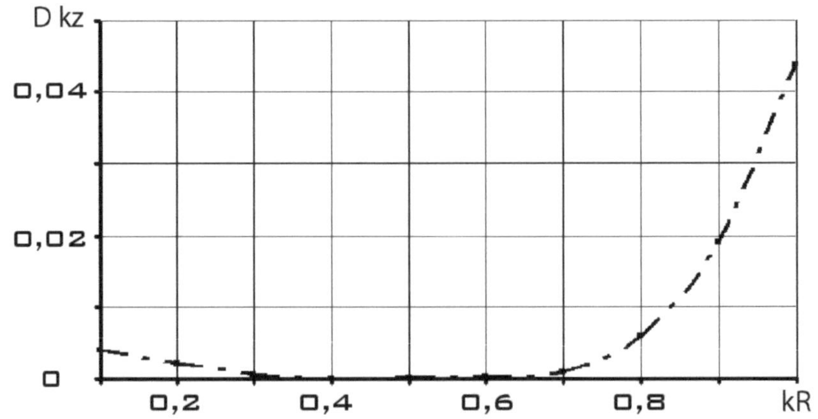

Figure 5. Sample dispersion of values kz_i in dependence on kR.

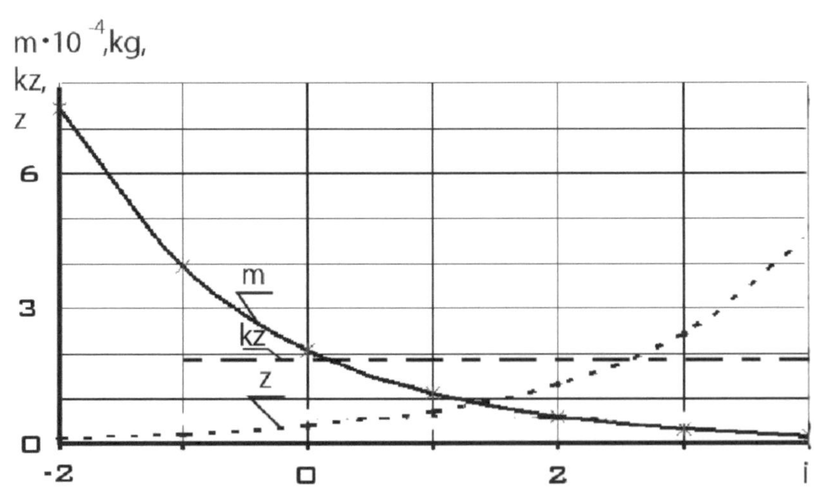

Figure 6. Changing dynamic of spacecraft's mass and generalized parameter when Eq. (11) is right taking into account changing of spacecraft's radius for iterations $-2 \le i \le 4$.

Consequently, we can confirm that constant scaling coefficients of central body's radius and generalized parameter z are found for viewed example of spacecraft. These coefficients provide invariance of appearance of dependence of constructive component of microaccelerations on time (**Figure 2**). The laws in Eqs. (8), (13) and (14) are realized simultaneously.

In this case, the fractal quality of constructive component of microaccelerations can be determined as *invariance of the form of time-dependence of microaccelerations' module whole scaling of spacecraft on generalized parameter z.*

Scaling coefficients among axes of graph of time-dependence of microaccelerations are equal to 0.25 among time-axis and one among microacceleration-axis. Thus, the graph does not change its appearance as well as values of microaccelerations. It is very important for providing of required microacceleration level in the area with technological equipment. Graph compression among time-axis is explained by fourfold increase of frequency of proper oscillations of a beam while double reduction of its length.

There is used the simple scheme of spacecraft to show scaling property without hard computing in mentioned case. The same situation would be for more complicated scheme which is close to real spacecraft.

Scaling coefficients of mass km and radius kR are determined accordingly to the chosen scaling coefficient of length $kl = 0.5$ in such a way as generalized parameter z was changed according to the law:

$$z_i = kz_i \cdot z_0, \tag{15}$$

where z_0 is value of spacecraft's generalized parameter while $i = 0$, i.e., for basic spacecraft, which is the base for the future different types of spacecraft.

Application of another changing law of the length in place of Eq. (8) also does not change anything fundamentally. In this case, scaling coefficient of time-dependence of microaccelerations among time-axis will differ from 0.25.

It is worth to mention that only constructive component of microacceleration field has this property. Neither metastable even more nor random components according to classification [8] meets the fractal property. Metastable component is determined by the sphere of cosmic space where spacecraft operates. The component forms the part of microacceleration field, which is caused by outer disturbing factors. This disturbance, such as aerodynamic drag and light pressure, can change when spacecraft is scaled and can remain unchanged, such as magnetic disturbance. That is why metastable component has no scaling property.

The same thing is for random component. Appearance of emergency situations, failure of different equipment or random oscillations of basic parameters of outer disturbing impact on spacecraft is not connected with its scaling. On the other hand, probability of hitting of micrometeorite to spacecraft is connected with square of its outer surface and will change with scaling. Thus, we can claim that random component has no scaling property.

However, the remarks relatively to metastable and random components do not reduce the value of the found property of constructive component of microacceleration field. Actually, just constructive component is determined already in early developmental stages of spacecraft. This component characterizes dynamical features of spacecraft's constructional layout and takes into account the algorithms and ways to orient its orbital motion. Only constructive component can be influenced by using of the total power of different methods to provide and control microacceleration level in the area with technological equipment. This power can be significantly strengthened by application of fractal analysis as well as classical.

Scaling property of constructive component of microacceleration field allows application of methods of fractal analysis for designing of technological spacecraft as well as for correct data processing of microacceleration measurements, finding of failures of gauges and recovery of lost measuring data. It is very hard to do all of these only on the base of classical methods of analysis. The perspectives of addition of classical methods described by fractal property of constructive component of microaccelerations are mentioned below.

The launchings of spacecraft with specialized technological purposes happen with big intervals on the modern stage of space technologies' development. For instance, spacecraft "FOTON – M" No 3 was launched in 14 September 2007 and the following spacecraft "FOTON – M" No 4—only after almost 7 years in 19 July 2014. Thus, it is evident that it is impossible to arrange space experimental batch manufacturing only on the base of specialized space technic. One of the perspective ways to solute the problem is the wide application of small spacecraft. One start of carrier rocket of middle class is able to provide launching of tens small spacecraft to their orbits. Series-produced homotypic spacecraft can provide meeting of similar conditions for realization of concrete gravity-sensitive process. Continuous production can be organized using project launchings with small spacecraft as additional load. However, only application of small spacecraft would significantly limit production value. Limits on mass would not allow placement on spacecraft's board more than 100 kg of technological equipment and materials for production.

The approach, which includes advantages of spacecraft with specialized technological purposes and small spacecraft, is necessary to meet requirements of experimental batch production in space. This approach should integrate opportunities of these ways of space technic's development: regularity of launchings and significant scales of production while maintenance of equal favorable conditions for realization of gravity-sensitive processes. It is evident that such task statement leads to necessity to create line of spacecraft from different classes which meet these equal conditions. The scaling property of constructive component of microaccelerations should be applied to solute this task. For this purpose, the mathematical model of microaccelerations should be built using fractal function. Fractal model of microaccelerations worked out in the articles [15, 16] on the basis of cosine part of Weierstrass-Mandelbrot function in Eq. (1) can be example of such model. Behavior of an average value of microaccelerations as a random process was compared with behavior of cosine part of Weierstrass-Mandelbrot function with usage of statistical analysis. At that parameters of the function were identified with spacecraft's characteristics. So, functional fractal dimension D is identified with moment from orientation engine of spacecraft and scaling parameter b—with generalized parameter z.

Such line approximately can be presented in the form of scheme shown in **Figure 7** [13], where "NIKA-T" is the base spacecraft and actually created space objects are used.

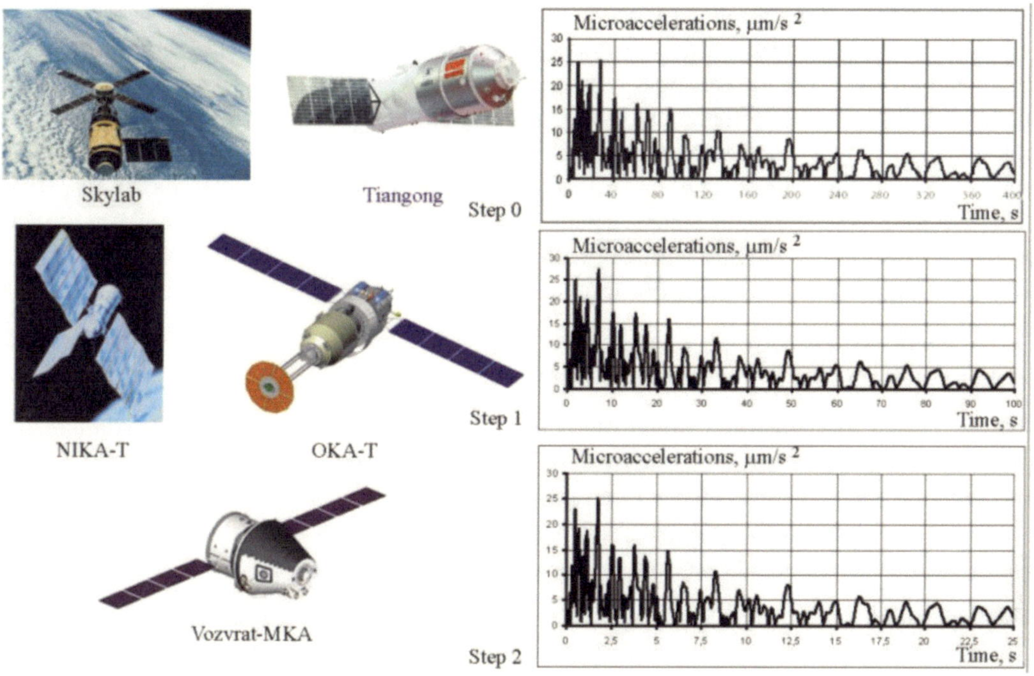

Figure 7. Line of spacecraft from different classes, which realize equal microacceleration level in the area with technological equipment.

The line developed using fractal property would guarantee a correspondence of microacceleration levels on the spacecraft from different classes. It would allow conducting of the same process on the spacecraft and taking a step toward full-scale experimental batch production in space.

Described property also can be effectively applied for study of microacceleration field of inner environment of spacecraft. Correctness of the gauges' run after launching overloads can be estimated by comparison of experimental data with results of on-board measurements. There is worked out an algorithm of measuring equipment efficiency check on the basis of fractal property of constructive component of microaccelerations in Ref. [13]. The algorithm is effective when constructive component dominates in microacceleration field of inner environment. Significant differences between the fractal model and measuring data can be explained by following factors:

- There is no dominance of constructive component in microacceleration field.
- An additional source of microaccelerations presents, for instance, as the result of emergency situation.
- The gauge is failed.

Thus, checking of efficiency of all gauges after launching overloads is possible. It is unachievable by classical methods of analysis.

For instance, failed gauge was found in the result of such checking. It is possible to avoid significant mistakes while forming of model of microacceleration field of inner environment of spacecraft by neglect of measuring data from the gauge. However, accuracy would be reduced along with loss of measurement channel from failed gauge. In this case accuracy can be raised using fractal analysis. Well-known method of majority control [13] can be applied for replacement of the measurement channel by fractal mathematical model of microaccelerations, which was used for measuring equipment efficiency check. Undoubtedly, accuracy of such estimation would be lower in comparison with valid effective measurement channel. On the other hand, it would be higher than without application of majority control.

One of the possible algorithms of majority control with the aim to find out failed gauge and to smooth results of measuring data processing is shown in **Figure 8** [2].

At that, fractal mathematical model plays the role of additional measurement channel. The presence of significant statistical differences between measurements with additional channel and without ones can be evidence of failed gauge. If the differences will be considered to be random, we can claim that all gauges run correctly. It increases the reliability of measuring data which leave much to be desired on the present developmental stage.

Thus, invention of scaling property of constructive component of microacceleration field which is similar to self-affine property of fractal functions allows application of methods of fractal analysis for creation of a line of spacecraft from different classes. At that, these spacecraft should realize the equal microacceleration level in the area with technological equipment.

The application of the scaling property of constructive component of microaccelerations gives the ability to select the row of values of generalized parameter z, which provides invariance of the form of time dependence of microaccelerations. The generalized parameter will allow a choice of corresponding inertia-mass characteristics of spacecraft and creation of the family of spacecraft (from orbital space stations to small spacecraft) which realize the identical microacceleration level.

Such result could not be achieved by classical methods because the fractal property is the base of supposed approach to creation of line of spacecraft from different classes. At that, it is worth to note that application of fractal analysis gives a significant practical result. Actually, creation of spacecraft line which realizes equal microacceleration level will have to be allowed in the beginning of pilot batch production in space already on the present developmental stage of space-rocket technic. The potential and advantages of spacecraft from all classes included in the line created on the base of scaling property of microaccelerations will be applied in the production. It is a large step forward in comparison with separated irregular experimental investigations with technological purposes, which are conducted in separated spacecraft today. On the other hand, creation of line of spacecraft on the basis of scaling property of microaccelerations is one of a few ways which lead to pilot batch-space-production on the present stage.

In spite of the main aim of designing of the spacecraft line, fractal model of microaccelerations, built, for example, on the basis of Weierstrass-Mandelbrot function, combined with classical methods of analysis allows solution of other tasks. These tasks include increasing of reliability

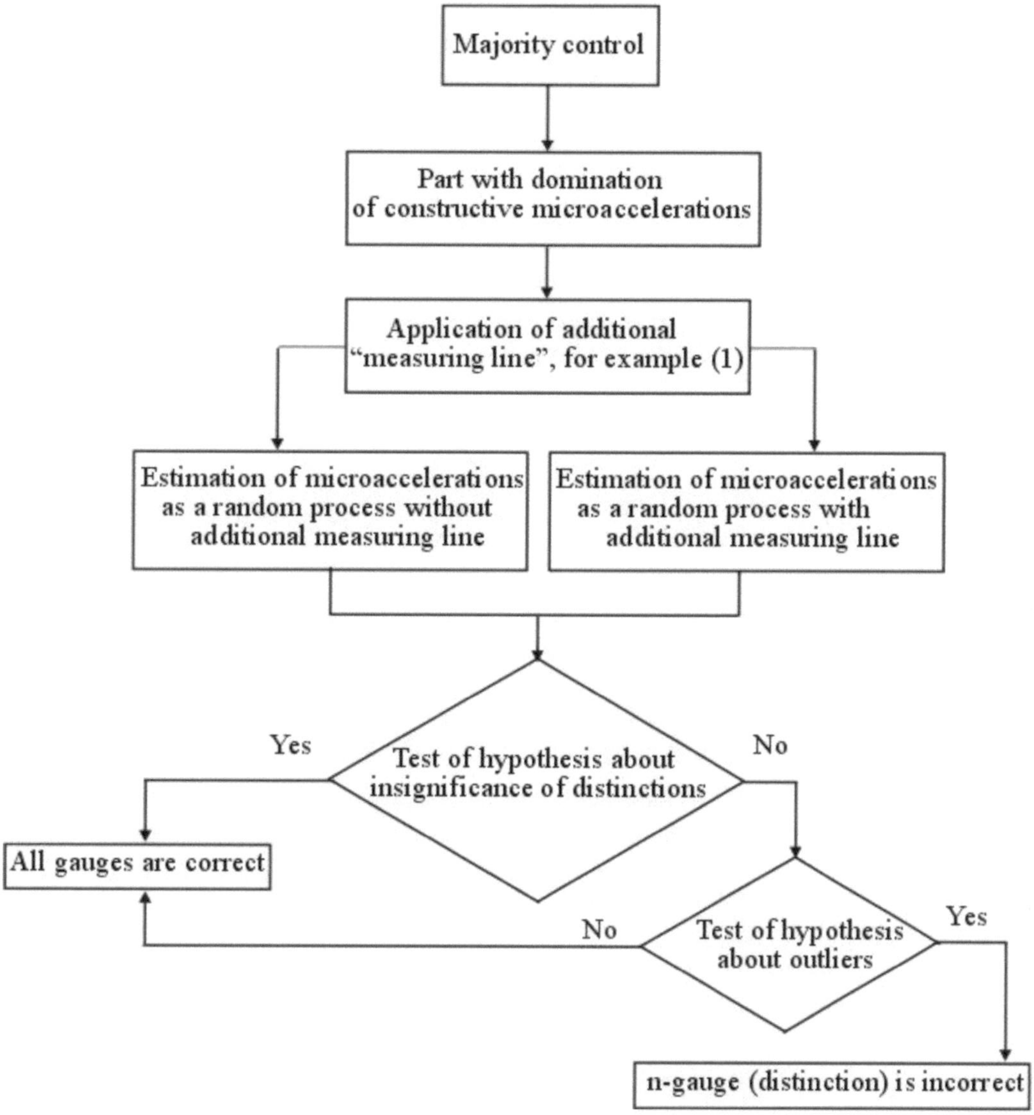

Figure 8. Algorithm to realize majority control over gauges during exploitation stage.

of data from on-board measuring equipment for microaccelerations; increasing of efficiency of control over microaccelerations during conducting of gravity-sensitive technological processes and also taking the decisions about possibility and advisability to realize some or another gravity-sensitive process on the board of concrete spacecraft.

We can claim that fractal analysis significantly widens opportunities of classical methods for conducting of majority control of microaccelerations, for recovering of lost measuring data and also for taking the decisions about realization of active control over microaccelerations on the exploitation stage of spacecraft.

Evidently, significant complications of spacecraft's model are necessary for effective application of fractal analysis and achieving required accuracy of estimations. The model described in the chapter was chosen only for approval of existence of scaling property of constructive component of microacceleration field and its illustration.

Author details

Andrew V. Sedelnikov and Ksenia I. Potienko*

*Address all correspondence to: potienko97@mail.ru

Samara University, Samara, Russia

References

[1] A.V. Sedelnikov, The problem of microaccelerations: from comprehension up to fractal model, (Moscow, Russian Academy of Sciences: The Elected Works of the Russian school, 2012), p. 277.

[2] A.V. Sedelnikov, K.I. Potienko, How to estimate microaccelerations for spacecraft with elliptical orbit, Microgravity Sciences and Technology, 28 (1) (2016) 41–48.

[3] R.G. Owen, D.I. Jones, A.R. Owens, A.A. Robinson, Integration of a microgravity isolation mount within a Columbus single rack, Acta Astronautica, 22 (1990) 127–135.

[4] V.V. Sazonov, S.Yu. Chebukov, V.I. Abrashkin, A.E. Kazakova, A.S. Zaitsev, Analysis of low-frequence microaccelerations onboard the foton-11 satellite, Space Research, 39 (4) (2001) 419–435.

[5] I.A. Babushkin, A.F. Glukhov, V.A. Demin, E.A. Zilberman, G.F. Putin, Measurement of inertial microaccelerations with the use of convection sensors, Journal of Surface Investigation: X-Ray, Synchrotron and Neutron Techniques 3 (1) (2009) 142–147.

[6] G.F. Putin, A.F. Gluchov, I.A. Babushkin, D.A. Zavalishin, M.Yu. Belyaev, A.I. Ivanov, V.V. Sazonov, Study of microaccelerations onboard the international space station with the DAKON-M convection sensor, Space Research 50 (5) (2012) 346–342.

[7] K.I. Potienko, Basic development stages of the algorithms applied to recover lost microacceleration data and check efficiency of measuring equipment on the board of space laboratory, American Journal of Aerospace Engineering, 3 (1) (2015) 10–16.

[8] A.V. Sedelnikov, Classification of microaccelerations according to methods of their control, Microgravity Sciences and Technology 27 (3) (2015) 245–251.

[9] A.V. Sedelnikov, On the fractal properties microaccelerations, Microgravity Sciences and Technology 24 (5) (2012) 345–350.

[10] A.V. Sedelnikov, D.P. Molyavko, K.I. Potienko, Study of over-assessment of microaccelerations when using a beam-model of elastic elements, International Review of Aerospace Engineering 9 (1) (2016) 9–12.

[11] A.V. Sedelnikov, A.A. Kireeva, Alternative solution to increase the duration of microgravity calm period on board the space laboratory, Acta Astronautica 69 (6–7) (2011) 480–484.

[12] A.I. Belousov, A.V. Sedelnikov, K.I. Potienko, Study of effictive application of electric jet engine as a mean to reduce microacceleration level, International Review of Aerospace Engineering 8 (4) (2015) 157–160.

[13] A.V. Sedelnikov, The usage of fractal quality for microacceleration data recovery and for measuring equipment efficiency check, Microgravity Sciences and Technology 26 (5) (2014) 327–334.

[14] A.V. Sedelnikov, A problem of choosing a generalized parameter of flexible spacecraft structures to construct a fractal model of microaccelerations, Russian Aeronautics 51 (1) (2008) 84–86.

[15] A.V. Sedelnikov, Fractal assessment of microaccelerations at weak damping of natural oscillation in spacecraft elastic elements. I, Russian Aeronautics 49 (3) (2006) 329–331.

[16] A.V. Sedelnikov, Fractal assessment of microaccelerations at weak damping of natural oscillation in spacecraft elastic elements. II, Russian Aeronautics 50 (3) (2007) 322–325.

Chapter 6

Specific Emitter Identification Based on Fractal Features

Janusz Dudczyk

Additional information is available at the end of the chapter

http://dx.doi.org/10.5772/67894

Abstract

If we take into consideration the fact that the radar signal recognition and identification process is an integral part of contemporary combat operations, the importance of the fractal analysis increases significantly. For this reason, the fractal analysis is used in the process of radar sources identification on the contemporary battlefield. Radar Signal Recognition (RSR) with the use of classical methods, that is based on statistical analysis of basic measurable parameters of a radar signal, such as Radio Frequency (RF), Amplitude (A), Pulse Width (PW) or Pulse Repetition Interval (PRI) is not enough to carry out the distinction process of particular copies of the same radar type. Only by this approach, the identification process of particular copies in a set of the same type emitters can be carried out. As a result, it is possible to maximize Correct Identification Coefficient (CIC) in the final stage of the recognition process, which is realized in Electronic Warfare (EW) systems. One of the most important elements of the whole recognition and identification process, which is realized in ELectronic INTelligence (ELINT) battlefield system, is building a measurement data vector, then a radar's metrics and the same database. This approach is called Specific Emitter Identification (SEI).

Keywords: fractal feature, pattern of radar, Specific Emitter Identification (SEI), radar vector parameters, ELINT system, Radar Signal Recognition (RSR), Correct Identification Coefficient (CIC)

1. Introduction

Developing an innovative method for generating distinctive features extracted from radar signals in order to achieve explicit identification is a main goal in the process of Specific Emitter Identification (SEI). As a result, it is possible to maximize Correct Identification Coefficient (CIC) and identify particular copies of radars of the same type in ELectronic INTelligence (ELINT) system on the contemporary battlefield. The presented achievements in this chapter concern methods and techniques adjusted to electromagnetic emitter source recognition and identification with particular reference to the systematic approach, that is, from the acquisition process, through initial data transformation, main transformation, classification and identification and building a model in the DataBase (DB). Only a comprehensive approach can be coherently fitted into a harmonious whole of all the ELINT recognition systems' processes on the contemporary battlefield. As a result, this approach may contribute to an increase in effectiveness of activity by minimization of time, which is necessary for the decision process realized in Electronic Warfare systems.

Generally, it is possible to distinguish the main task, which has a significant contribution to the development of recognition and identification of radar signals, that is, developing innovative methods for generating distinctive features of radar signals [1–5]. As mentioned above, the task is connected with developing a method for setting the structure of basic measurable parameters of a radar signal in the form of time-frequency-formalized Pulse Descriptive Word (PDW) [5, 6], developing and implementing in SEI process, an advanced Hierarchical Agglomerative Clustering Algorithm (HACA) based on the 'bottom up' agglomerative formula, which makes it possible to receive dendrograms of hierarchical clustering for pulse repetition intervals and their distinctive analysis [1], applying the inter-pulse modulation analysis to extract distinctive features [7], developing an innovative histogram procedure to build PRI decomposition [2] and finally developing an innovative method for defining a transformation attractor of radar signals measurement clusters [3, 4]. This approach is called Specific Emitter Identification.

2. Classic recognition of radar signals

As a general rule, the systems aiming at acquisition, analysis and recognition of radar signals on the contemporary battlefield are autonomous systems, which are made of electronic recognition devices doing ELINT tasks. These systems accomplish complex procedures in the scope of emission acquisition, analysis, transformation and radar emission recognition with the range of wavelengths 0.5–18 GHz, long-term data archivization and full synthesis and fusion of information. These systems, in most cases, are made of the following subsystems and modules, that is, radar signal acquisition subsystem, radar signal processing and analysis subsystem, database management subsystem and communication module[1] between particular subsystems.

Accomplishing basic tasks in the designed systems of radar signal acquisition, analysis and recognition on the contemporary battlefield is based on:

- automated searching for and detecting[2] electromagnetic emitters (in the range of wavelengths 0.5–18 GHz);

[1] Communication module between particular subsystems and components of acquisition system and radar signal analysis on the contemporary battlefield is not the subject of this chapter.

[2] In the aspect of tasks connected with searching for and detecting signals and measuring their parameters, the specialist equipment of the radar signal acquisition, analysis and recognition system makes it possible to accomplish the tasks above on the land, from the air or the sea and may be adjusted to a plane, helicopter, a ship or installed in a special container case.

- automated parameter measurement of the detected emitters;
- analysis of measured signal parameters in a thick electromagnetic environment (a few thousand or more pulses) and initial measurement data processing (initial parametric selection and/or signal reduction);
- main signal processing (extraction of basic features and estimation of basic measurable radar signal parameters i.e. signal radio frequency, amplitude, pulse width, pulse repetition interval) on the basis of statistical functions concerning estimating e.g. average values of parameters, class models and hypothesis verification;
- radar signal archivization in files with measurement data in the DataBase and updating procedure as well as DataBase structure modification;
- radar types recognition and classification on the basis of radar signals received from them by comparing signal parameters with the model in BD in the shortest time possible and using expert's knowledge in the process of emitter sources recognition, classification and location;
- visualization of signals, measurement data clusters, results of recognition and other data and distinctive information presented in a tabular and graphic form;
- depict results of recognition on a digital map and automatically adjusting a unit to the area;
- possibilities of import and analysis of measurement data from electronic recognition, which is received from other sources and data measurement and recognition information fusion;
- creating a simulation software to generate warfare scenarios, test correctness of emission sources classification, recognition and location procedures and to estimate the effectiveness of the system and trainings for operators.

The analysis of radioelectronic situation on the contemporary battlefield and long-term radar signal measurements makes it possible to admit that during the process of signal recognition and classification, there is a phenomenon of penetrating ranges of radar parameters, many ranges of particular parameters for single radars, different types of emissions (constant, pulse, interrupting), complexity of sounding pulses and specific work properties (signal polarization depending on the weather), decrease in frequency of repetition and top pulse as the beams go above the horizon or the change of top power in case of e.g. weather condition changes. The general block diagram of radar signal acquisition, analysis and recognition is presented in **Figure 1**.

The radar recognition system (see **Figure 1**) is only able to recognize and classify particular types of radars. A definitely more advanced one is the recognition process understood in a 'narrow sense'. Its aim is to identify these signals, thus their emission sources as well. I deal with the recognition understood in a 'narrow sense', which concerns identification of particular radar copies of the same type depending on the detail level.

The process of distinguishing the radar emission source even 'a single copy' is the exact identification of radar signal source in the aspect of SEI. Thus, applying in the acquisition system, the

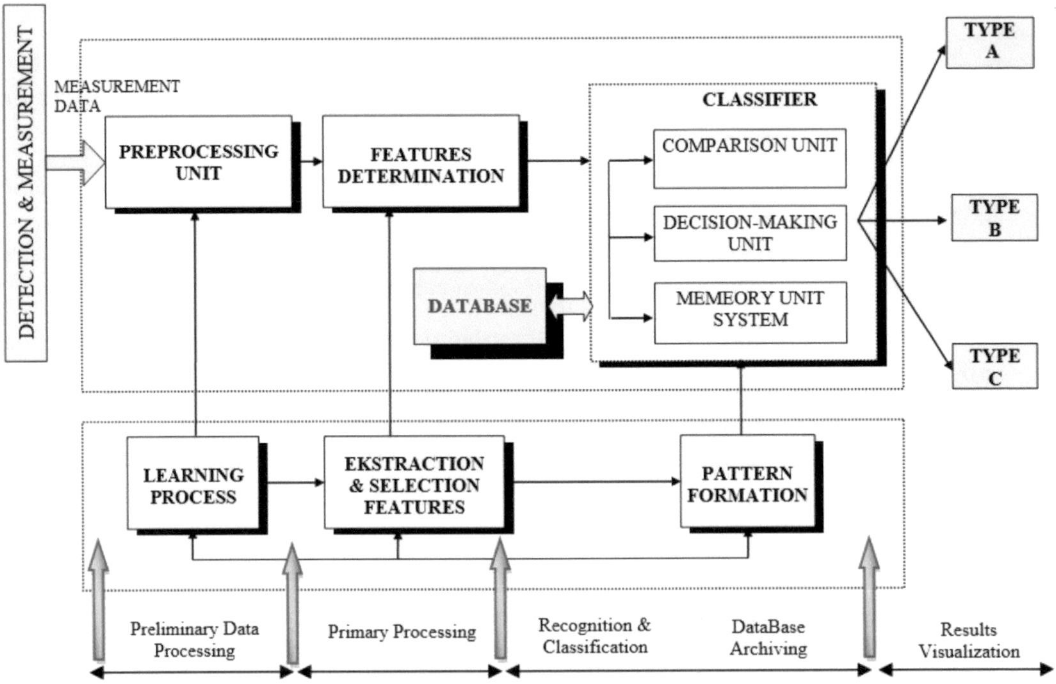

Figure 1. The process of information processing in the radar recognition system.

analysis and recognition of radar signals, inter-pulse analysis of signal parameters (defining the type of inter-pulse modulation and estimation of parameters' modulation), intra-pulse analysis (defining the type and parameters of intra-pulse modulation on the basis of a single pulse), applying innovative methods for generating distinctive features, using fast-decision identification algorithm and advanced DataBases prepared as a result of modelling entity relationships and using AI[3] devices is an immanent specificity of ELINT systems and makes it possible to identify particular copies of radar emitters with the use of the above dedicated methods for generating distinctive features' signals in Electronic Warfare (EW) systems. As concerns contemporary used ELINT class systems, data classification and recognition techniques are currently developed fields of science, it is not possible to formulate optimal model of their structure and maximize the recognition and classification function as well as the identification function.

In **Figure 2**, a block diagram of the acquisition, analysis and identification system, including the subsystems that implement modern and advanced methods for generating distinctive fractal features, is presented.

[3]AI (Artificial Intelligence), a term of artificial intelligence, which in BD systems is realized on the basis of using artificial neural networks (so-called AI bionic trend) and expert systems, based on predicate calculus (so-called AI pragmatic trend).

http://dx.doi.org/10.5772/67894

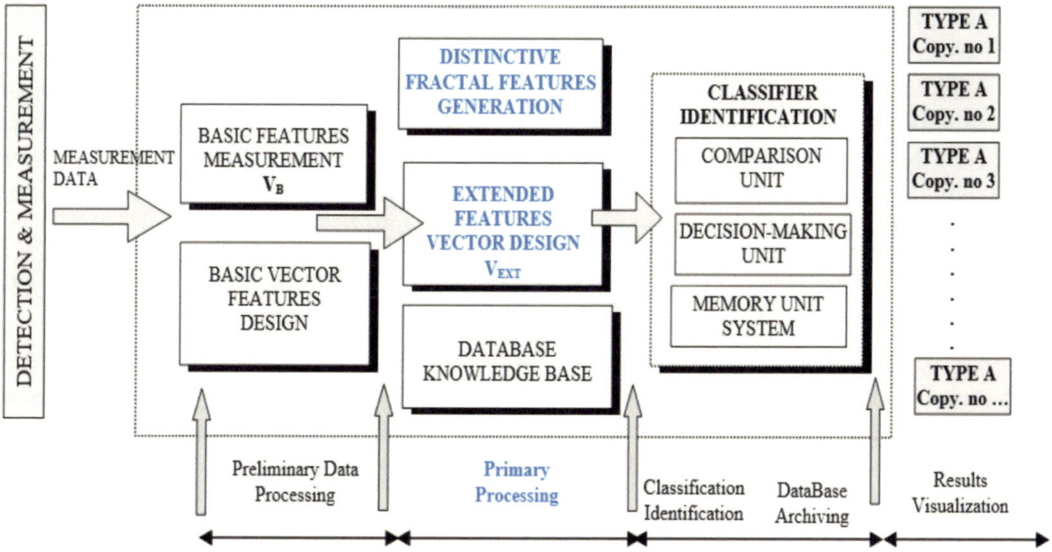

Figure 2. The process of distinctive fractal features generation in the RSR system.

3. Innovative method for generating distinctive features based on fractal analysis

RSR with the use of classic techniques, which are based on the statistical analysis of basic measurable parameters, such as radio frequency, signal amplitude, pulse width, and pulse repetition interval, is completely not sufficient for SEI process to carry out the process of distinguishing particular radar copies of the same type.

For this reason, on the stage of initial data processing, a method for defining the structure of basic measurable parameters of a radar signal in the form of formalized time-frequency Pulse Description Word was developed. These vectors are input data into the further process of generating distinctive features, in the main processing stage [3, 4]. As mentioned above, the PDW vector is a formalized data structure of record type, where particular fields consist of frequency parameters and time parameters of radar signal according to Eq. (1), where Nr(k) is the number kth of the pulse, $t_p(k)$ is the time of appearing kth pulse in [μs], A(k) is the amplitude of kth pulse, PW(k) is the width of kth pulse in [μs], PRI(k) is the Pulse Repetition Interval of kth pulse in [μs], RF(k) is the Radio Frequency of kth pulse in [MHz], n is the number of pulses in the record of these which are qualified to the analysis while k is the number of pulses in the measured sample.

$$\mathbf{PDW} = \begin{bmatrix} \mathrm{Nr}(1) & t_p(1) & A(1) & \mathrm{PW}(1) & \mathrm{PRI}(1) & \mathrm{RF}(1) \\ \mathrm{Nr}(2) & t_p(2) & A(2) & \mathrm{PW}(2) & \mathrm{PRI}(2) & \mathrm{RF}(2) \\ \cdots & \cdots & \cdots & \cdots & \cdots & \cdots \\ \mathrm{Nr}(k) & t_p(k) & A(k) & \mathrm{PW}(k) & \mathrm{PRI}(k) & \mathrm{RF}(k) \\ \cdots & \cdots & \cdots & \cdots & \cdots & \cdots \\ \mathrm{Nr}(n) & t_p(n) & A(2) & \mathrm{PW}(n) & \mathrm{PRI}(n) & \mathrm{RF}(n) \end{bmatrix} \tag{1}$$

The effect of the further transformation of PDW vector is Basic Signal Vector $\mathbf{V_B}$, whose fields are of particular signal frequency and time parameters, according to Eqs. (2)–(4). The time parameters of the vector $\mathbf{V_B^{PRI}}$ are as follows: minimum pulse repetition interval value PRI_{min}, average pulse repetition interval value PRI_{EV}, maximum pulse repetition interval value PRI_{max}, the number of values of pulse repetition interval $nPRI$, the number of values of average pulse repetition interval $nPRI_{EV}$, the minimum value of pulse width PW_{min}, the average value of pulse width PW_{EV} and the maximum value of pulse width PW_{max}.

$$\mathbf{V_B^{PRI}} = [PRI_{min}, PRI_{EV}, PRI_{max}, nPRI, nPRI_{EV}, PW_{min}, PW_{EV}, PW_{max}] \tag{2}$$

The frequency parameters of the signal vector $\mathbf{V_B^{RF}}$ are defined according to Eq. (3) and are as follows: the minimum value of the signal radio frequency RF_{min}, the average value of the radio frequency for the period RF_{EV}, the maximum value of the radio frequency in the period of changes RF_{max}, the number of values of the radio frequency nRF and the number of average radio frequencies in the cycle of changes nRF_{EV}.

$$\mathbf{V_B^{RF}} = [RF_{min}, RF_{EV}, RF_{max}, nRF, nRF_{EV}] \tag{3}$$

The vector $\mathbf{V_B}$ of the final structure presented according to Eq. (4) consists of parameters concerning information about the accuracy measurements of: radio frequency $sigRF$, pulse repetition interval $sigPRI$ and pulse width $sigPW$.

$$\mathbf{V_B} = [\mathbf{V_B^{PRI}}, \mathbf{V_B^{RF}}, sigRF, sigPRI, sigPW] \tag{4}$$

These parameters are the base for defining the brackets of acceptable changes of radar signal, that is, RF, PRI and PW are used in the estimation process of effectiveness of the Fast decision Identification Algorithm (FdIA), described in [6]. The process of developing the signal vector $\mathbf{V_B}$ also undergoes the process of implementation and automation in the stage of initial data

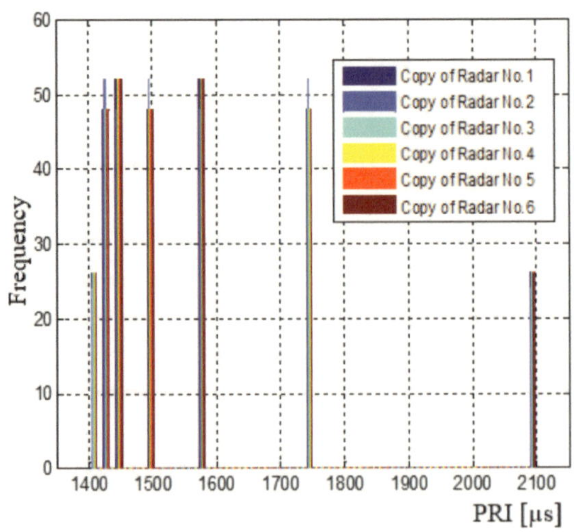

Figure 3. PRI histogram for six copies of the same type of radars marked by colours.

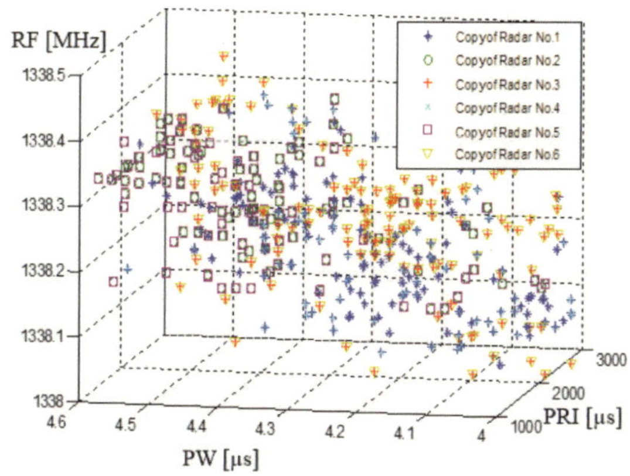

Figure 4. 3D graphic depicting of PW, PRI and RF for six copies of the same type of radars.

processing and the main data processing by ELINT system (see **Figure 2**). During the analysis, in total hundreds of radar samples coming are carried out. The received record collections (e.g. six copies of the same type of radars) with basic measurable parameters of PDW are presented in the form of a graph with basic measurable parameters, that is, RF, PW and PRI in **Figures 3** and **4**. **Figure 3** presents the PRI histogram of six tested radar copies in an overall depiction. **Figure 4** presents a 3D graph of RF, PRI and PW parameters from six copies, in an overall depiction as well.

Figures 5 and **6** present 3D depicting of radio frequency and pulse width for three selected copies of the same type of radars received with the use of 'mesh' function in the MatLab software.

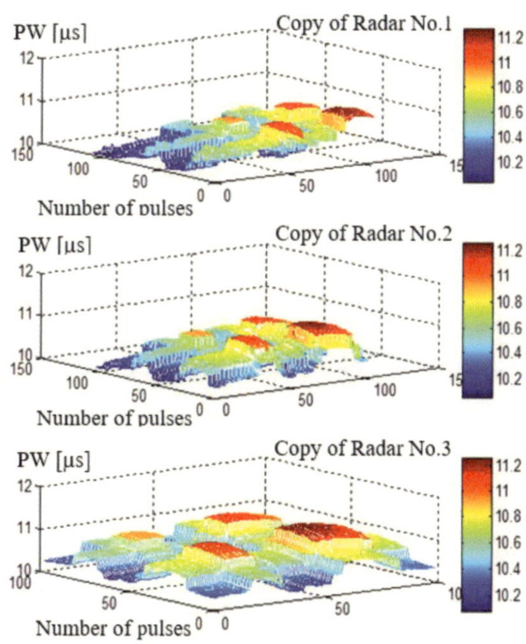

Figure 5. 3D graphic depicting of PW for three selected copies of the same type of radars.

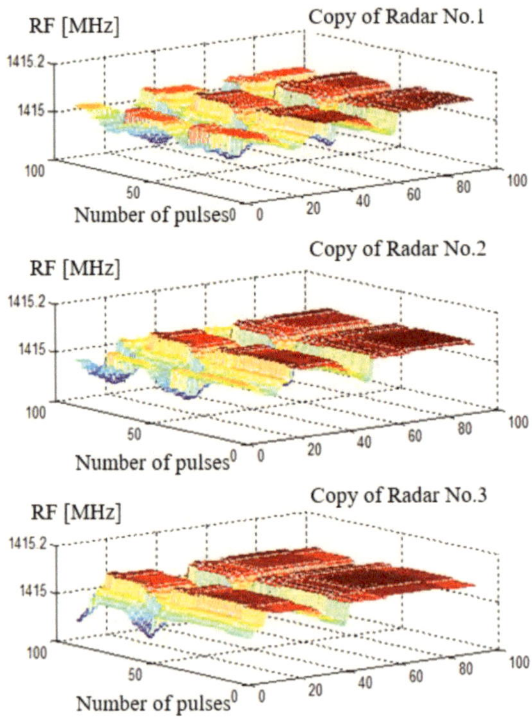

Figure 6. 3D graphic depicting of RF for three selected copies of the same type of radars.

On the basis of the recordings and initial analysis to further process of identification, only these copies were admitted whose basic measurable parameters, that is, RF, PW and PRI were much the same—see **Figures 5–8**. **Figures 5–8** present the biggest similarity of the radar signal parameters which those sources generated.

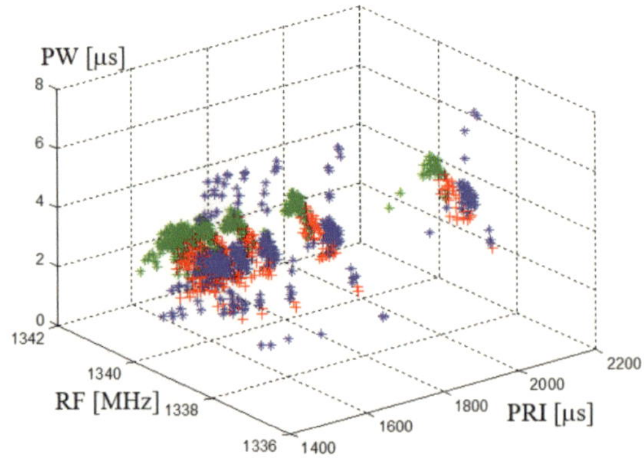

Figure 7. 3D graphic depicting of RF, PRI and PW for three selected copies of the same type of radars marked by three shades of gray.

http://dx.doi.org/10.5772/67894

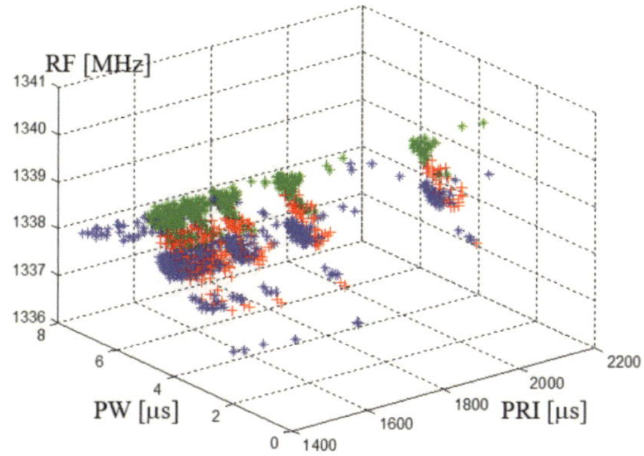

Figure 8. 3D graphic depicting of PW, PRI and RF for three selected copies of the same type of radars marked by three shades of gray.

4. Defining the transformation attractor and distinctive features extraction

One of the ways to increase the number of details of definition is specific identification of electromagnetic emitter sources SEI which extracts distinctive features in the process of signal transformation. The distinctive features may be a result of the received transformations of measurement data sets. New data sets will have fractal features which will make it possible to define clearly the source of radar emission. The fractal features and the theory of fractals is adopted by researchers, especially in the field of SAR (*Synthetic Aperture Radar)* image transformation [8, 9], acoustic signal transformation and the analysis of radar signals. New possibilities of Digital Signal Processing (DSP) in Frequency Modulated Continuous Wave (FMCW) radar and fractal image compression is a promising brand new compression method [10, 11]. It should be noted that the identification of emitter sources based on classical methods of the analysis of basic parameters is currently inefficient. The methods of SEI [12, 13] should be used in order to identify, more precisely, a radar copy of the same type.

4.1. An attractor of transformation

The easiest way to make fractals is by using a set of affine transformations, which are contractions or narrowing transformations. In this case, the set of affine transformations is Iterated Function System (IFS). A recording of radar signal was made. Further frequency values, for which the recording was made, correspond to particular measurement points. By transforming the sets of measurement points in the scope of their symmetry or left-side/right-side asymmetry, what was received was the attractor of transformation which can be a fractal in a special case. As a result, the attractor of generalized measurement function appeared, which was the result of the procedure of SEI described here. While doing the analytical procedure of defining the attractor of measurement function, right-side measurement vectors $\mathbf{p}^{\mathrm{r}}$ and left-side ones $\mathbf{p}^{\mathrm{l}}$ with the beginning in the particular point of reference f_0, so that $\mathbf{p}^{\mathrm{r}} = [p_1^{\mathrm{r}}, p_2^{\mathrm{r}}, \ldots, p_N^{\mathrm{r}}]^T$ and $\mathbf{p}^{\mathrm{l}} = [p_1^{\mathrm{l}}, p_2^{\mathrm{l}}, \ldots, p_M^{\mathrm{l}}]^T$ were assigned. In order to define the desirable selective features, the

$T : \mathbf{p}^r \rightarrow \mathbf{t}$ transformation was done. In this transformation, **t** is the image of the $\mathbf{p}^r$ vector in the form of a vector with coordinates corresponding to the $\mathbf{p}^l$ vector. For the transparent record of the transformation above with the use of vectors $\mathbf{p}^r$ and $\mathbf{p}^l$, the mapping was written in the Euclidean plane, that is, $T : E^1 \rightarrow E^2$. In the issue, which is considered here, these transformations are linear mappings, so they can be written in the matrix form as $\mathbf{t} = T(\mathbf{p}^p, \mathbf{A})$, in which **A** is the matrix of a given transformation. Depending on the received symmetry or asymmetry (right/left-hand) of measurement points, they will create different dispersion graphs. An example of right-side and left-side asymmetry dispersion graph is presented in **Figures 9** and **10** and precisely described in the work of Dudczyk [13]. The number of measurement points is chosen empirically and is a double value of the maximum filter's width of the IF frequency from the superheterodyne receiver, which is used in the measurement procedure, that is, 40 MHz. As a result of this assumption, the critical original number of measurement points is as follows N = M = 80.

Depending on the received symmetry or asymmetry (right/left-side) of measurement points, it is possible to create dispersion graphs. Measurement points presented in **Figure 11**, transformed and depicted together, form the so-called measurement function $K(f_n)$. **Figure 11** shows the

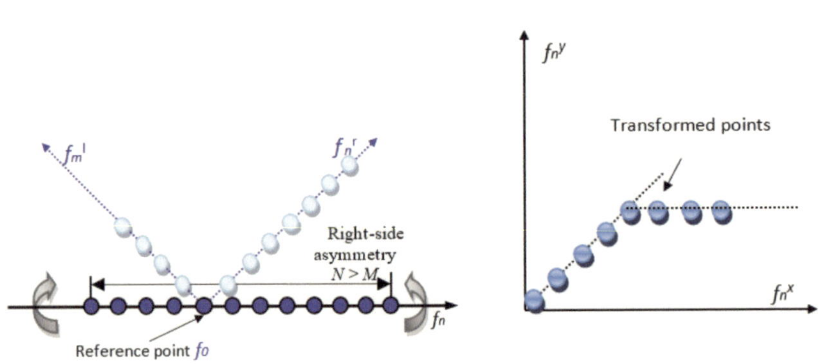

Figure 9. Depiction of transformation of measurement points in a 2D Euclidean space for N > M, that is, the range of right-hand asymmetry.

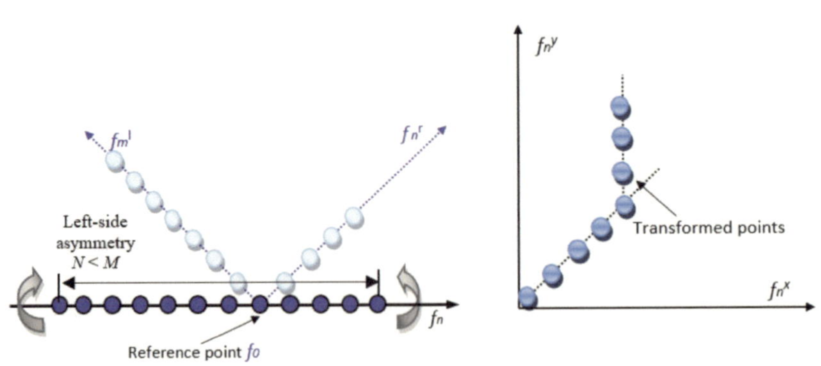

Figure 10. Depiction of transformation of measurement points in a 2D Euclidean space for N < M, that is, the range of left-hand asymmetry.

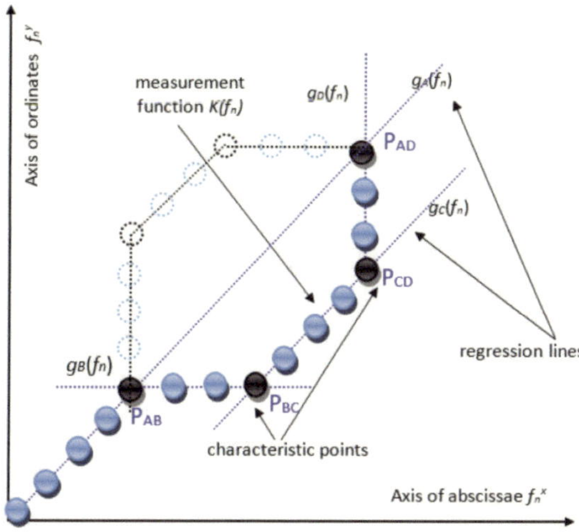

Figure 11. A graph of measurement points dispersion after transformation—attractor of transformation.

coordinate plane, where an abscissa (the value of x) is marked as a f_n^x and an ordinate (the value of y) is marked as a f_n^y.

On the basis of distinctive streaks that were formed, such hypothesis can be proposed: functions $g_A(f_n)$, $g_B(f_n)$, $g_C(f_n)$ and $g_D(f_n)$belong to the class of linear functions, in which$g_A(f_n)$, $g_B(f_n)$, $g_C(f_n)$and $g_D(f_n)$ will be the regression lines for the streaks formed through the measurement points [14]. Linear equation of regression for the presented case is defined with the following equation $\mathbf{g}(f_n) = \boldsymbol{\alpha} \cdot f_n + \boldsymbol{\beta}$, in which $\boldsymbol{\alpha}$ can be expressed as a vector $[\alpha_A, \alpha_B, \alpha_C, \alpha_D]^T$ and $\boldsymbol{\beta}$ can be expressed as a vector $[\beta_A, \beta_B, \beta_C, \beta_D]^T$ and $\mathbf{g}(f_n)$ can be expressed as a vector $[g_A(f_n), g_B(f_n), g_C(f_n), g_D(f_n)]^T$. To define the value of $\boldsymbol{\alpha}$ and $\boldsymbol{\beta}$, Eq. (5) should be minimalized.

$$E[f_n^Y - \boldsymbol{\alpha} \cdot f_n - \boldsymbol{\beta}]^2 = \min \tag{5}$$

$$\begin{cases} \dfrac{\partial}{\partial \alpha} E[f_n^Y - \boldsymbol{\alpha} \cdot f_n - \boldsymbol{\beta}]^2 = -2E[(f_n^Y - \boldsymbol{\alpha} \cdot f_n - \boldsymbol{\beta}) f_n] \\ \dfrac{\partial}{\partial \beta} E[f_n^Y - \boldsymbol{\alpha} \cdot f_n - \boldsymbol{\beta}]^2 = -2E[(f_n^Y - \boldsymbol{\alpha} \cdot f_n - \boldsymbol{\beta})] \end{cases} \tag{6}$$

After comparing the calculated derivatives Eq. (6) to zero, appears the system of normal equations in which after replacing the expected values with particular moments of equation systems, the following equation can be written:

$$\begin{cases} \boldsymbol{\alpha} \cdot m_{20} + \boldsymbol{\beta} \cdot m_{10} = m_{11} \\ \boldsymbol{\alpha} \cdot m_{10} + \boldsymbol{\beta} = m_{01} \end{cases} \tag{7}$$

in which m_{10} and m_{01} are sample *1st* moments, m_{20} is sample *2nd* moment and m_{11} is mixed sample *1st* moment. After further transformations, the regression equation is as follows:

$$g(f_n) = \frac{\mu_{11}}{\mu_{20}} \cdot f_n + \left(m_{01} - \frac{\mu_{11}}{\mu_{20}} m_{10} \right) = \boldsymbol{\alpha}_{21} f_n + \boldsymbol{\beta} \tag{8}$$

where

$$\boldsymbol{\alpha}_{21} = \left[\frac{\mu_{11}^A}{\mu_{20}^A}, \frac{\mu_{11}^B}{\mu_{20}^B}, \frac{\mu_{11}^C}{\mu_{20}^C}, \frac{\mu_{11}^D}{\mu_{20}^D} \right]^T = [\alpha_A, \alpha_B, \alpha_C, \alpha_D]^T \tag{9}$$

$$\boldsymbol{\beta} = \left[m_{01}^A - \frac{\mu_{11}^A}{\mu_{20}^A} m_{10}^A, m_{01}^B - \frac{\mu_{11}^B}{\mu_{20}^B} m_{10}^B, m_{01}^C - \frac{\mu_{11}^C}{\mu_{20}^C} m_{10}^C, m_{01}^D - \frac{\mu_{11}^D}{\mu_{20}^D} m_{10}^D, \right]^T = [\beta_A, \beta_B, \beta_C, \beta_D]^T \tag{10}$$

and μ_{11} means mixed *2nd* central moment and μ_{20} means *2nd* central moment. As a result of further transformations, four linear regression equations were given. The particular equation system given by the regression equation allows to calculate characteristic points of coordinates. Examples of four characteristic points presented in **Figure 11** in the form of black points, such as (P_{AB}, P_{BC}, P_{CD}, P_{AD}), were formed. Then, with the use of characteristic points of coordinates, the measurement function $K(f_n)$ was formed.

4.2. Distinctive fractal features extraction

As a result of further transformations, four equations of linear regression were received. Then, it is possible to draw a measurement function $K(f_n)$ in the form of a product k degree given $k + 1$ characteristic points, defined by the Lagrange's polynomial formula in accordance with Eq. (11), where a_k, a_{k-1}, …, a_0 are characteristic parameters of measurement function $K(f_n)$, as shown in **Figure 12**.

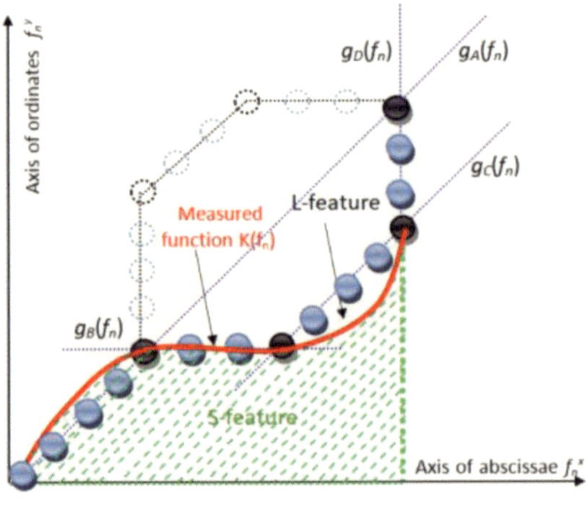

Figure 12. The image of measured function according to Lagrange polynomial and fractal features extraction.

$$K(f_n) = a_k f_n^k + a_{k-1} f_n^{k-1} + a_{k-2} f_n^{k-2} + \cdots + a_0 \tag{11}$$

The formalized notation of the measurement function $K(f_n)$ allows to extract distinctive features through defining the space area under the measurement function and the arc length of the function, which appeared for the SEI process. The feature S is the value of the space area of a closed surface expanding from the generalized measurement function $K(f_n)$ in the bracket $\langle f_n^{\min}, f_n^{\max}\rangle$, respecting Eq. (12).

$$S = \int_{f_n^{\min}}^{f_n^{\max}} K(f_n) df_n = \int_{f_n^{\min}}^{f_n^{\max}} (a_k f_n^k + a_{k-1} f_n^{k-1} + a_{k-2} f_n^{k-2} + \cdots + a_0) df_n \tag{12}$$

Simultaneously, the arc length of the generalized measurement function $K(f_n)$ as the second distinction feature of the radar emission source will be represented through the arc length L of the function $K(f_n)$ in the brackets $\langle f_n^{\min}, f_n^{\max}\rangle$, respecting Eq. (13).

$$L = \int_{f_n^{\min}}^{f_n^{\max}} \left[1 + \left(\frac{\partial K(f_n)}{\partial f_n}\right)^2\right]^{\frac{1}{2}} df_n = \int_{f_n^{\min}}^{f_n^{\max}} \left[1 + \left(k a_k f_n^{k-1} + (k-1) a_{k-1} f^{k-2} + \ldots + a_1\right)^2\right]^{\frac{1}{2}} df_n \tag{13}$$

According to Eqs. (12) and (13), it is possible to extract two additional distinctive features, that is, the length of measurement function and the value of area which is included under this function. The presented method of features extraction makes it possible to estimate numerical surface areas under the measurement functions (feature S) and the distance of arc of these functions (feature L). Then the vector of basic measurable parameters of radar signal was extended with two additional features.

Given in that way, two additional features expand the $\mathbf{V_B}$ vector of the basic features of radar signal measurable parameters, such as PW, PRI and RF. And these features are a good separation measure in the SEI process. The way of defining these two additional features and using them in the process of identification of the radar copies of the same type was presented in further part of this chapter.

5. Fractal of generalized measurement function

Generalization of the method of radar signal identification on the basis of the transformation fractal is defining the generalized measurement function $\hat{K}(f_n)$ going through all particular characteristic points P_n, in which $n = 0,1,\ldots,k_{gr}$. **Figure 13** presents the fractal character of the measurement function received as a result of the transformation of the set of measurement points.

The generalized measurement function $\hat{K}(f_n)$ preserves the character of not decreasing function in a particular bracket $\langle f_n^{\min}, f_n^{\max}\rangle$ and out of definite character, it shows prediction

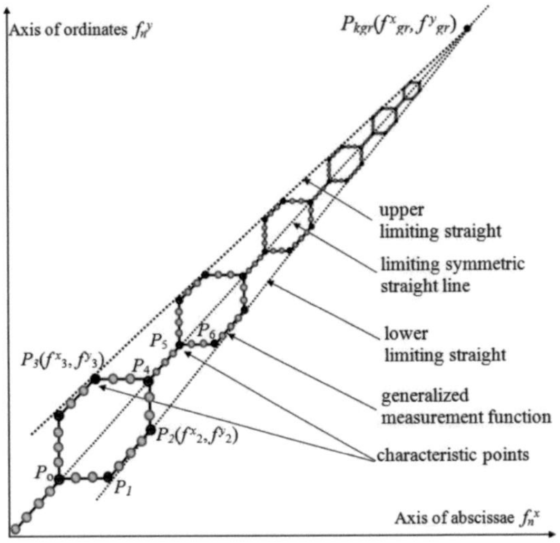

Figure 13. The fractal character of generalized measurement function.

features. Simultaneously, $\hat{K}(f_n)$ is located in an area that is mapped by the lower limiting straight $F_L(f_n)$ and upper limiting straight $F_U(f_n)$ and is symmetric relatively to the symmetrical limiting straight $F_S(f_n)$ according to Eqs. (14)–(16), in which f_2^x, f_3^x, f_{gr}^x are the abscissae of characteristic points and f_2^y, f_3^y, f_{gr}^y are the ordinates of characteristic points. **Figure 13** presents a fractal character of the generalized measurement function which has the form of a contraction mapping.

$$F_L(f_n) = \left[\frac{f_{gr}^y - f_2^y}{f_{gr}^x - f_2^x}\right] \cdot (f_n - f_2^x) + f_2^y \tag{14}$$

$$F_U(f_n) = \left[\frac{f_{gr}^y - f_3^y}{f_{gr}^x - f_3^x}\right] \cdot (f_n - f_3^x) + f_3^y \tag{15}$$

$$F_S(f_n) = \left[\frac{f_{gr}^y}{f_{gr}^x}\right] \cdot f_n \tag{16}$$

It should be mentioned that the received shape of the measurement function (according to **Figures 11** and **12**) is an individual model of a radar emission source. 'An individual model' means 'lines on the fingers' of the radar which make a clear identification possible.

6. Results of analysis

To compare the received results with other RSR methods based on, for example, Fast decision Identification Algorithm of emission source pattern described in Ref. [6], or out-of-band radiation of radar devices described in Ref. [13], or inter-pulse modulation described in Ref. [7] and intra-pulse analysis of a radar signal shown in Ref. [15] or method based on data modelling presented in Ref. [16], Correct Identification Coefficient (CIC) is set according to Eq. (17), where n_{B-P} is the number of correct comparisons of basic features' vectors $\mathbf{V_B}$ (presented according to Eq. (4)), to extended vectors $\mathbf{V_{EXT}}$ (with two additional features L and S) in a particular class, where N is the number of all comparisons divided by the number of test collections.

$$\mathrm{CIC} = \frac{n_{B-P}}{N} \tag{17}$$

The number of n_{B-P} correct comparisons is set according to Eq. (18), where γ_i^j function assigns to a pair of vectors $(\mathbf{V}_{\mathrm{B}}^i, \mathbf{V}_{EXT}^j)$ the value which equals '1' if i = j, or the value which equals '0' if i $\neq$ j. The example of CIC received values are presented in the following part of this chapter.

$$n_{B-P} = \sum_{i=1}^{I}\sum_{j=1}^{J}\gamma^{i\,j} \tag{18}$$

The process of identification was made on the basis of length measurement and the decision about the criterion of minimal distance classification. A correctness estimation of tests with particular class were Mahalanobis, Euclidean and Hamming distances (metrics) [17, 18]. The criterion of classification was the criterion of 'the nearest neighbour', which was used as one of the basic threshold criteria [19]. In order to assess the quality of the classification/identification process, the Correct Identification Coefficient was defined.

According to Eqs. (8) and (9), it was possible to extract two additional distinctive features, that is, the length of measurement function and the value of area which is included under this function. The results were presented in **Figures 14–16**.

Also, the received estimation results are presented in **Figures 17–19**. Appropriately crossed columns and lines of each $\mathrm{V_B}$ vector and extended vectors $\mathrm{V_{EXT}}$ present the degree of their similarity defined by the distance value. The less value of this distance means the bigger similarity of $\mathrm{V_B}$ vector to the extended vectors. Also, in **Figures 15–17**, there are minimum values of the distance marked with a red dotted ellipse.

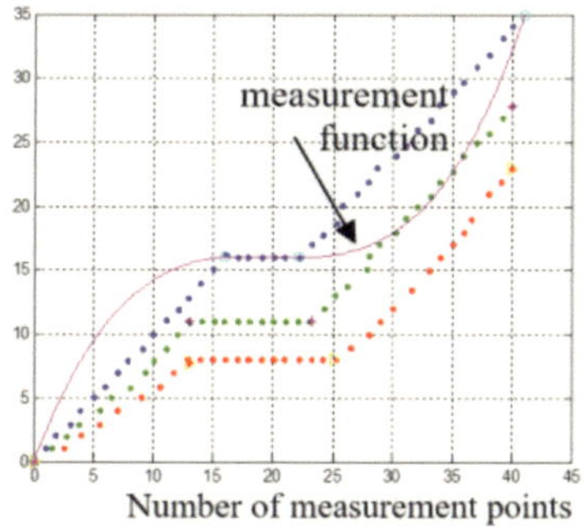

Figure 14. An attractor of transformation for Copy of Radar No. 1.

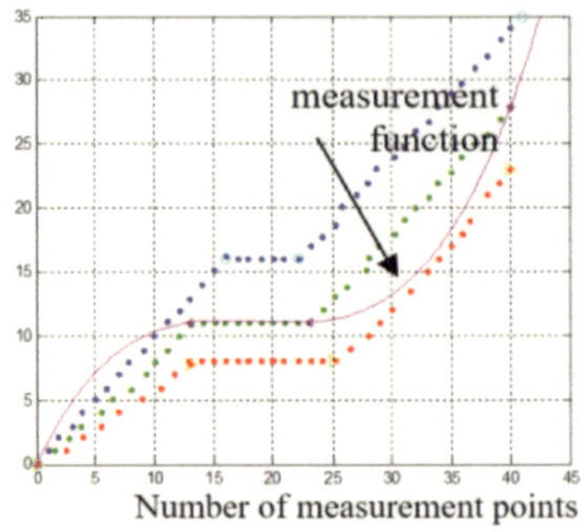

Figure 15. An attractor of transformation for Copy of Radar No. 2.

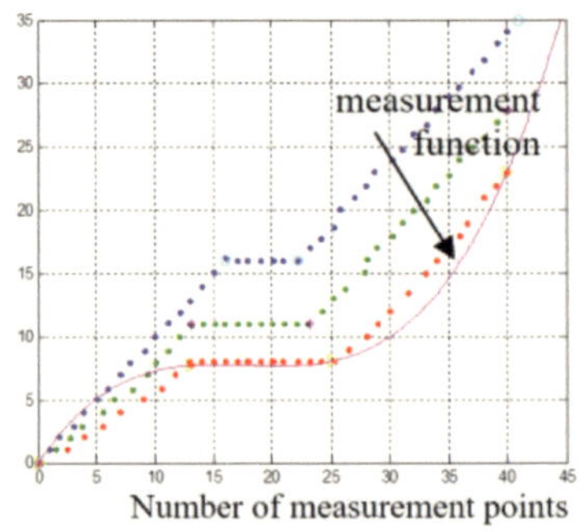

Figure 16. An attractor of transformation for Copy of Radar No. 3.

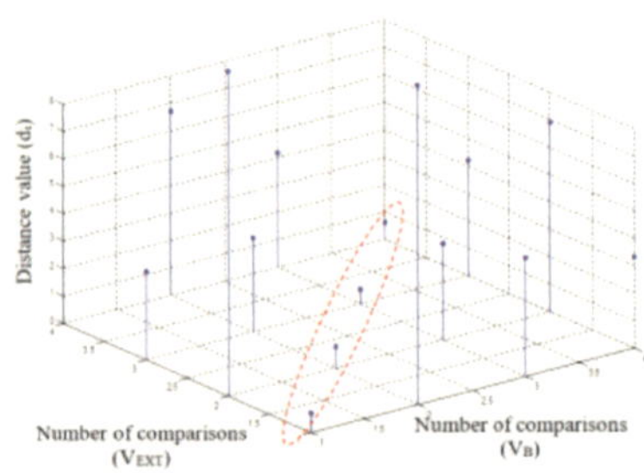

Figure 17. The values of Mahal distances for Radar Copy No. 1.

A correctness estimation of tests with particular class were Mahalanobis, Euclidean and Hamming distances (metrics) [21]. The SEI estimation results are presented in **Figures 20–22**.

According to the SEI methods listed in this chapter, the received RER results are as follows: the use of out-of-band radiation described in the work of Dudczyk [13] and the CIC value for RSR is about 90%. The method based on fractal features described in the work of Dudczyk and Kawalec [3], and the CIC value is 91.6% for Mahalanobis metric and 96.7% for Euclidean and Hamming metrics. Very similar RER results are received in the work of Dudczyk and Kawalec

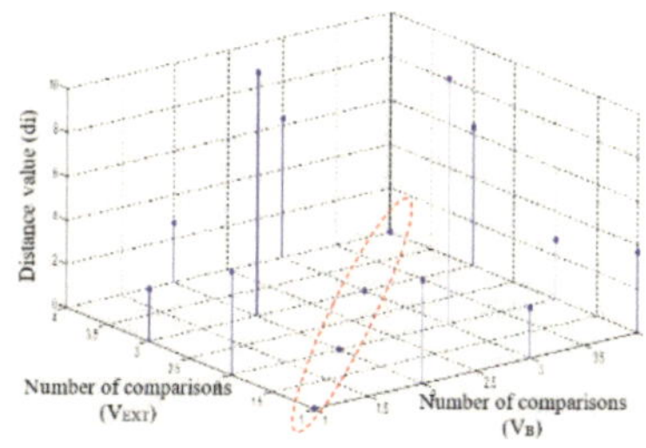

Figure 18. The values of Mahal distances for Radar Copy No. 2.

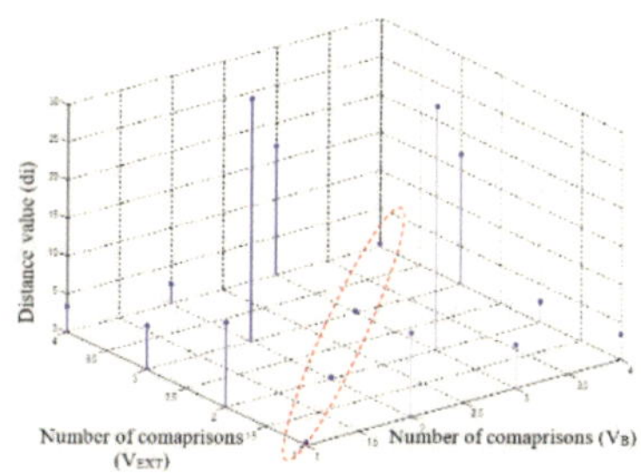

Figure 19. The values of Mahal distances for Radar Copy No. 3.

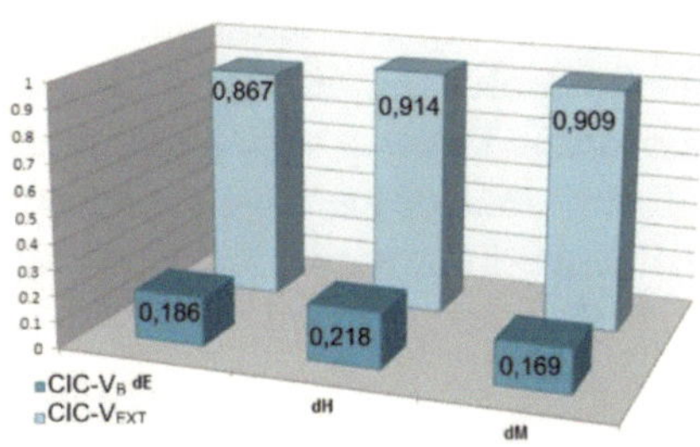

Figure 20. The values of CIC for Radar Copy No. 1.

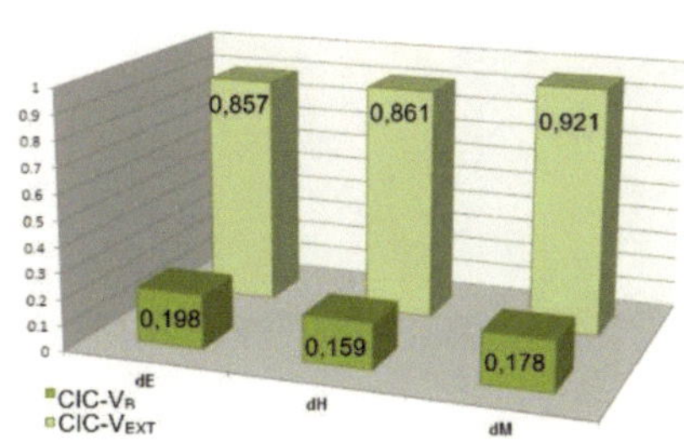

Figure 21. The values of CIC for Radar Copy No. 2.

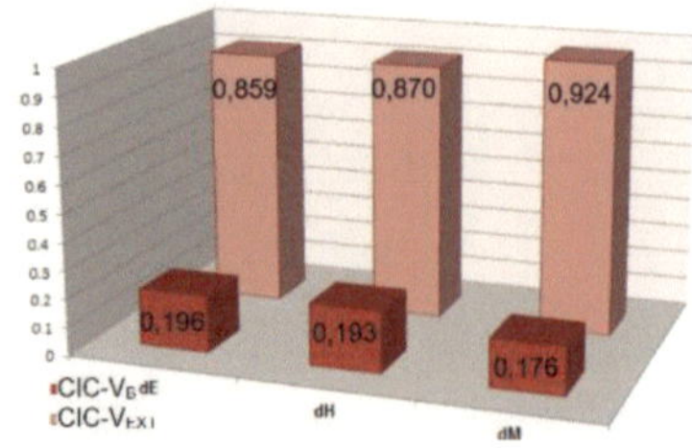

Figure 22. The values of CIC for Radar Copy No. 3.

[4], where RSR is also based on the analysis of fractal features. The method based on inter-pulse analysis described in the work of Dudczyk et al. [7] increases the CIC coefficient up to 70%, and the method based on intra-pulse analysis described in the work of Kawalec and Owczarek [15] makes it possible to receive RSR results reaching 90% level. Data modelling applied to RSR and identification is presented in the work of Kawalec and Owczarek [16]. In this work, the value of CIC equals 98%. In the work of Dudczyk and Kawalec [6], the Fast Identification Algorithm for RER is presented. This algorithm is parameterized in three stages by implementation of three different ways to define the similarity degree of the signal vector to the pattern in the database. Based on this algorithm, the CIC value is 63%. In order to depict it, in **Figure 23**, there have been presented CIC values.

The presented method of features extraction makes it possible to estimate numerical surface areas under the measurement functions (feature *S*) and the distance of arc of these functions (feature *L*). Then, the vector of basic measurable parameters of radar signal was extended with two additional features. Given in that way, two additional features expand the vector of the basic features of radar signal measurable parameters, such as PW, PRI and RF, are a good separation measure in the SEI process. The way of defining these two additional features and using them in the process of identification of the radar copies of the same type was presented. The features S and L are distinctive information for good separating measure in the SEI process. Simultaneously, as a result of transformations in collections of measurement points,

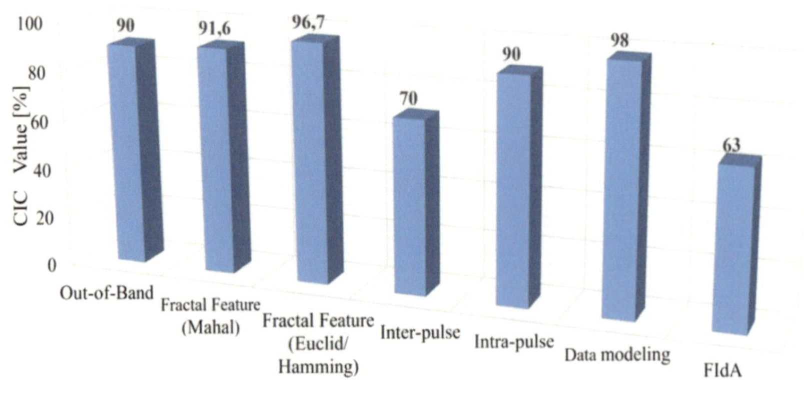

Figure 23. Graphic illustration of CIC values for other RER methods.

the transformation attractor of the generalized measurement function is received. The received attractor is used later on to optimize the SEI process.

It needs to be emphasized with full conviction that referring to the works above, during the SEI procedure, the same recordings of a few hundred radar signals coming from the same type of radars are used. Only by this approach, it is possible to compare the received results. It needs to be emphasized that the RSR methods listed in this chapter differ from each other as concerns the test procedure, the compilation level, calculation time and algorithm complexity. However, the main difference is that in the process of generating distinctive features, it is possible to achieve different distinctive features from a radar signal. In that way, a quasi-optimum radar signal pattern appeared.

7. Conclusion

Radar signal recognition with the use of classical methods, that is, based on statistical analysis of basic measurable parameters of a radar signal, such as radio frequency, amplitude, pulse width or pulse repetition interval, is not enough to carry out the distinction process of particular copies of the same radar type.

The received measurement data have a significant influence on the SEI process of radar, in which it is aimed to receive very high level of radar signal identification. Ultimately, signal source identification, which is 100%, should be characterized by the maximization of explicitness of RER procedure. It is not a trivial matter to achieve such a result. It is also known that stochastic context-free grammars (SCFG) appear promising for the recognition and threat assessment of complex radar emitters in radar systems, but the computational requirements for learning their production rule probabilities can be very onerous [20]. As shown in Ref. [21], a self-organizing map and the maximum likelihood gamma mixture model classifier and adopted Bayesian formalism are too complicated for direct analytical use in automatic radar recognition. The presented SEI method based on fractal features is realized on the basis of MatLab software package and received vectors are recorded in a dedicated database for ELINT system. The received CIC value indicates that there has been a noticeable rise in the radar signal correct identification. Comparing the received results of the identification process with other methods, it may be admitted that the presented method makes it possible to increase the value of CIC. In order to increase the CIC coefficient value, in further works on RSR use in SEI process, a common similarity matrix should be defined. This matrix should include the complexity of algorithms which are used in the RER method, estimation time, the requirements of the equipment platform and other requirements, which are significant in the process of quality estimation of a particular method. Thus, it will be possible to count automatically the similarities between vectors of basic measurable parameters for different radar copies of the same type. For ELINT systems working in real conditions, on a contemporary battlefield, the automation of RSR process and explicit identification of every single emitter in real time (with minimum time burden) are currently primary challenges for ELINT specialists

This chapter highlights the fact that the RSR process described here is a complex problem. It is also generally known that a number of aspects such as defining the DataBase, the method which creates the pattern, classification process and identification process, the criteria which are used and the method to calculate the correct identification coefficient, are currently a great challenge for researchers and, for the time being, there are no optimal solutions to them. Many solutions are still a mystery in this subject and because of the fact that they are a matter of current EW field and specific programme-device applications, they cannot be published. All attempts to implement such solutions to ELINT systems and electronic warfare should be optimized to a particular device not to overload the SEI system.

Author details

Janusz Dudczyk

Address all correspondence to: j.dudczyk@wb.com.pl

WB Electronics S.A., R&D Department, Poznanska, Poland

References

[1] J. Dudczyk. Radar emission sources identification based on hierarchical agglomerative clustering for large data sets. Journal of Sensors. 2016;**2016**(2016):1879327. doi:10.1155/2016/1879327

[2] J. Dudczyk, A. Kawalec. Specific emitter identification based on graphical representation of the distribution of radar signal parameters. Bulletin of the Polish Academy of Sciences, Technical Sciences. 2015;**63**(2):391–396. doi:10.1515/bpasts-2015-0044

[3] J. Dudczyk, A. Kawalec. Fractal features of specific emitter identification. Acta Physica Polonica. 2013;**124**(3):406–409. doi:10.12693/APhysPolA.124.406

[4] J. Dudczyk, A. Kawalec. Identification of emitter sources in the aspect of their fractal features. Bulletin of the Polish Academy of Sciences, Technical Sciences. 2013;**61**(3):623–628. doi:10.2478/bpasts-2013-0065

[5] J. Dudczyk, A. Kawalec. The use of fractal features extracted from radar signals in the process of specific identification. Przegląd Elektrotechniczny. 2014;**R.90**(11/2014):212–215. doi:10.12915/pe.2014.11.54

[6] J. Dudczyk, A. Kawalec. Fast-decision identification algorithm of emission source pattern in database. Bulletin of the Polish Academy of Sciences, Technical Sciences. 2015;**63**(2):385–389. doi:10.1515/bpasts-2015-0043

[7] J. Dudczyk, J. Matuszewski, A. Kawalec. Specific emitter identification based on an inter-pulses modulation of radar signals. Przegląd Elektrotechniczny. 2016;**R.92**(9/2016):267–271. doi:10.15199/48.2016.09.64

[8] F. Berizzi, G. Bertini, M. Martorella. Two-dimensional variation algorithm for fractal analysis of sea SAR images. IEEE Transactions on Geoscience and Remote Sensing. 2006;**44**(9):2361–2373. doi:10.1109/TGRS.2006.873577

[9] M. German, G.B. Bénié, J.M. Boucher. Contribution of the fractal dimension to multiscale adaptive filtering of SAR imagery. IEEE Transactions on Geoscience and Remote Sensing. 2003;**41**(8):1765–1772. doi:10.1109/TGRS.2003.811695

[10] B. Świdzińska. Fractal compression using random encoding algorithm. Bulletin of the Polish Academy of Sciences, Technical Sciences.1998;**46**(4):525–532. ISSN 0239-7528

[11] A. Wojtkiewicz, M. Nałęcz, K. Kulpa, R. Rytel-Andrianik. A novel approach to signal processing in FMCW radar. Bulletin of the Polish Academy of Sciences, Technical Sciences. 2002;**50**(4):347–359. ISSN 0239-7528

[12] K.I. Talbot, P.R. Duley, M.H. Hyatt. Specific emitter identification and verification. Technology Review Journal Spring/Summer 2003:113–133.

[13] J. Dudczyk. Applying the radiated emission to the radio-electronic devices identification. Dissertation thesis, Department of Electrical, Military University of Technology, Poland; 2004, (in Polish).

[14] Z. Hellwig. Theory of probability and mathematical statistics. Warsaw: Scientific Press PWN; 1998.

[15] A. Kawalec, R. Owczarek. Radar emitter recognition using intrapulse data. In Proceedings of the NordSec 2005—The 10th Nordic Workshop on Secure IT-Systems, pp. 444–457, Warsaw, Poland, 17–19 May 2004.

[16] A. Kawalec, R. Owczarek, J. Dudczyk. Data modelling and simulation applied to radar signal recognition. Molecular and Quantum Acoustics. 2005;**26**:165–173.

[17] R.O. Duda, P.E. Hart, D.G. Stork. Pattern classification (2nd ed.). New York: John Wiley & Sons; 2000.

[18] K. Fukunaga. Introduction to statistical pattern recognition (2nd ed.). New York: Academic Press; 1990.

[19] C.T. Zahn. Graph-theoretical methods for detecting and describing gestalt clusters. IEEE Transactions on Computers. 1971;**1**:68–86.

[20] G. Latombe, E. Granger, F. Dilkes. Fast learning of grammar production probabilities in radar electronic support. IEEE Transactions on Aerospace and Electronic Systems. 2010;**46**: 1262–1289.

[21] K. Copsey, A. Webb. Bayesian gamma mixture model approach to radar target recognition. IEEE Transactions on Aerospace and Electronic Systems. 2003;**39**:1201–1217.

Chapter 7

Application of Fractal Dimension in Industry Practice

Vlastimil Hotař

Additional information is available at the end of the chapter

http://dx.doi.org/10.5772/intechopen.68276

Abstract

Today, industrial production lines commonly use off-line and automatic on-line quality monitoring and control. Monitoring and control units analyse data from a production process, and analysis should be able to obtain reliable information that correspond with the character of the data obtained. The character of the data set obtained from production processes or from products can be highly structured in all industrial areas. Structured surface, complex time series (topologically one dimensional signals), difficulty to describe dividing curves are much more common than it can be expected. For this kind of data set, a powerful tool for analysis of complexity —fractal geometry (especially a fractal dimension) should be used. The fractal dimension with a combination of statistical tools is an interesting and powerful tool for complex data quantification, for tracing the source of poor quality, production optimization and investigating the source of instability of production process subsystems in industrial applications. The methodology for evaluation of complex and irregular data was developed and applied in industrial practice. This methodology searches appropriated parameters for a complex evaluation of data. Only the chosen parameters are used for a complete analysis of the data in order to reduce processing time.

Keywords: fractal geometry, fractal dimension, statistic tools, monitoring, control, complex data

1. Introduction

A demand for objective measurement and control methods for materials, processes and production processes stems from continuously increasing pressure from competitors to improve the quality of products. However, description of many complex and irregular structures (e.g. defects, surfaces, cracks, signals from dynamic processes) is almost impossible by conventional methods. The application of fractal geometry, which is successfully used in science, appears to be a powerful approach. The industrial application of the fractal dimension (FD) is generally experimental [1, 2], but application to production is possible and brings benefits. Currently, there are tools to monitor three basic data format types: topological one-dimensional signals (analysis of time series for control systems, roughness from a surface roughness tester), dividing curves (application for the corrugation test of flat glass and surface roughness—iron aluminides in comparison with carbide-nickel steel in contact with molten glass), and digitalized photos (evaluation of 2D pictures of surface defects).

It is possible to compare statistical tools and FD, but the comparison should be done carefully. To claim that the FD is better than statistics and vice versa is impossible, because the FD provides added information about the character of data set description and does not substitute standard tools. Moreover, the FD should not be used separately for the reason that the dimension does not give all the information about data set captures. FD using together with added parameters (statistics, topology, spectral analysis, etc.) brings benefits and is recommended.

2. Description of one-dimensional signals

Dynamic subsystems can be found in many production processes, and they have a strong influence on the production. Data measured from production sensors contain mentioned dynamic effect. Therefore, the time series are structured and the typical statistical data evaluations are not often sufficient.

Also roughness from a surface roughness tester is in many cases complex, and using standard roughness parameters may be not satisfactory.

To analyse one-dimensional signals, we use statistical methods, power spectral analysis, and an estimate of the FD [3, 4]. The estimation of the FD is calculated using the rescaled range method [5] and the box counting method from an "iso-set" [6], also compass counting and EEE method can be used.

2.1. Rescaled range analysis

The rescaled range analysis (R/S) represents method for estimating the FD of self-affine fractals and uses statistical tools. The method is based on an analysis of a changed interval of time series. Consider an interval of time series, length w. Within this interval, one can define two quantities:

R(w), the range taken by the values of y in the interval (vector y—the vertical axis of the time series—represent the time series without a sampling interval x—the horizontal axis of the time series). The range is measured with respect to a trend in the interval, where the trend is estimated simply as difference between maximum and minimum of the interval. This subtracts the average trend in the interval.

S(w), the standard deviation of the first differences Δy of the values of y within the interval. The first differences of the y's are defined as the differences between the values of y at some location x and y at the previous location on the x axis:

$$\Delta y(x) = y(x) - y(x - \Delta x) \tag{1}$$

where Δx is the sampling interval, that is, the interval between two consecutive values of x.

The rescaled range $R/S(w_{RS})$ is defined as follows:

$$R/S(w_{RS}) = \langle \frac{R(w_{RS})}{S(w_{RS})} \rangle \tag{2}$$

where w is the interval length and the angled brackets $<R(w_{RS})>$ represent the average of a R (w_{RS}) value numbers. The self-affinity incurs that one expects the range taken by the values of y in an interval of length w to be proportional to the interval length to a power equal to the Hurst exponent H, that is,

$$R/S(w_{RS}) = \mathrm{w}_{\mathrm{RS}}^{\mathrm{H}} \tag{3}$$

For a given window length w_{RS}, the input series in a number of intervals of length w_{RS} are then subdivided, $R(w_{RS})$ and $S(w_{RS})$ in each interval are measured and $R/S(w_{RS})$ is calculated as the average ratio $R(w_{RS})/S(w_{RS})$, see Eq. (1). Mentioned process is then reiterated for several window lengths. Logarithms of $R/S(w_{RS})$ are subsequently plotted versus the logarithms of w_{RS}. Considering the self-affine trace, this plot follows a straight line with a slope equals to the Hurst exponent H. FD of the time series can be calculated from the relationship between the Hurst exponent H and the FD:

$$D_{RS} = 2 - H \tag{4}$$

where R/S dimension D_{RS} denotes the FD estimated from the Rescaled Range analysis. The R/S dimension has value from 1 to 2, and Hurst exponent has value from 0 to 1. More about the method can be found in Ref. [5].

2.2. Box dimension from "Iso-set"

An "iso-set" is constructed from the time series as shown in **Figure 1** that contains zeros and ones, and the FD is estimated from this set. The "iso-set" (as time series) can be generated by two basic ways, where the ones represent crossings of a pre-selected threshold. The "iso-set" is generated by setting of suitable thresholds and marking the time at which the time series cross these thresholds (**Figure 1**). The threshold values can be perceptually dependent on the time series average (a floated threshold value, in **Figure 1**) or can be pre-selected fixed values.

The FD of the "iso-set" is estimated by using box counting method that is described in Section 4.4. The principle of the box dimension method used for "iso-set" is given in **Figure 1**. (The box size r_B=t_s.b.) Starting from box size r_B = t_s (t_s is sampling time interval), the number of boxes that contain a crossing is recorded. The box size is then increased by the factor b and the procedure continues until the entire "iso-set" is contained in one single box. This is illustrated for the factor b = 2. The box dimension D_B is determined from the central slope of the

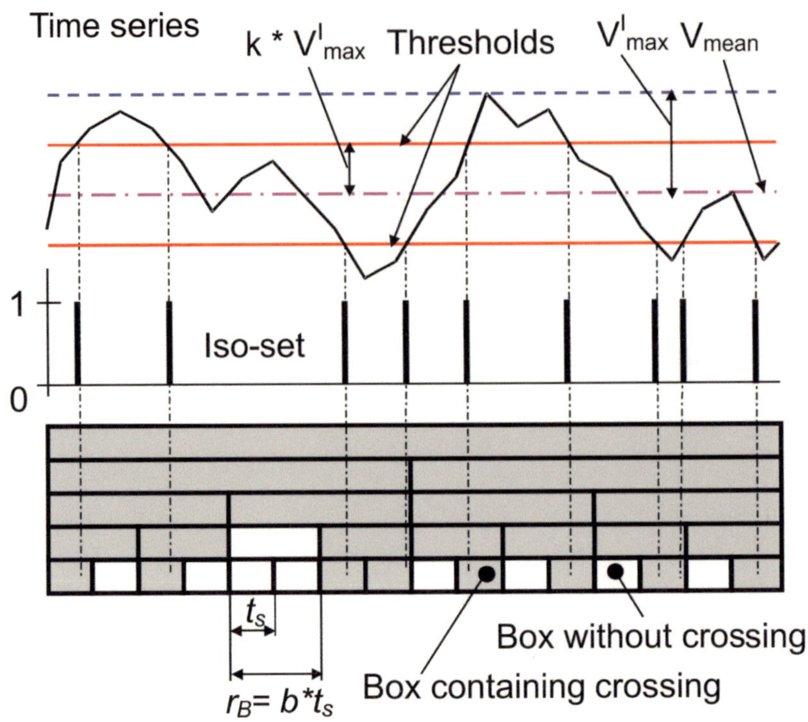

Figure 1. Construction of iso-set by marking each time at which trace crossed the chosen threshold values and a principle of box dimension computing.

regression line of the Richardson-Mandelbrot plot (logarithmic dependence between $\log_2 N(r)$ and $\log_2 r_B$). For more information about the method, please see Refs. [6–8].

2.3. Compass counting

The estimated Compass Dimension expresses the degree of complexity of the profile. A compass method [5, 9, 10] is based on measuring of the profile (curve) using different ruler sizes (**Figure 2A**) according to the equation:

$$L_i(r_i) = N_i(r_i).r_i \tag{5}$$

where L_i is the length in i-step of the measurement, r_i is the ruler size and N_i is the number of steps needed for the measurement.

If the profile is fractal, and hence the estimated FD is larger than the topological dimension, then the length measured increases as the ruler size is reduced. The logarithmic dependence between $\log_2 N(r_i)$ and $\log_2 r_i$ (Richardson-Mandelbrot plot) is shown in **Figure 2B**. The Compass Dimension is then determined from the slope s of the regression line:

$$D_C = 1 - s = 1 - \frac{\Delta \log_2 L(r)}{\Delta \log_2 r} \tag{6}$$

http://dx.doi.org/10.5772/intechopen.68276

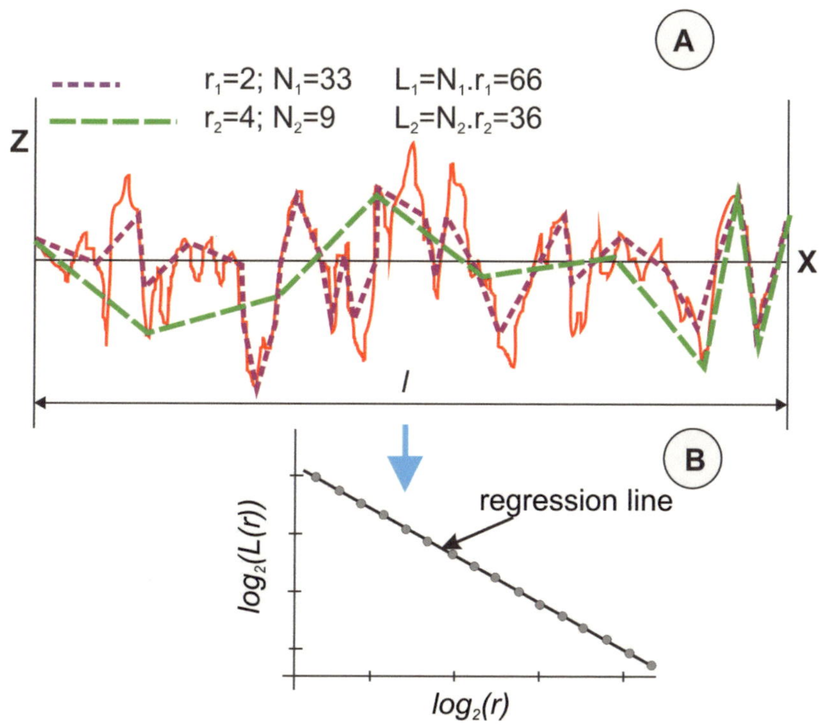

Figure 2. Estimation of the fractal dimension by the compass method.

2.4. Relative and proportional length

The rate of signal (profile) deformation might be evaluated by its relative length L_R. This fast and reliable method measures the ratio of the profile length l_{PIXEL} (solid line in **Figure 2A**) using the smallest ruler (1 pixel) r_{PIXEL} and the length of the projection l (**Figure 2A**):

$$L_R = \frac{l_{PIXEL}}{l} \tag{7}$$

Another similar approach is to compute the proportional length of the profile L_P. The proportional length is the ratio of the profile length measured with a defined ruler l_r (e.g. dashed line in **Figure 2A**) and the length measured with the maximum ruler $l_{r\ max}$ (the length between the first and the last point of the profile):

$$L_P = \frac{l_r}{l_{rmax}} \tag{8}$$

2.5. EEE method

The method is based on length evaluation of a curve (signal). The curve is defined by measured values and they are isolated points $x_1, x_2, \ldots, x_n$ in the range $y(x_1), y(x_2), \ldots, y(x_n)$. The dots represent local extremes (minima and maxima). Unnecessary extremes are classified with a

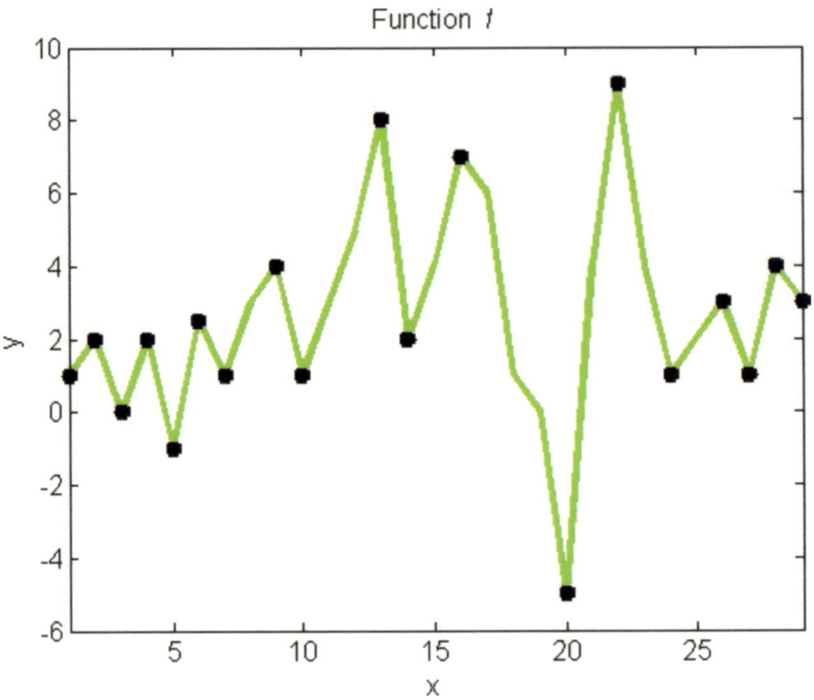

Figure 3. Function *f* and its local extremes.

defined rule on the curve, and a new simplified function is defined by the remaining points. For the next classification, the new function is used.

An example of a function defined by points and connected into the linear by parts function *f* is in **Figure 3**. The function *f* is in the first step purged of points which are not local extremes using the rule:

First, the difference proportion of the dependent variables y to the independent variables x between neighbouring points x_i, x_{i+1} is determined from function *f*:

$$\Delta f(x_i, x_{i+1}) = \frac{\Delta y(x_i)}{\Delta x_i} = \frac{y(x_{i+1}) - y(x_i)}{x_{i+1} - x_i} \tag{9}$$

Second, irrelevant points, concretely points where the difference Δf has the same sign, are eliminated. Remained points (marked by black dots in **Figure 3**) are considered as the local extremes. Extremes are supplemented by first and last points. The prepared points form a simplified function g, **Figure 4**. A relative length of the function g is computed, and the result is saved. Based on the absolute length of the function the relative length, $L_{R\ 1}$ (Section 2.4) is evaluated from point to point and divided by the length of its x axis projection.

The elimination of insignificant extremes procedure is used to the simplified function g. The procedure applies functions formed from function g minima and maxima, **Figure 5**, the functions g_{min} and g_{max} are extended by the g function's first and last points, as it was

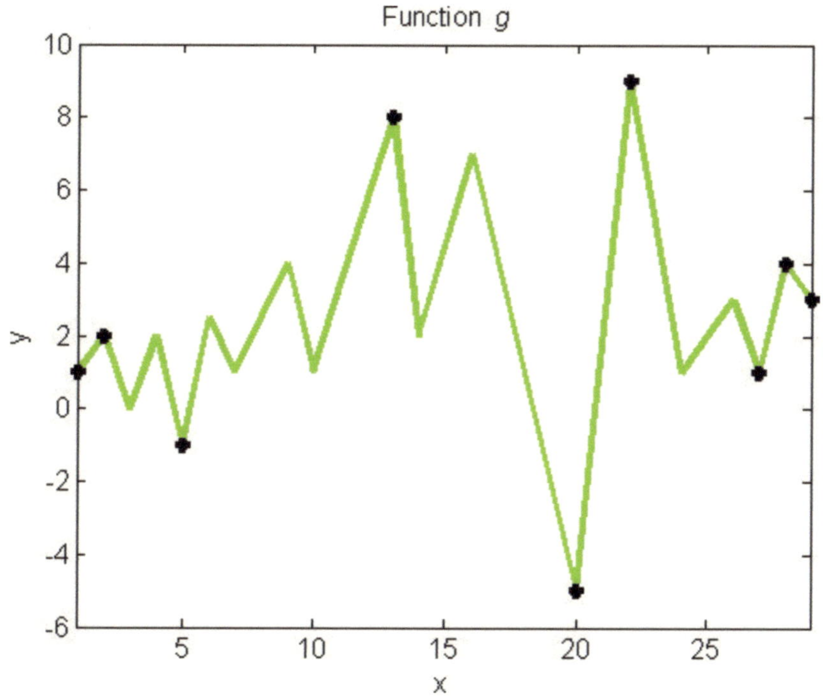

Figure 4. Simplified function g and its local extremes.

mentioned above. The function g_{max} is formed by the g function's maxima (dotted line in **Figure 5**). There are local maxima presented in this function by black dots in **Figure 5**. The definition of local maxima using above-mentioned rule, but only maxima are used (obtained minima are forgotten). The function g_{min} is generated from function's minima (dashed line). Obtained local minima are presented as black dots using above-mentioned rule; however, only minima are used (obtained maxima are forgotten).

The local minima and maxima of the functions g_{max} and g_{min} are used for the generation of the function g_{red}, **Figure 6**. In this function, again local minima and maxima are defined using the rule (**Figure 6**, black dots). These final local extremes of the function g (**Figure 4**, black dots) and the first and last points from the function g define the function h, **Figure 7**. The relative length $L_{R\ 2}$ of function h is computed, and the result is saved.

The similar procedure is subsequently used for the new simplified function h. Global extremes and the first and last point (**Figure 7**, black dots) define the function k. The function is formed from the global minimum and maximum of all functions (f, g, h), and therefore the analysis is stopped. All functions are represented in **Figure 8**.

The analysis steps j are plotted versus g, h, k, function's computed relative lengths $L_{R\ j}$ (Section 2.4.) **Figure 9**. The dependence between the relative lengths $L_{R\ j}$ and steps of elimination j is computed by a sufficient regression function. Concretely, it can be represented by a regression line (**Figure 9**), a quadratic function or a hyperbolical function. Parameters of those regression functions are used for the function f determination. (Logarithmic axis use is not beneficial.)

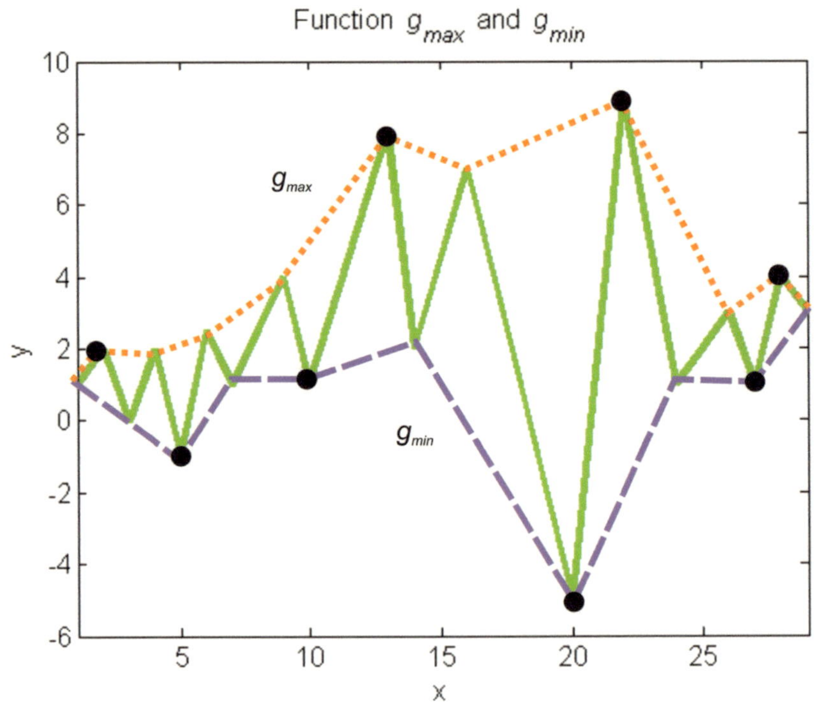

Figure 5. Functions g_{max} and g_{min} generated from local extremes of function g.

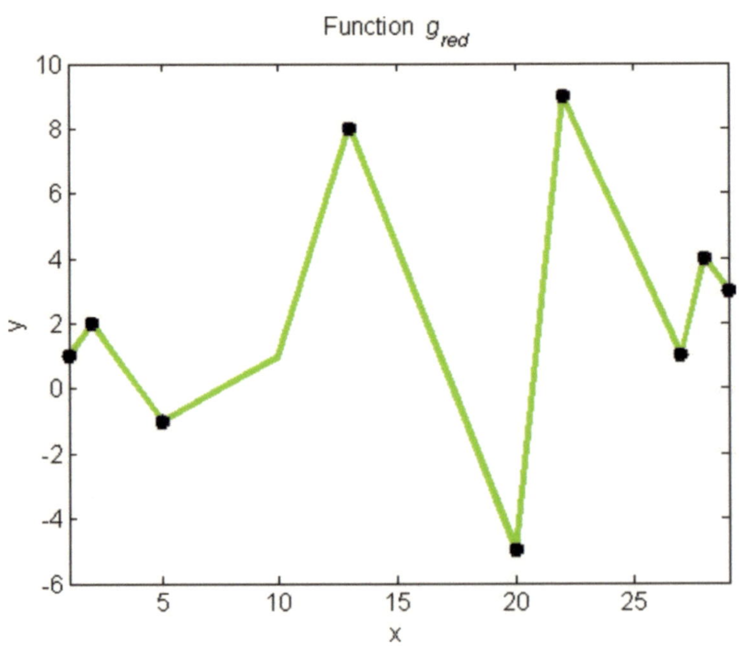

Figure 6. Function g_{red} generated from maxima of function g_{max} and from minima of function g_{min}.

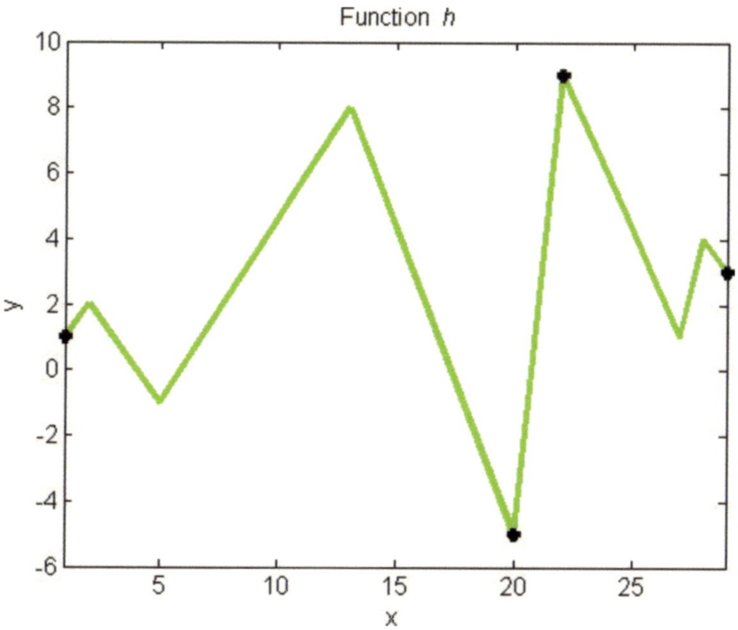

Figure 7. Simplified function *h* and its local extremes.

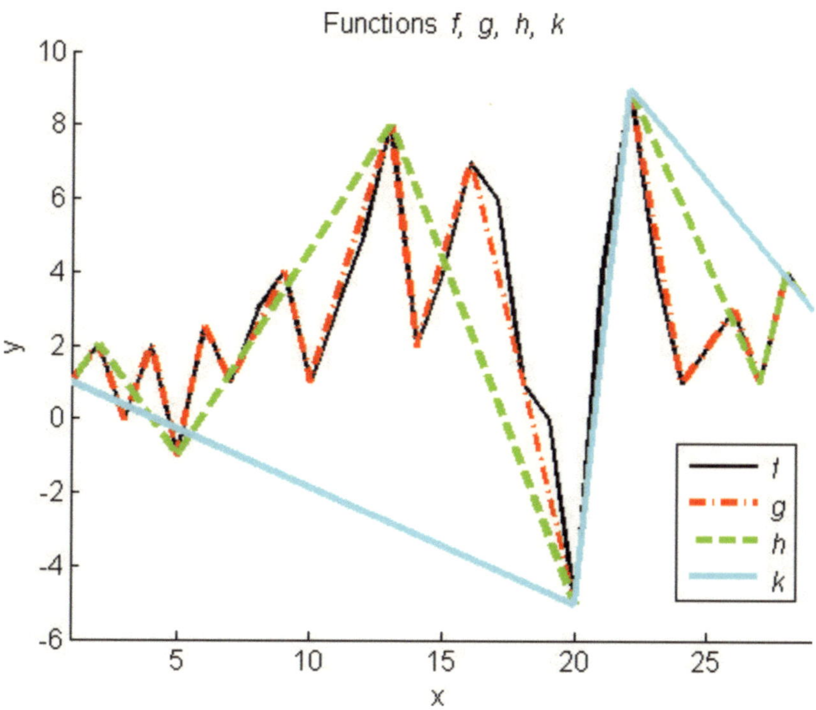

Figure 8. Function *f* and simplified functions *g, h, k*.

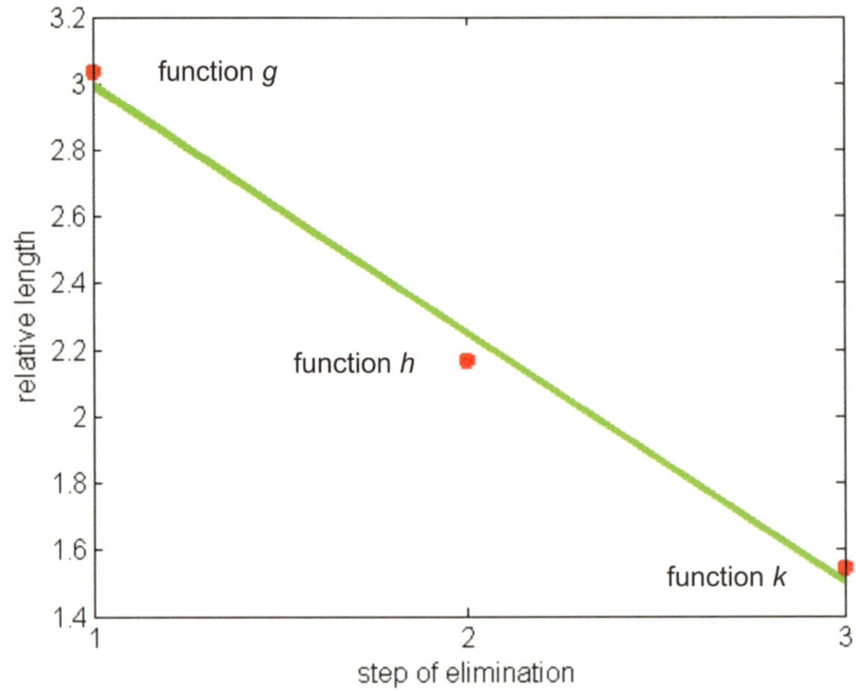

Figure 9. Plot of the relation between steps of analysis and relative length, regression line.

Signal (time series) obtained from simulation of fractional Brownian motion using Cholesky-Factorization of the related covariance matrix (FBM) was used to test the developed method. An example of testing signal is generated using the input Hurst coefficient $H = 0.4$, in **Figure 10**. The coefficient represents the character of signal and can achieve value 0 to 1 (lower coefficients generate more complex functions, and further information can be found in Refs. [5, 10]). The dependence between relative lengths and elimination steps is shown in **Figure 11**.

A standard process to estimate the FD using the dependence between $\log_2 L_{R\ j}$ and $\log_2 j$ was tested. However, such representation was not beneficial as in this method the rulers are not used with different lengths. For that reason, the length of rulers cannot be used in the plot.

The regression function, which fits the most to describe a relation of relative lengths $L_{R\ j}$ and the number of steps j, is estimated by the hyperbolic regression model:

$$L_{R\ j} = \frac{d}{j + a} + b \tag{10}$$

This might be calculated by parameters d and a. Parameter b is always set as $b = 1$. Parameter a needs to be computed numerically with application of an error function.

For verification of the EEE method, over 900 simulated time series from FBM were used with the Hurst coefficient between 0.1 and 0.95. The dependence between Hurst coefficients and the average value of parameters a and d is shown in **Figure 12**.

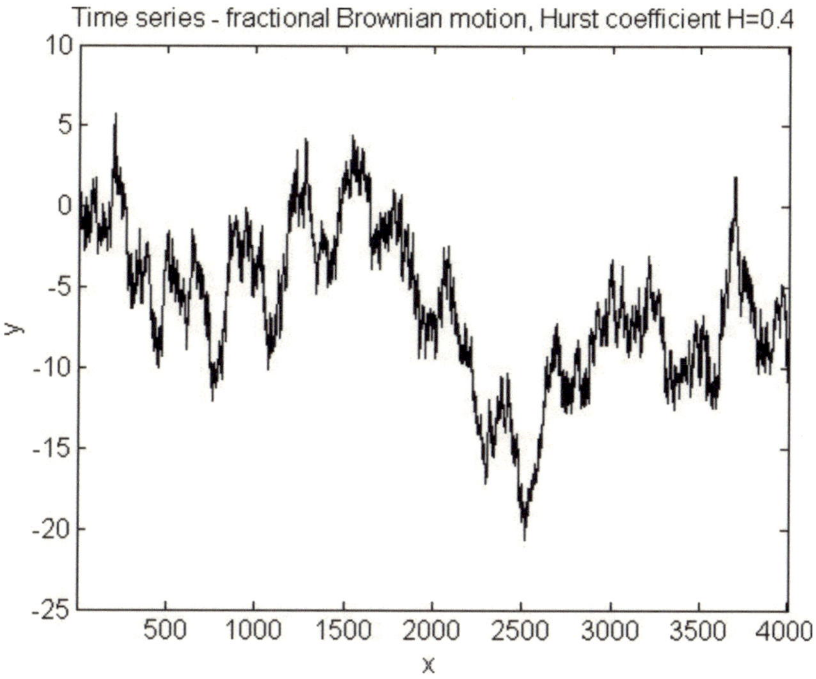

Figure 10. Simulation of the time series using fractional Brownian motion.

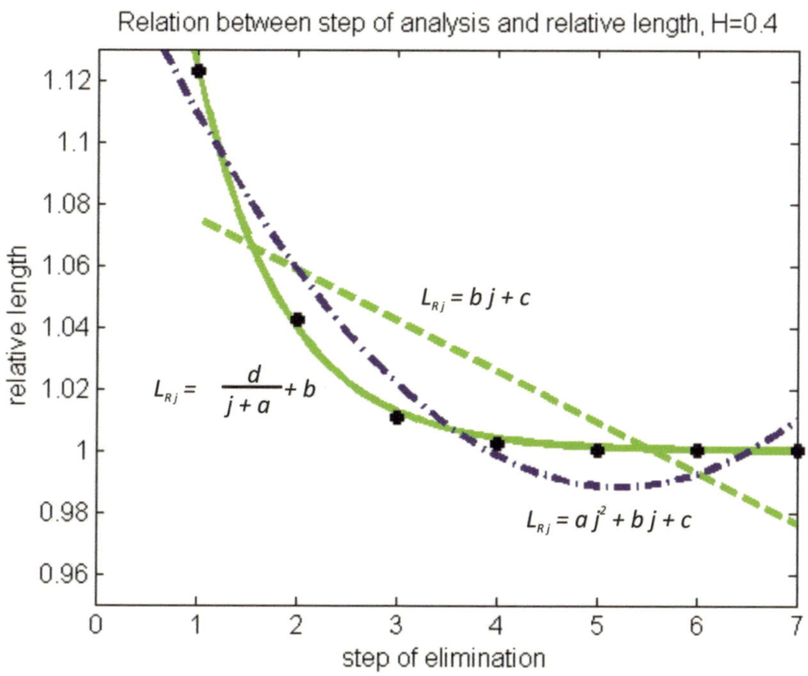

Figure 11. Relation between number of steps of analysis and relative length for the time series FBM, H=0.4.

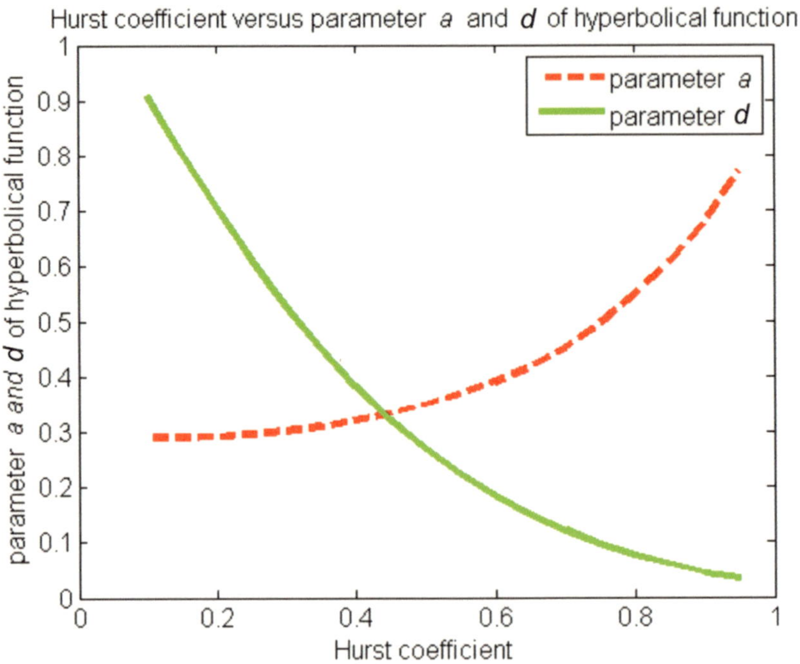

Figure 12. Hurst coefficient versus average value of parameters *a* and *d* of hyperbolic function.

EEE dimension for self-affine fractals can be estimated on base of parameter *d*:

$$D_{EEE} = 1 + |d| \tag{11}$$

2.6. Example of time series results

There are numerous possible applications of a fast and accurate description of time series (signals) from production process sensors using fractal geometry and statistical analyses.

Figure 13 represents results of three time series analysis from a glass tank: temperature of a tank main arch. The time series are from one position of production processes in different times. Box and R/S dimensions of the time series are written under the pictures. The complexity is estimated by one number, larger number represents higher complexity. First time series has relatively the largest complexity, and third is relatively the smoother (**Figure 13**).

Figure 14 shows results of a tank siege analysis, time series were obtained during defined time, and whole signals were analyzed using the standard deviation, the R/S dimension, and the box dimension. The average temperature indicates a temperature profile of the tank siege and an implicit temperature profile of the molten glass in contact to the tank siege. The box dimensions were computed for threshold values $k = \{0.2, 0.5, 0.7\}$, see **Figure 14**. From positions 5 and 6, the FD decreases and between positions 6 and 8 rises. The large dimension indicates a smoother time series, and the higher dimension represents a complex time series. Glass melt is in permanent movement in the tank. This movement is important for a good quality of glass melt in the end of the glass tank. However, this movement is impossible to monitor by

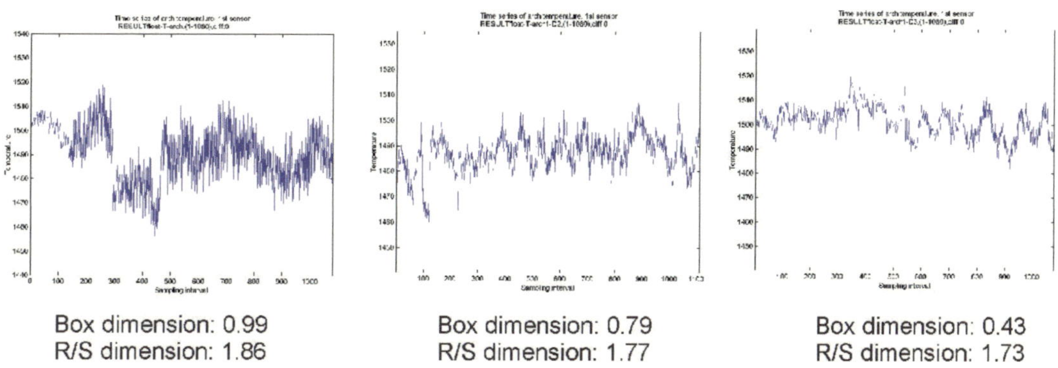

Figure 13. Time series of glass tank main arch temperatures from a first sensor T1 and these box and R/S dimension in different time.

standard measurements. The character (complexity) of time series represents changes in temperature. These changes are caused by the movement of molten glass. The decrease of the complexity occurs where a change of longitudinal glass currents is expected. This shows that fractal analysis can be used for the detection of molten glass currents [3, 8] (**Figure 14**).

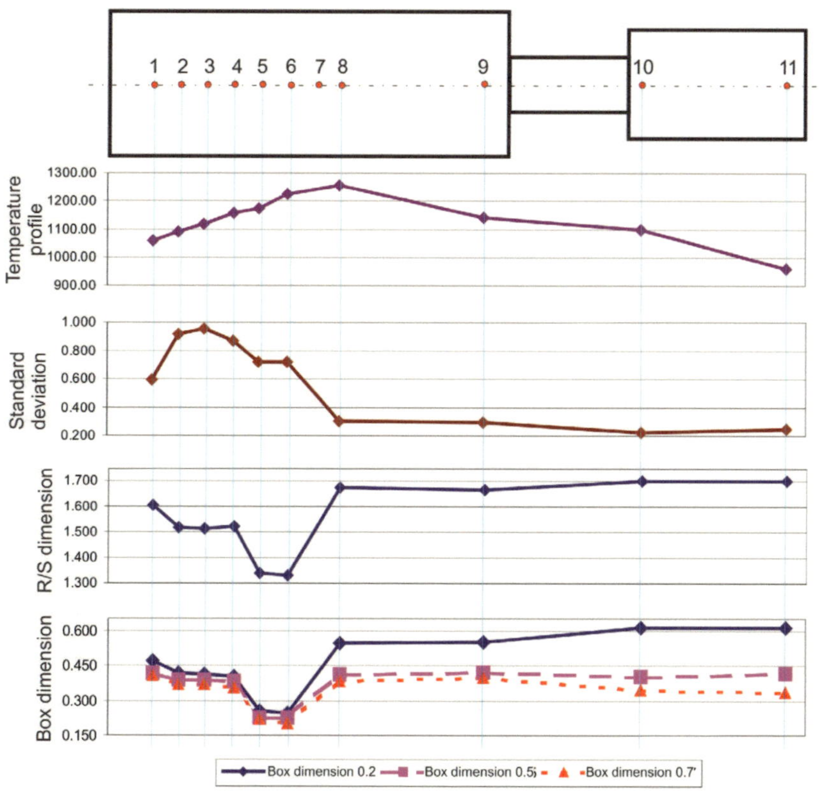

Figure 14. Results of glass tank siege time series analysis.

2.7. Examples of roughness analysis results

The raw data reading (unfiltered reading) from a surface roughness tester presented by a curve is labelled as a profile. Obtained parameters can be divided into three groups:

- *parameters of frequency* describing surface profile spacing parameter and corrugation frequency characterization (e.g. *Sm*—Mean Spacing),
- *parameters of amplitude* describing depth characterization (*Ra*—Average Roughness, *Std*—Standard Deviation, *Rz*—Mean Roughness, *Rt*—Maximum Roughness, Depth, etc.).
- *parameters of complexity and deformation* describing FD estimation by Compass Dimension (*Dc*), by Profile Proportional Length (L_P), EEE method, or Profile Relative Length (L_R).

Mentioned statistic parameters, amplitude, and frequency are widely used in industry. Surface profile parameters, such as Maximum Roughness, Average Roughness, Mean Spacing, and Mean Roughness Depth, are defined by ISO 4287-1997 [12] standard. Complexity and deformation parameters were chosen on the basis of previous experiences.

For better result comparison, the dimension is multiplied by 1000 (D_{C1000}).

We analysed 14 surfaces produced by five different processes and in different conditions, **Table 1**. **Figure 15** shows 28 samples (with 14 surfaces) [13]. The analysed structures were chosen so as to be different and to cover the most common surfaces in industrial practice. Chosen samples were made purposely from identical material. This allows us to subsequently ignore material properties and to analyse the change of technological parameters and the influence of used technology.

Sample	Technology of surfaces production
1	Polished surface to maximum gloss
2	Ballotini (glass beads) blasting, grain size F120 (mean diameter 0.109 mm)
3	Corundum blasting, grain size F36 (mean diameter 0.525 mm)
4	Corundum blasting, grain size F12 (mean diameter 1.765 mm)
5	Electro-erosion machining 29A
6	Electro-erosion machining 42A
7	Electro-erosion machining 54A
8	Sandpaper, K400
9	Emery cloth, 120
10	Emery cloth, 80
11	Vertical milling machine, milling cutter 20 mm, 120 rpm, feed 30 mm/min
12	Grinding wheel, 98A 60 J 9 V C40
13	Grinding wheel, 96A 36P 5V
14	Vertical milling machine, milling cutter 20 mm, 120 rpm, feed 240 mm/min

Table 1. Correlation coefficients of selected parameters.

Figure 15. Analysed samples with machined surfaces.

Measurement was realized on a surface roughness tester Mitutoyo SV 2000. Parameters were set as follows: stylus measuring speed: 0.5 mm/s; positioning: 2 mm/s; traverse range: 50 mm; linearity of traverse: 0.3 μm/50 mm. Standard type of stylus with a 60° angle with a measuring force: 0.75 mN was used. The length of measurement is 4800 μm, and the sampling interval is 0.5 μm.

All samples (two samples with the same surface) were measured in nine positions, each position in three directions, *x*, *y*, and transversely. All data obtained in the form of unfiltered

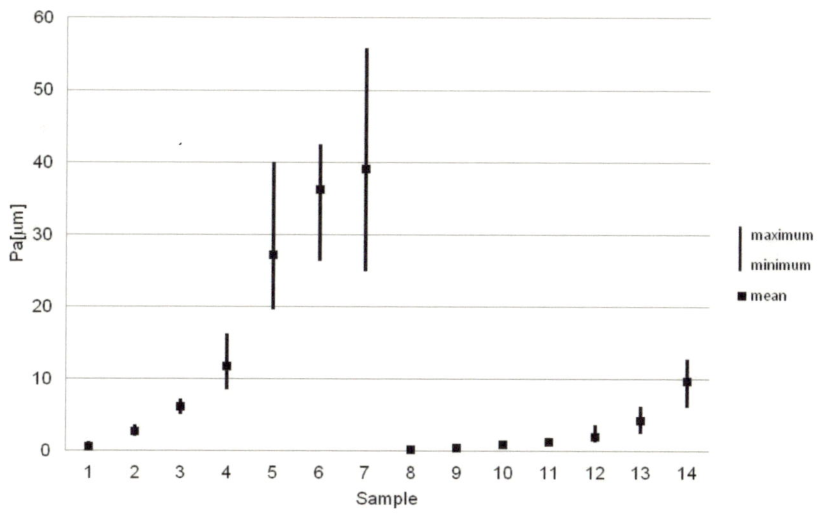

Figure 16. Results of *Pa* parameter.

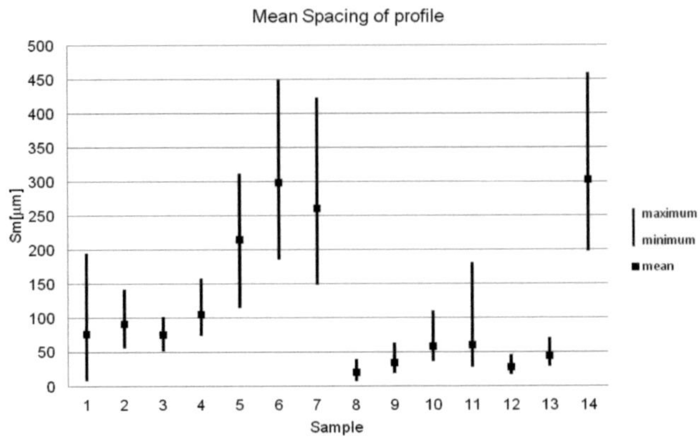

Figure 17. Results of *Sm* parameter.

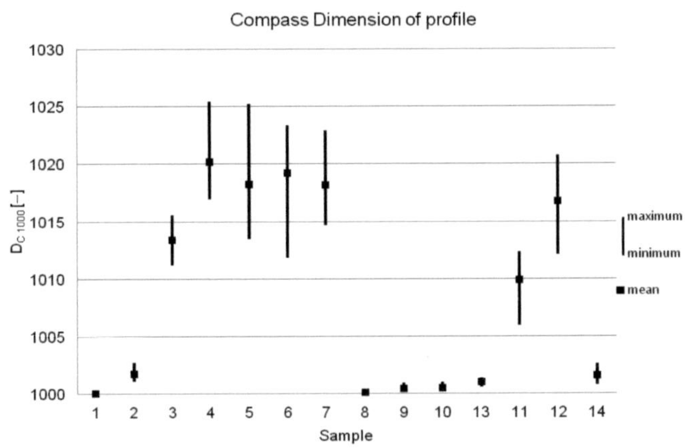

Figure 18. Results of FD estimation, $D_{C\ 1000}$.

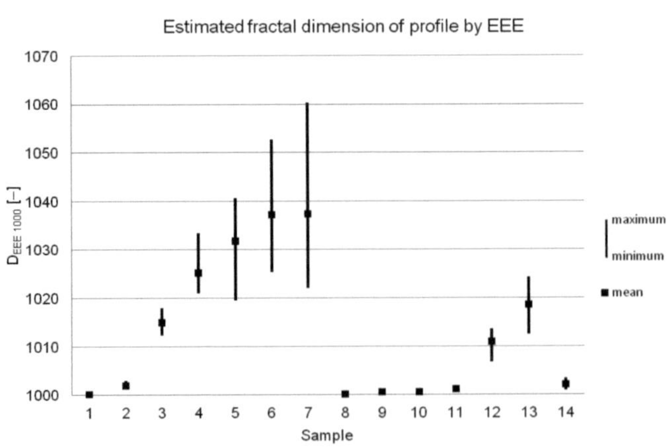

Figure 19. Results of FD estimation, $D_{EEE\ 1000}$.

profiles were used for analyses. Matlab platform was used for a data evaluation, and necessary software tools were developed.

Based on a linear correlation from obtained parameters (Section 3.1) and simplifying results, we can specify suitable parameters for evaluation of these types of data: Average Roughness, *Pa* (*parameter of amplitude*, **Figure 16**), Mean Spacing, *Sm* (*parameter of frequency* **Figure 17**), Compass Dimension, $D_{C\ 1000}$ (*parameter of complexity and deformation*, **Figure 18**). Diverse information from data can be provided by these three parameters. EEE dimension (*parameter of complexity and deformation*) is depicted in **Figure 19**. More information can be found in Ref. [13].

3. Classification of topological one-dimensional dividing curve

The research focuses on the application of the methodology for a quantification of metal surface changes and on an objectification of corrugation test for flat glass. Tools for analysis (the estimation of FD and statistical tools) are in the principle the same as for the topological one-dimensional signals, described in Section 2. The following text shows examples of FD application in practice.

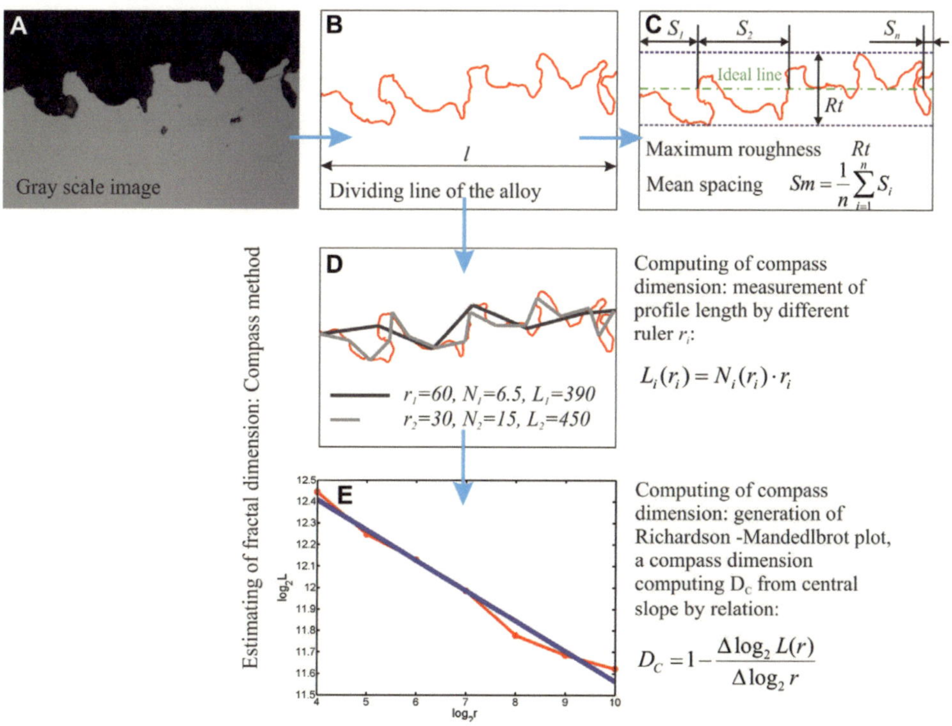

Figure 20. Image analysis of dividing curves between alloy and glass: A—gray scale image from light optical microscopy, B—evaluated dividing curve between the alloy and glass (boundary curve), C—parameter of amplitude (Rt—maximum roughness) and parameter of frequency (Sm—mean spacing), D—computing of compass dimension, E—compass dimension D_C computed from slope.

3.1. Example: surface roughness changes after corrosion tests

Quantification of metal surface changes after exposition on air or in a glass melt is important for objective comparison of materials corrosion resistance. Example is showed in relatively new materials: iron aluminides [14], compared with currently used chrome-nickel steels in contact with molten glass. The sample roughness changed during an interaction of metal surface with molten glass, and the effect of disruption can be evaluated after the end of the corrosion test. The obtained roughness of metal surface was quantified using FD and statistical tools.

First, a digital camera (in a light optical microscope) takes a photograph of a metal surface profile from metallographic sample (five photographs from one sample), **Figure 20A**.

	Fe-14Al-6Cr (iron aluminides)	Pt Sm D_{C1000}	Pt Sm D_{C1000}	EN X8CrNi25-21 (chrome-nickel steels)
Before test		0.60 11.60 1004.6	0.92 10.32 1012.8	
After test 48 h		1.33 10.46 1014.4	11.23 31.29 1040.6	
After test 72 h		1.52 11.24 1015.1	2.71 16.05 1046.2	
After test 168 h		1.80 13.98 1018.5	9.17 22.01 1102.82	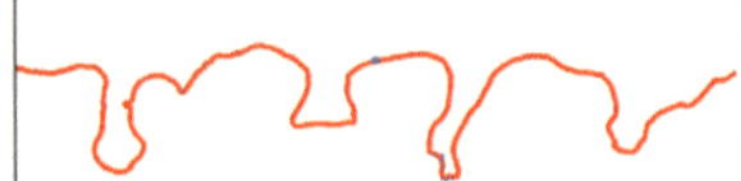

Figure 21. Examples of dividing curves chrome-nickel steel material and iron aluminide, after static glass melt effects in different temperatures and results of analyses (compass dimension multiplied by 1000, $D_{c\ 1000}$).

Second, a dividing curve is generated from obtained images by a software tool. Software exactly defines a curve between material alloys and its surrounding, **Figure 20B**. The generated dividing curve is evaluated by the FD (a compass dimension multiplied by 1000, $D_{C\ 1000}$), **Figure 20D** and **E**. The average standard deviation of all the curves (STD), average mean spacing(Sm), **Figure 20C**, and the average maximum roughness of all the curves (Rt), **Figure 20C**, are then described using statistics [15–17].

Examples of analysis are shown on materials: Fe-14Al-6Cr (iron aluminides) and EN X8CrNi25-21 (chrome-nickel steels). More information about tests and materials can be found in Ref. [17]. The dividing curves are depicted in **Figure 21**.

The evaluation of roughness parameters was carried out on 10 places for each sample (each alloy and time interval). The corrosion attack of the tested alloy may also be described by the roughness of the surface. Many parameter types can be used for a quantification of the metal roughness. Parameters can be divided into three groups: *parameters of amplitude, parameters of frequency, and parameters of complexity and deformation* (described in Section 2.7). In this field of research, a filtered profile is not being used. For this reason, the Average Roughness is called *Pa*, maximum Roughness is denoted *Pt* etc.

Graphs in **Figures 22–24** show the results of analysis for the dividing curves between alloys and glass. Average values were used in order to compare results.

The analysis using the developed methodology has two steps. The first step is a specification of appropriate parameters for fast and reliable analysis for data evaluation. Mentioned methodology

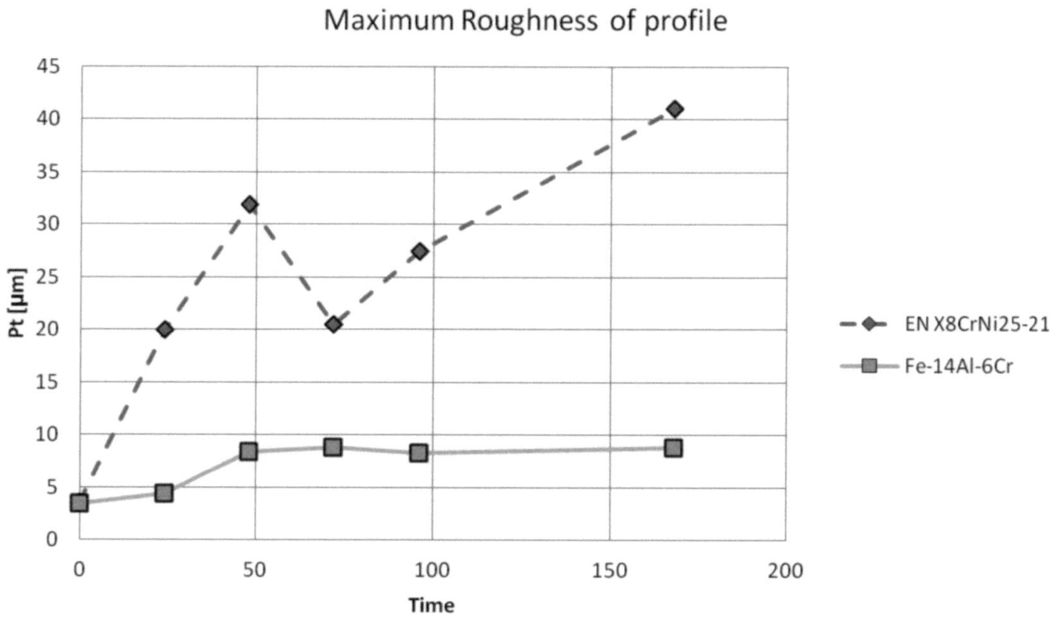

Figure 22. The average value of maximum roughness of profile, Pt as a function of time for corrosion in molten soda-lime glass.

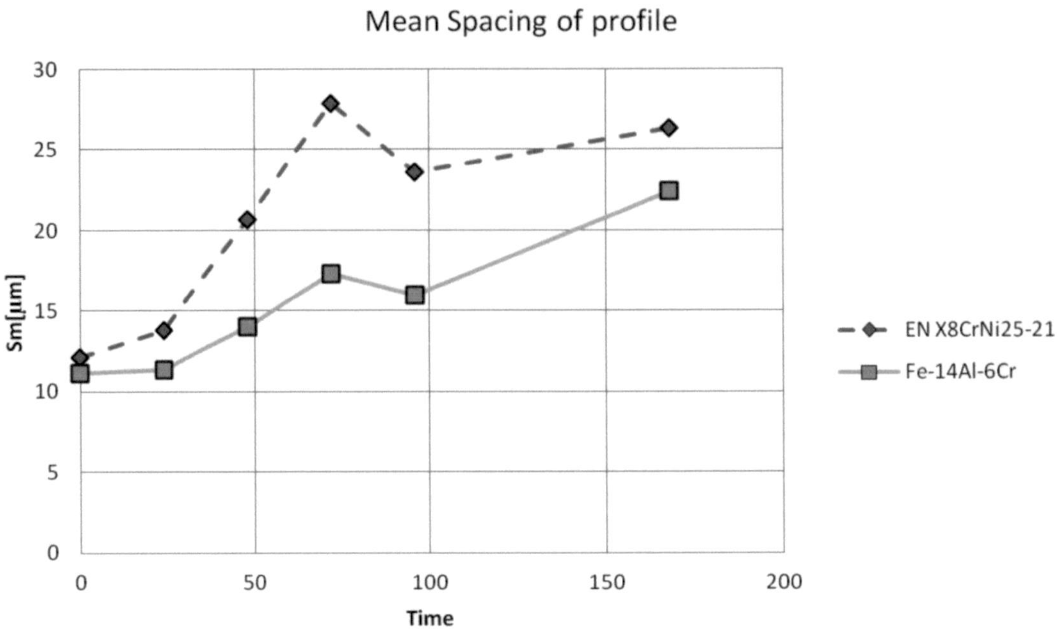

Figure 23. The average value of mean spacing of profile, Sm as a function of time for corrosion in molten soda-lime glass.

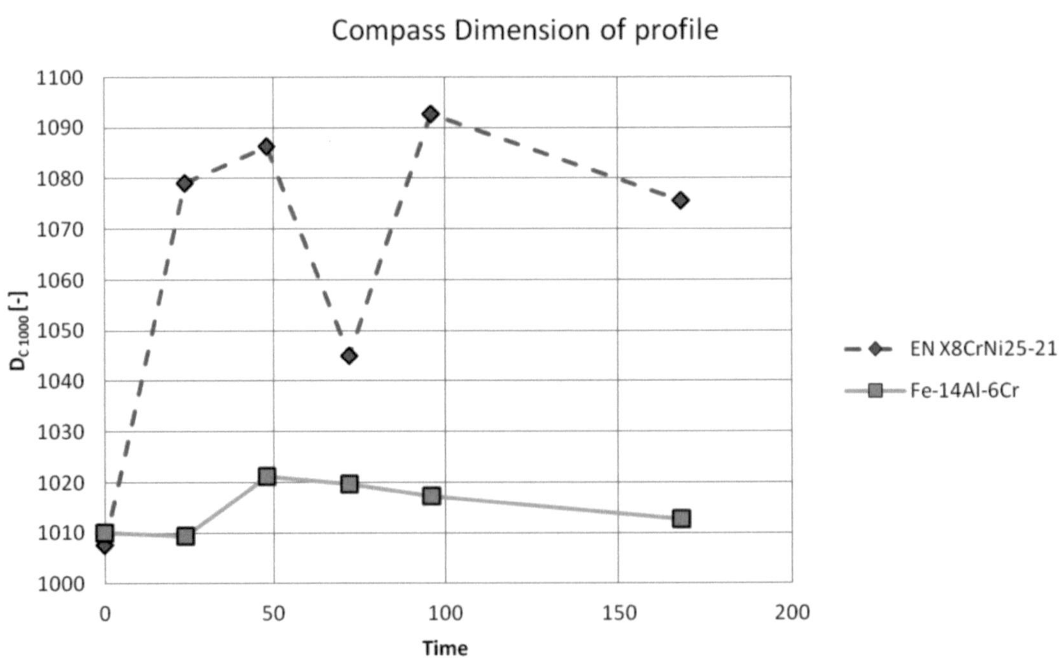

Figure 24. The average value of fractal dimension estimation, compass dimension of profile $D_{C\ 1000}$ as a function of time for corrosion in molten soda-lime glass.

	Relative length, L_R[-]	Compass dimension, $D_{C\ 1000}$ [-]	Mean spacing, *Sm* [um]	Average roughness, *Pa* [um]	Maximum roughness, *Pt* [um]	Standard deviation, *Std* [um]
Standard deviation, *Std* [um]	0.90	0.83	0.50	0.99	0.98	1
Maximum roughness, *Pt* [um]	0.88	0.85	0.47	0.96	1	
Average roughness, *Pa* [um]	0.91	0.82	0.49	1		
Mean spacing, *Sm* [um]	0.33	0.30	1			
Compass dimension, $D_{C\ 1000}$ [-]	0.92	1				
Relative length, L_R[-]	1					

Table 2. List of analysed samples with their production properties.

contains 22 parameters. However, only chosen parameters were used for a complete analysis in order to simplify the analysis and to reduce the processing time.

Some parameters linearly correlate with the others (they provide similar information about the data). To evaluate the parameters objectively, Pearson's correlation coefficients were computed, see **Table 2**. A correlation between chosen parameters is clearly visible (*Pa* and *Pt*). The L_R parameter correlates less with the parameters *Pa* and *Pt*, but still significantly. The $D_{C\ 1000}$ parameter correlates less with the parameters *Pa* and *Pt*, but correlates with the L_R parameter. The *Sm* parameter does not correlate with other parameters. On the base of a linear correlation from obtained parameters and simplifying results, we can specify suitable parameters for evaluation of these types of data: Maximum Roughness, *Pt* (*parameter of amplitude*), Mean Spacing, *Sm* (*parameter of frequency*), Compass Dimension, $D_{C\ 1000}$ (*parameter of complexity and deformation*). Diverse information from data can be provided by these three parameters

Second, it is possible to objectively describe a character of metal structure after corrosion attack. We are able to draw the conclusions:

Parameter of amplitude, Maximum Roughness *Pt* (**Figure 22**), shows deepness of corrosion attack of metal surface by glass melt. The average dividing curve deepness of alloy Fe-14Al-6Cr grows slowly up to 168 h because iron aluminide dissolved slowly and uniformly in the molten glass than EN X8CrNi25-21 [17–19]. The average dividing curve deepness of austenitic steel grows from raw state during attack to 48 h (**Figure 21**). After 48 h, it is apparent that corrosion protrusions penetrate less deeply into the surface of steel. It does not mean any increase of corrosion resistance, but probably a progress of corrosion attack is more uniform in this time period. This should be analysed by other methods, for example, the measurement of weight loss, chemical analyses, etc. The maximal deepness of corrosion attack is after 168 h.

Parameter of frequency, Mean Spacing *Sm* (**Figure 23**), shows surface profile spacing. The parameter describes corrugation frequency of the dividing curve after corrosion attack, how many wavelets can be observed on the surface. For both alloys grows the parameter up to 72 h, where the maximum is. The frequency is connected to the corrosion mechanism.

Parameter of complexity and deformation, Compass Dimension $D_{C\ 1000}$ (**Figure 24**), shows level of dividing curve complexity. In case of alloy Fe-14Al-6Cr, the average complexity of surface expressed as Compass Dimension grows from 24 to 48 h and then slowly decreases. On the other hand, the average surface complexity of alloy EN X8CrNi25-21 increases significantly from raw state during attack up to 48 h, then falls down, and grows to the maximum after 72 h. After 96 h, slight decrease of $D_{C\ 1000}$ was observed. The significant decrease in value of $D_{C\ 1000}$ after 48 h is probably related to more uniform a progress of corrosion. For understanding of the steel corrosion mechanism, it is necessary to carry out further analysis.

It is obvious that dividing curves of alloy Fe-14Al-6Cr after corrosion attack are smoother and less complex. It seems that the corrosion resistance of Fe-14Al-6Cr is higher than steel EN X8CrNi25-21. Austenitic steel showed corrosion protrusions due to probably intergranular corrosion (preferential attack of some phases at grain boundaries). However, this statement should be supported by structural and phase analysis.

3.2. Example: corrugation test

The optical test using a zebra plate is widely used measurements for mass production. The test is one of the many important measurements, and it is used in a wide range of situations: as a production control by manufacturers of float glass; as a quality control by glass processors on the glass they buy; as a production control of products (laminated glass, thermal treated glass, etc.); as a quality control by the final customer on the glass they buy.

The corrugation test (**Figure 25**) is focused on the reflection while another type of tests specialize on the passage of light though the glass (test of distortion). The test is based on the reflection of light off a glass sample sheet from a skew striped plate. The zebra plate is 1 × 2 m with 25 mm wide black strips at an angle of 45°. Glass sheet is laid on a Table 4 m distanced from zebra plate. An observer is distanced additional 4 m from the table. The observer subjectively evaluated the corrugation of reflected light based on a comparison with etalons. The quality of the sheet is classified using a rate dimensionless number from 1.5 to 3.5. The evaluation is carried out off-line in a dark-room. Samples of flat glass are obtained from an on-line production process, and they are cut from the whole width of production glass ribbon. **Figure 26** shows good and poor quality of glass sheets during the corrugation test. Using a small angle of observation caused the relatively extreme "distortion".

A measuring system, which consists of both hardware and software, has been developed (**Figure 27**) according to executed experiments. The system is an important intermediate stage for development of the on-line measurement and its calibration. It is also a suitable solution for processors and customers, without an expensive on-line measurement. The hardware of the system includes the zebra-pate (**Figure 27A**) and a table for the glass sheet sample (C) as in the standard subjective test. The operator is replaced by a scanning unit (digital camera with power adaptor, B), a control unit (PC, D), a connection between the digital camera and the control unit (E), a ball head (F), a camera support system (placed on a top-wall, G), and a system for gripping the table (H) to keep it in a defined place.

An operator lays a tested glass sheet on the table into the defined place, fills in a form in the Corrugation software and by pressing a button he starts the evaluation. The software initiates the communication with the camera and the image is captured. The image analysis starts after

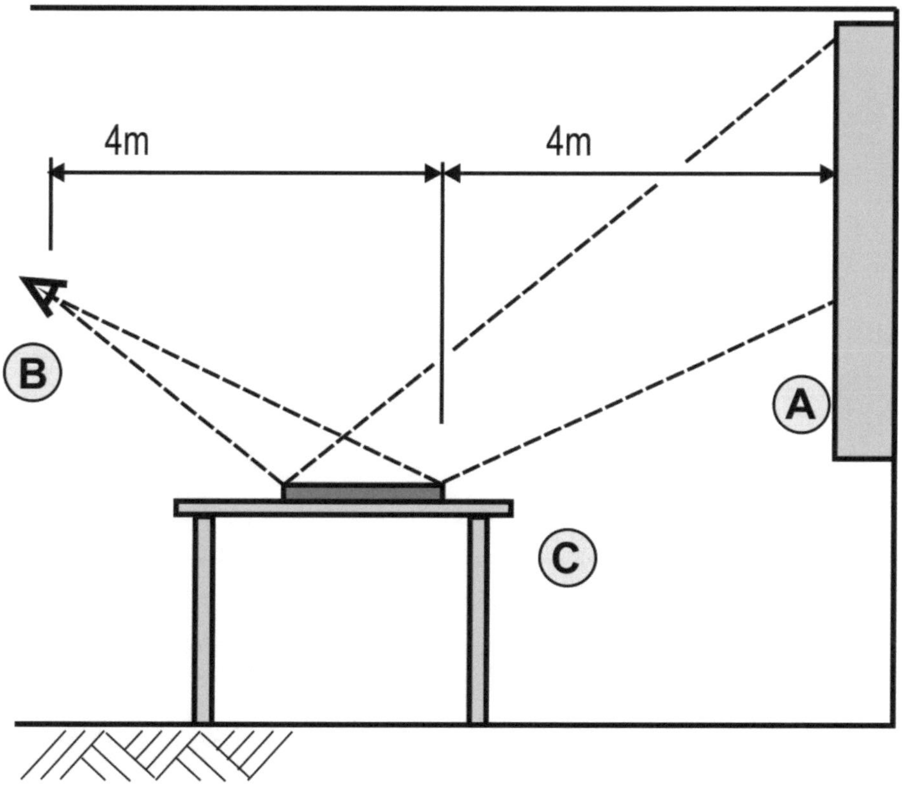

Figure 25. Scheme of corrugation test, A—zebra-plate, B—observer, C—table with window-glass sheet.

Figure 26. Good and poor quality of glass sheets with a specification of measured parts.

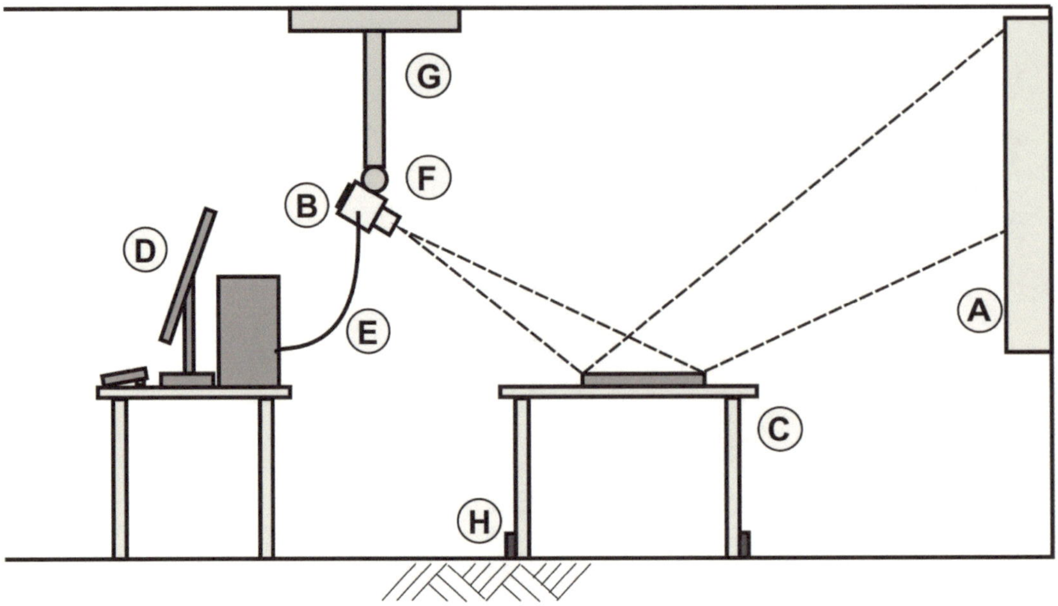

Figure 27. Scheme of measuring system for objective monitoring.

downloading the image from the camera to the computer. In a while, the operator can read the evaluated quality on the PC screen.

The analysis of a scanned image is used for the detection of a glass sheet position, for the necessary rotation and shifting, and for the generation of curves from detected boundaries between the light and dark areas of the sheet in the image. The curves are generated using a thresholding and a detection of contours. They are converted into sequences of points, where the axis X is defined from an ideal reflection, **Figure 28**. For curve obtaining, two thresholds are

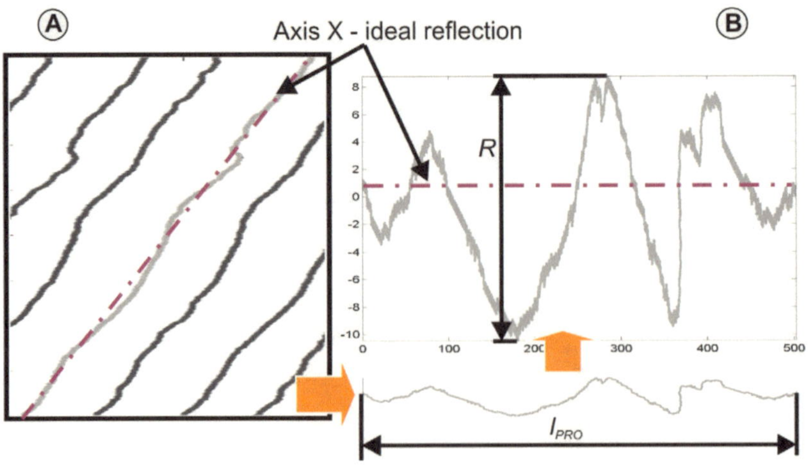

Figure 28. Conversion of the curve to a sequence of points and its evaluation using range R and relative length L_R.

used, as two different types of defects on the surface of the bottom part glass sheet can be found: a primary and a secondary corrugation **Figure 29**. (Different character of the reflected image can be found on heat treated glass and also the top and bottom parts can differ.) The methodology enables to describe both corrugations separately, and it evaluates the impact of the technological production parts to the quality of the production. The obtained curves are then evaluated by software tools: statistic; measuring the curve length; and estimating its FD.

Our research proved that only one parameter does not enable to describe all types of corrugation. We had to collect limited number of adequate parameters that evaluate completely the quality. An extensive analysis was performed to achieve the best conformity between a subjective evaluations and the evaluated quality using computed parameters. We used hundreds of measurements from one experienced operator with his specification of quality in the current scale. The results showed that three parameters (of the separated curves from the contours) are important to reliably assess the quality of flat glass:

- complexity (smoothness, used the compass dimension mentioned in Section 2.3);
- range of waviness from an ideal line;
- rate of deformation (using the relative length of the curves, Section 2.4).

All parameters are measured in pixels, and the parameters for one sample are defined as average values of all curves. Six quality parameters (the average values from two thresholds) are obtained and used for the final evaluation of the quality. The six quality parameters for each bottom part of a glass sheet obtained from the corrugation test were recalculated by means of weighted coefficients in accordance with the results obtained previously by subjective methodology. Six coefficients of the obtained parameters were defined using the particularly developed software for finding the smallest possible differences between the evaluated quality by means of the image analysis and with use of subjective monitoring by an experienced operator. An example of computed quality in relation to the quality of subjective assessment is showed in **Figure 30**.

Described system has been successfully tested for two years on a production line with the accuracy of 0.1–0.3 (in the used scale 1.5–3.5) in the latest version. Final results of over a thousand measurements from a real production process show a very good potential for the on-line application [3, 4, 8, 20]. The on-line corrugation test was implemented and showed

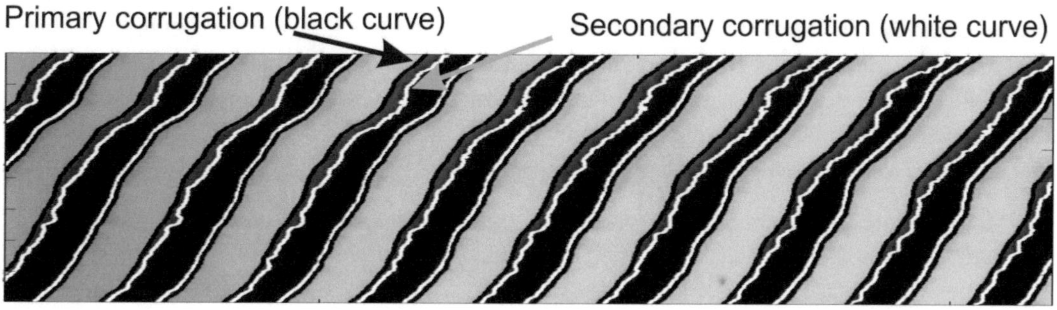

Figure 29. Primary and secondary corrugations of bottom part glass sheet.

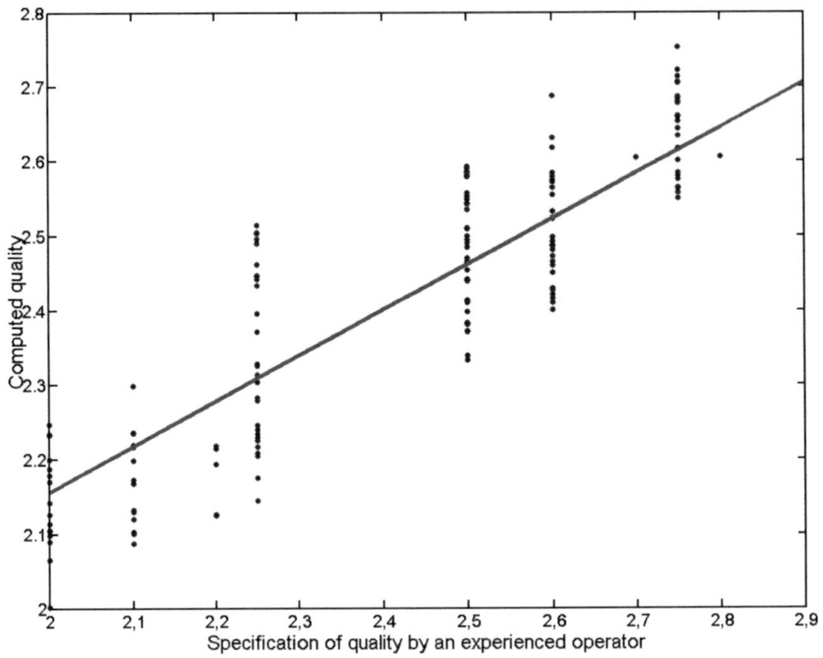

Figure 30. Relation between computed quality and the specification of quality by an experienced operator.

good stability of the evaluation methodology. Automatic measurement every few second is one of the many advantages. The system has been tested in a trial run.

4. Evaluation of 2D pictures with structure and defects

The explicit, objective, and automatic description of image complexity can be achieved by different methods, both statistical and using the FD. Only a few possibility examples are presented below. The analysis is done by evaluation of 2D images of surface defects—structures of the hole cracks in costume jewellery.

A digital image is represented as a matrix of pixels (or matrices for colour image, **Figure 31D**). Pixels can achieve different numbers, which depend on the format used for the digital images. The pixels are represented by numbers between 0 (black) and 255 (white) for the grey 8-bit palette bitmap, and the bitmap has only one matrix (brightness scale). **Figure 31C** shows two typical, poor quality surfaces of costume jewellery fissures. The cutting C-1 represents deep cracks, and C-2 represents a thin structure.

Matlab and HarFa software [7] were used for these experimental evaluations. A methodology for analysis of the pictures was developed based on: histogram evaluation, percentage of black pixels, percentage of large defects, the FD.

In practice, the methodology of surface structure description can be divided to five steps:

- Sample preparation—the costume jewellery is cut, as the structure must be visible, **Figure 31A**.

- Image acquiring—cracks in costume jewellery hole. An electronic microscope was used, **Figure 31B**.
- Software preparation, **Figure 31C** (separation of image parts needed for analysis)
- Image analyses.
- Evaluation of results.

4.1. Histogram

An evolution of a bitmap structure is possible by the histogram statistical description, **Figure 32**. Modus, median, average, range, standard deviation, and other statistic tools can be used easily. However, they are not applicable for surface structure of costume jewellery holes.

A potentially applicable method is the histogram cut off on 5% level. The image is described by a single number. It computes a 90% width of all the histogram pixel values from an average image value. However, this method is highly sensitive to shades that can possibly occur in the hole cracks. This easy method allows describing of all the defects, shades, cracks and structure collectively.

4.2. Thresholding

Following analyses are dealing with thresholding procedure. Thresholding is a technique that transforms grey or colour image into a binary one (black and white). For example, the binary image can be determined from the grey 8-bit palette bitmap, where black are all pixels which fulfil specific criteria, for example, $0 \leq \text{black} \leq 35$ and all the other pixels become white

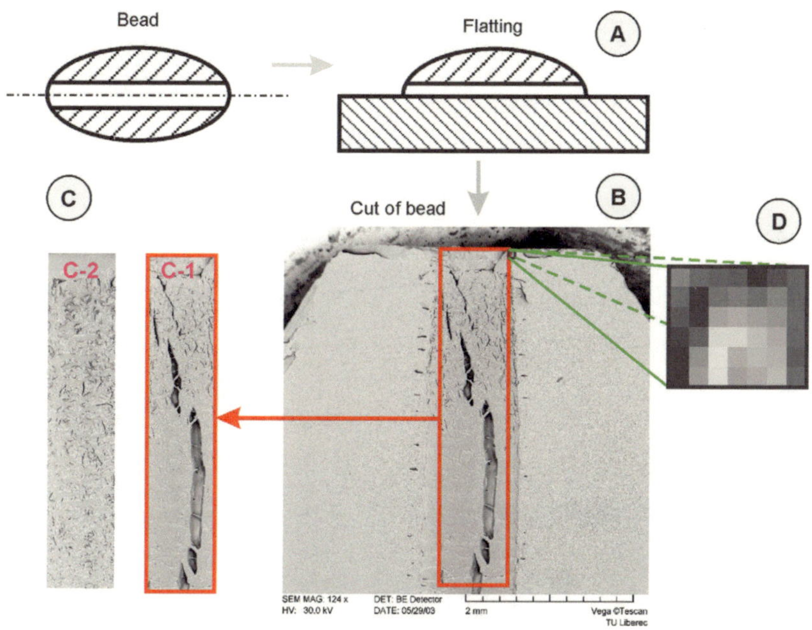

Figure 31. Preparation of samples, taking photographs, software preparation.

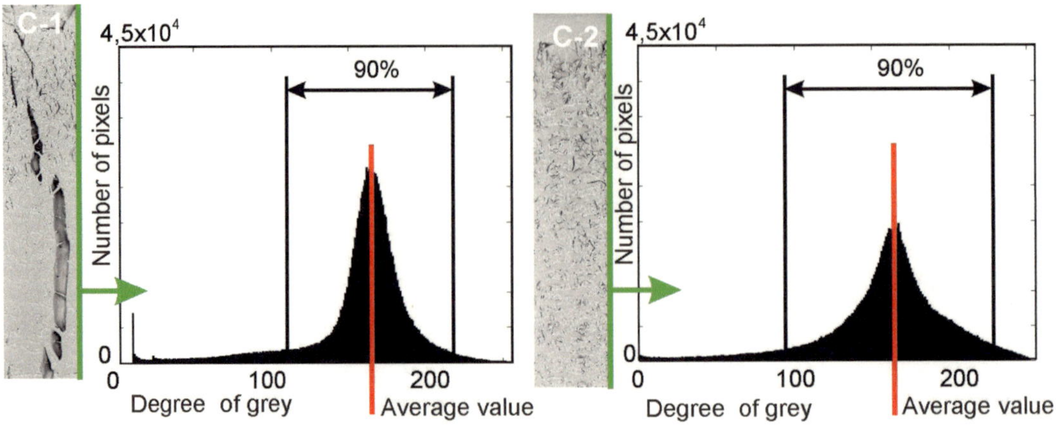

Figure 32. Histogram cut off on 5% level.

(35 < white ≤ 255), **Figure 33**. It means, that all pixels lesser than or equal to the threshold 35 are black and greater than 35 are white. We used the threshold 35. (More than one threshold can be used or the technique for matrixes of colour images can be used too.)

If the thresholding procedure is used for all thresholds of the grey 8-bit palette image, 256 binary images are obtained. Some of the image analyses can be done for all of obtained binary images. If a single number classifies one binary image, a spectrum of dependence between single number and threshold is given (e.g. **Figure 33**).

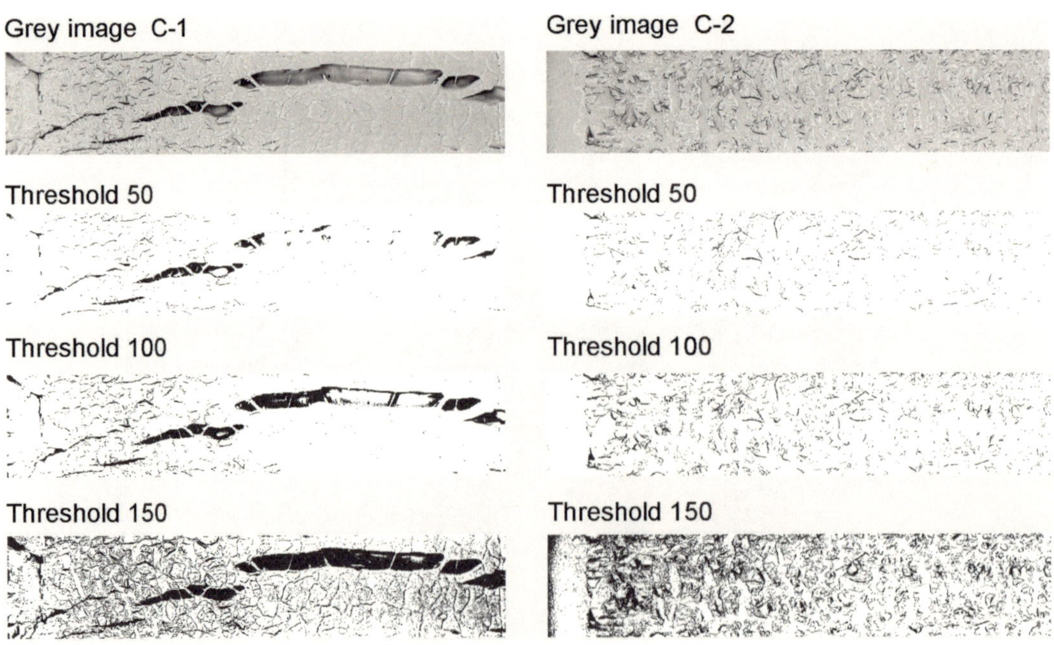

Figure 33. Binary images obtained by thresholding of grey images (8-bit palette bitmap).

4.3. Percentage of deep cracks

The method uses binary image obtained by thresholding and computes percentage of pixels with neighbouring pixels of the same value. The analysis searches black pixels (value 0) in a binary image, which have five or more neighbouring black pixels. **Figure 34** shows part of boundary crack. Black pixel in **Figure 34A** has five neighbours, and in **Figure 34B** has 8. The method is suitable for a detection of relatively large and single cracks and defects that are represented by black pixels with five or more neighbouring pixels.

A spectrum of dependence between percentage of black pixels with five or more neighbouring black pixels of binary images and thresholds is showed in **Figure 35A**. Thresholds from 10 to 50 are the most suitable for detection of large hole cracks in costume jewellery, **Figure 35B**. For the threshold 50, the cutting C-1 has more single cracks and defects, numerically: $T_{50\ C\text{-}1}$=3.17% than the cutting C-2, numerically: $T_{50\ C\text{-}2}$=0.8%.

4.4. Box dimension

The method again uses binary image obtained by thresholding. The box counting method [5, 7, 9, 10] estimates FD of a structure in an image. This estimation is a single number called the box dimension D_B. The box counting method works by laying meshes of different sizes *r* and then counting numbers of boxes *N* needed to cover a binary image (**Figure 36A**) completely (**Figure 36B, C**). The power law allows us to determine a number of boxes *N(r)* necessary to cover the structure:

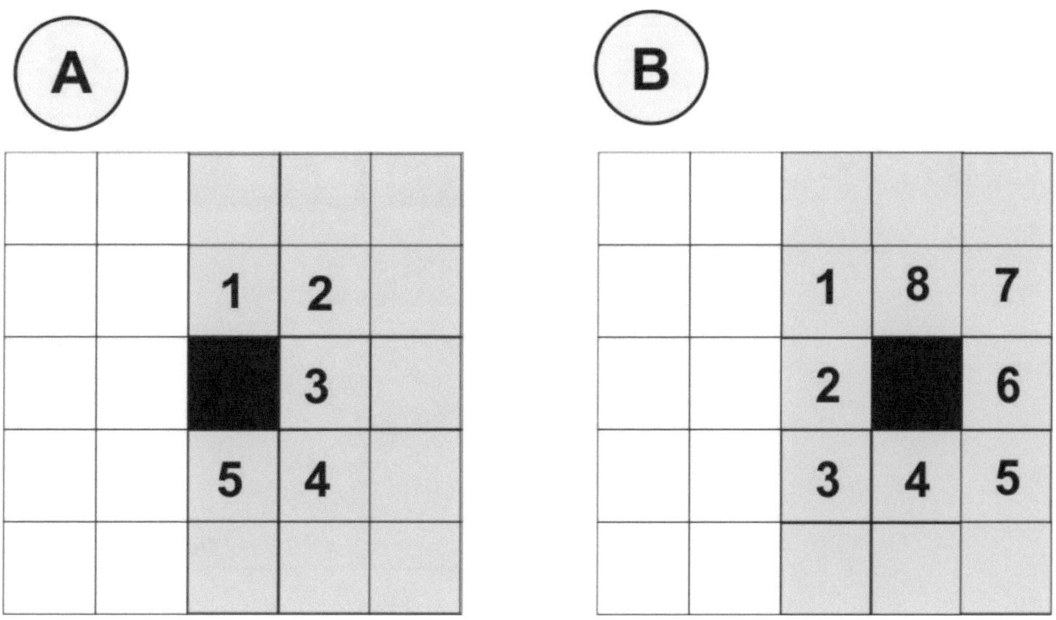

Figure 34. Pixels on boundary crack.

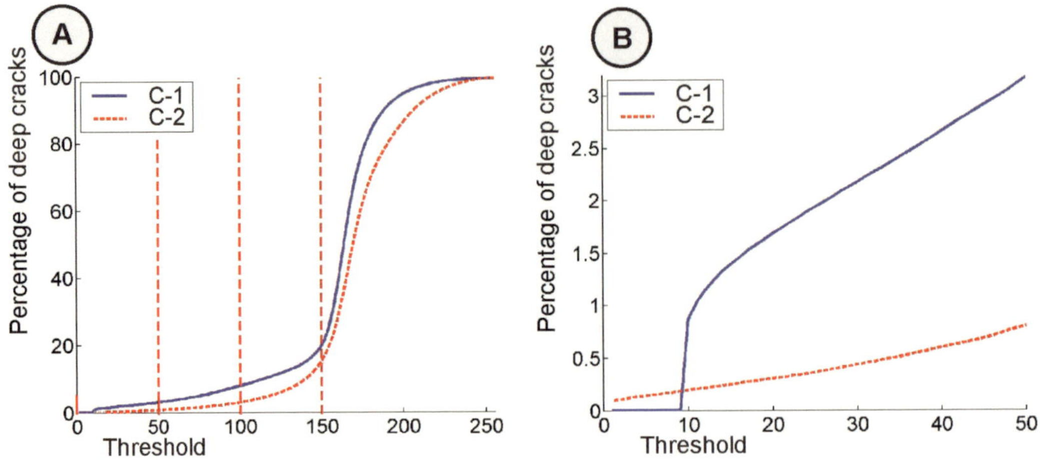

Figure 35. Percentage of black pixels with five or more neighbouring black pixels spectrum.

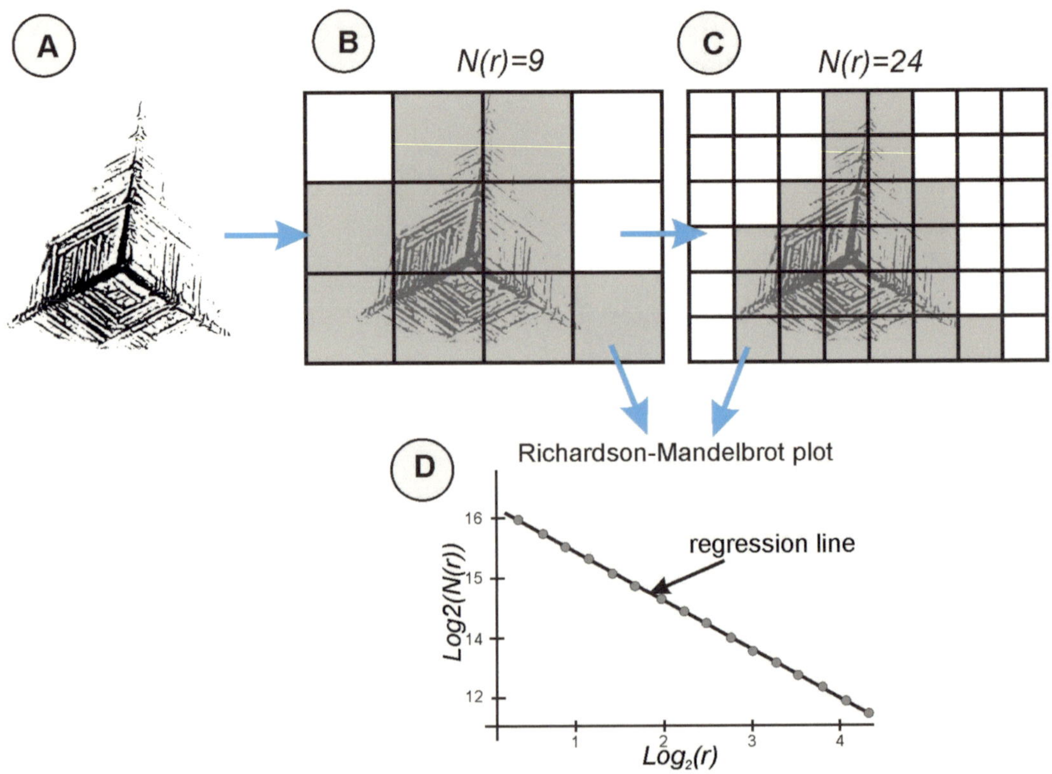

Figure 36. Box counting method.

$$N(r) = const. \cdot r^{-D_B} \tag{12}$$

where D_B is the box dimension. A relation between $\log_2 r$ and $\log_2 N(r)$ is known as Richardson-Mandelbrot plot (**Figure 36D**). The box dimension can be determined by slope s of the regression line in **Figure 36D**:

$$s = D_B = -\frac{\Delta \log N(r)}{\Delta \log r} \tag{13}$$

The software HarFa [7] is used for the analysis and software tools developed in Matlab make data evaluation. The box dimension is multiplied by 1000 for a better confrontation.

The fractal spectrum of the cuttings C-1 and C-2 is shown in **Figure 37**. The box dimensions over threshold 150 are similar, as a shadow influence is significant over this value. Results of analysis for threshold 120 are:$D_{B\_C\text{-}1}$=1429,6 (C-1) a $D_{B\_C\text{-}2}$=1562,4 (C-2), where the higher value represents more complex structure in the image. The cutting C-2 is more structured than the C-1 and box dimension quantifies the structures.

4.5. Example: detection of surface structure

The structures of the hole cracks in costume jewellery were analysed. Methodology of the image analysis, mentioned in the previous chapter, analysis was developed based on: histogram evaluation, percentage of black pixels, percentage of large defects, the FD. Only the last two analyses are suitable for describing these kinds of structures.

The percentage of large defects is suitable for the detection of individual, relatively large cracks and defects, **Figure 38**.

For this reason, many analyses for the FD estimation can be used. However, the most suitable for these kinds of data and structures is the box counting method (in **Figure 39**). For research purposes, the dimension was multiplied by 1000.Although an automatic classification of glass

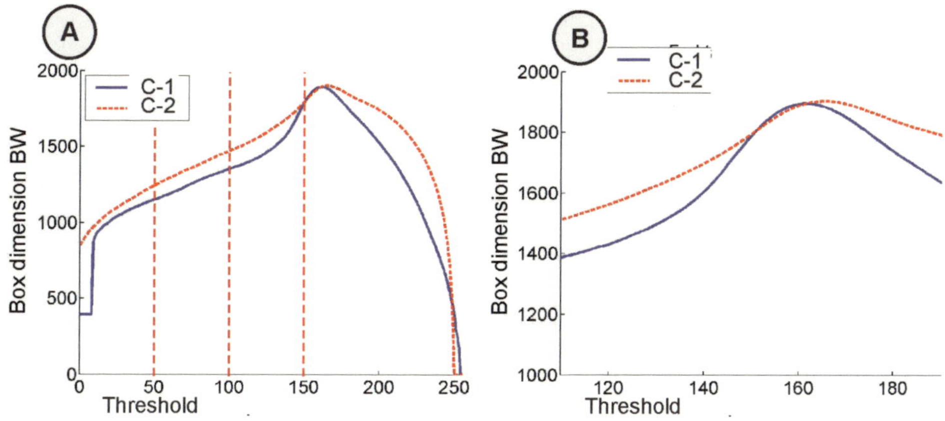

Figure 37. Fractal spectrum.

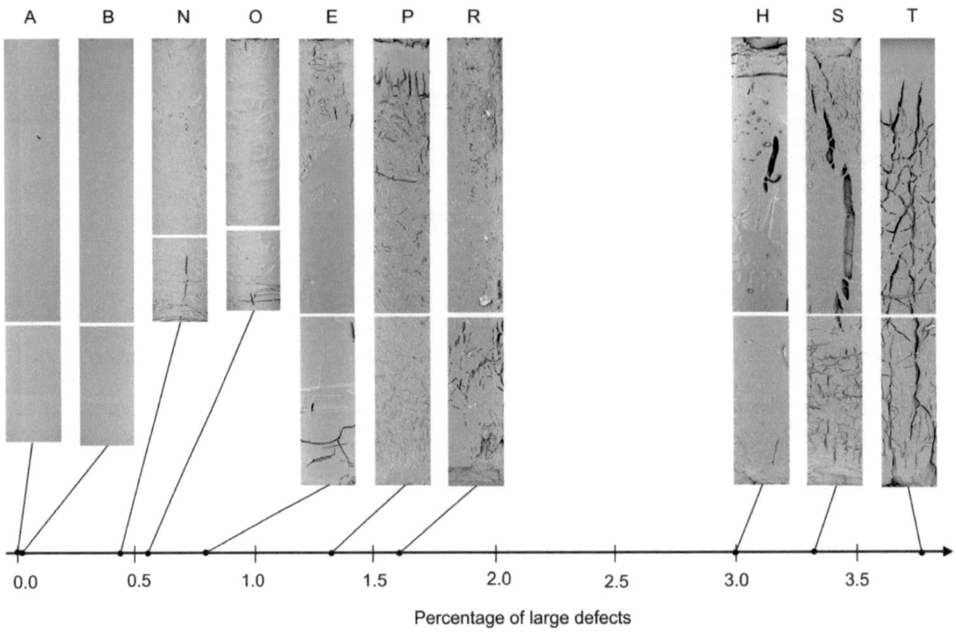

Figure 38. Computation of the percentage of pixels with neighbouring pixels of the same value—percentage of large defects.

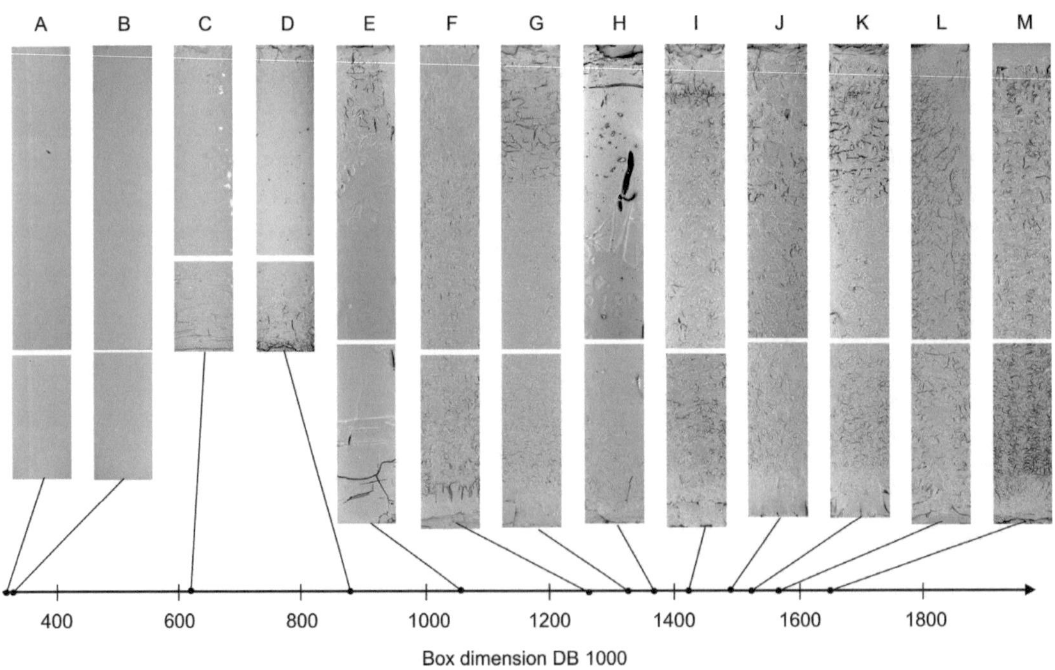

Figure 39. Description of glass defects using the box dimension D_B 1000 (dimension multiplied by 1000), percentage of large defects and automatic classification.

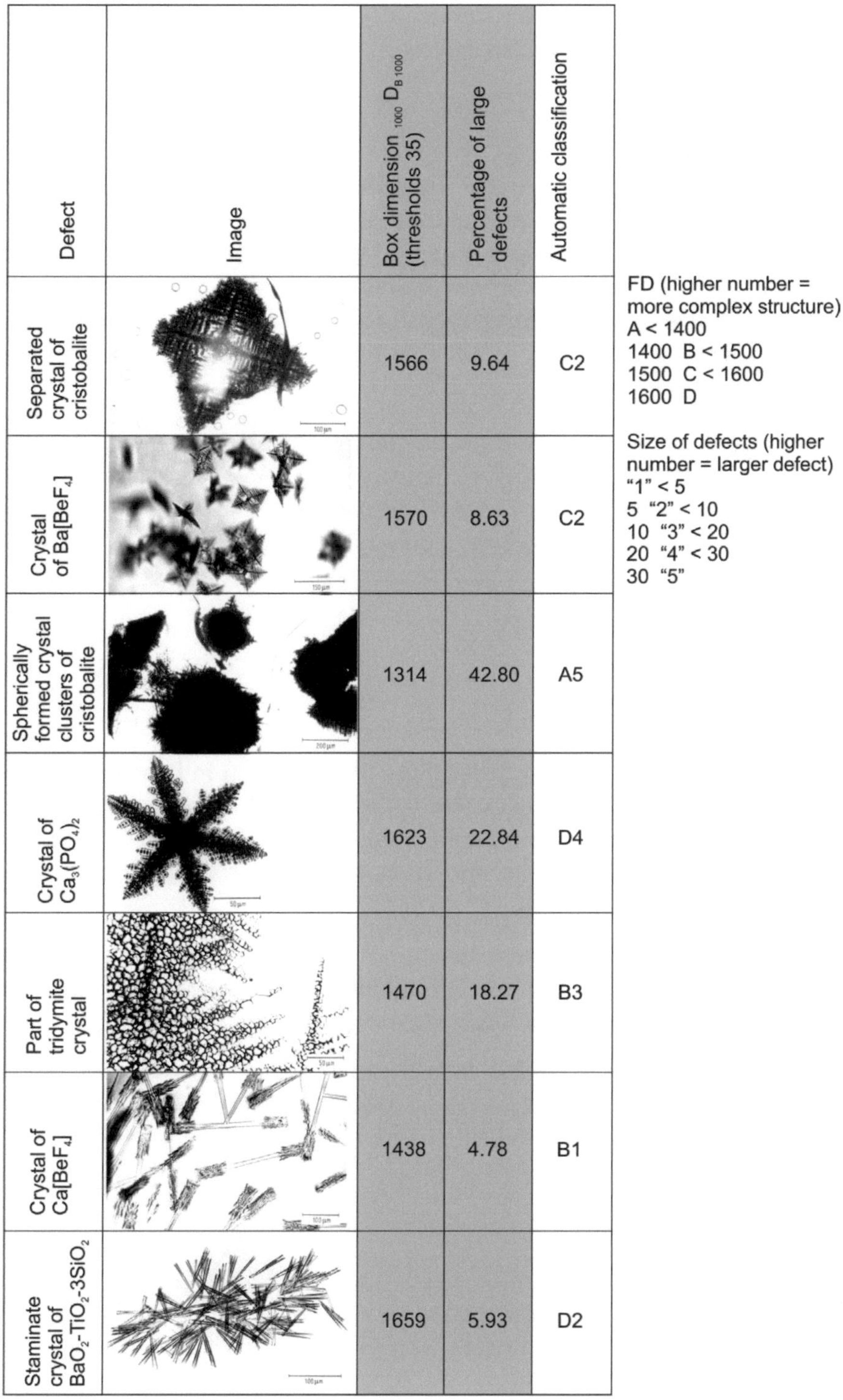

Figure 40. Description of glass defects using the box dimension, percentage of large defects and automatic classification.

defects is from the laboratory, presented example shows the possibilities of this method, its sensitivity to the different shapes of glass defects.

4.6. Example: defects classification

Automatic defect classification is another example of the FD application. A classification mark of obtained images can be done automatically and can be used for monitoring and control of production system. Software HarFa [7] was used for these experimental evaluations. The developed methodology was also based on: histogram evaluation, percentage of black pixels, percentage of large defects, the estimating of FD. On the base of our research, the last two analyses were chosen as suitable for describing these kinds of structures, **Figure 40**.

An automatic classification of glass defects can most easily be realized by using the defined boundaries for the analysis results. The simulated classification is on the right side of **Figure 40**. Presented example shows possibilities of this method and its sensitivity to the different shapes of glass defects. However, it does not solve problems with on-line obtaining of the images.

5. Conclusion

The examples of used methods show possibilities of application in industry and production laboratories. Structured surface, complex time series (topologically one dimensional signals), difficulty to describe dividing curves are much more common than it can be expected. The FD is then important for an objective quantification of complexity and should be use as an additional tool for the overall analysis.

The FD is widely used in science; nevertheless, industrial applications are rather rare. Data analysis using the FD has great potential in combination with statistical and other measurements in industry. However, FD cannot substitute standard analysis tools.

Although the text is focusing to a glass industry, emphasis was put on general application possibilities, where obtained knowledge, methodology and principles for product and industrial data evaluation more widely in industry might be used [11].

Acknowledgements

The results of this project LO1201 were obtained with co-funding from the Ministry of Education, Youth and Sports as part of the targeted support from the programme entitled "Národní program udržitelnosti I".

Author details

Vlastimil Hotař

Address all correspondence to: vlastimil.hotar@tul.cz

Technical University of Liberec, Liberec, Czech Republic

References

[1] Levy VJ, Lutton E, Tricot C, editors. Fractals in Engineering. 1st ed. USA: Springer-Verlag; 1997

[2] Levy VJ, Lutton E, editors. Fractals in Engineering: New Trends in Theory and Applications. 2nd ed. Germany: Springer; 2010

[3] Hotar V. Methodology of Industrial Data Description Using Fractal Geometry (in Czech language) [thesis]. Liberec: Technical University of Liberec; 2005. p. 124

[4] HotarV. Fractal geometry for industrial data evaluation. Computers & Mathematics with Applications. 2013;**66**:113–121

[5] EvertszCJG, PeitgenHO, VossRF, editors. Fractal Geometry and Analysis: The Mandelbrot Festschrift. 1st ed. Singapore: World Scientific Publishing Co. Pte. Ltd.; 1996

[6] Philpott D, Barnard HD, Bullen P. Some fractal measures used with the Noviscam project. In: Novak M, editor. ISQVPFD'2000; 3-4 April 2000; Bled. Lubljana, Slovenia: Faculty of Mechanical Eingeneering; 2000. pp. 61–70

[7] Zmeska O, Nezadal M, Buchnicek M, Sedlak O. HarFA and HarFA: e-journal [Internet]. 2002. Available from: http://www.fch.vutbr.cz/lectures/imagesci [Accessed: 25 January 2017]

[8] Hotar V, Novotny F. Some advanced analyse for quality monitoring. In: 21st International Congress on Glass; July 1–6, 2007; Strasbourg. Strasbourg: ICG; 2007. p. 91

[9] Mandelbrot BB. The Fractal Geometry of Nature. New York: W. H. Freeman and Co.; 1982

[10] Peitgen HO, Juergens H, Saupe C. Chaos and Fractals: New Frontiers of Science. New York: Springer-Verlag; 1992

[11] Hotar V, Novotny F, Chrpova E, Lang M. Application of fractal geometry in glass industry—Overview. In: Sirok B, editor. ISQVPFD'2002; February 4-5 2002; Bleb. Ljubljana, Slovenia: University of Ljubljana; 2002

[12] ISO 4287. Geometrical Product Specifications (GPS)—Surface Texture: Profile Method—Terms, Definitions and Surface Texture Parameters. Geneva: International Organization for Standardization; 1997

[13] Hotar V, Salac P. Surface Evaluation by Estimation of Fractal Dimension and Statistical Tools. The Scientific World Journal, vol. 2014, Article ID 435935; 2014. p. 10.

[14] Hotar A, Kratochvíl P. The corrosion resistance of iron aluminide $Fe_{28}Al_3Cr_{0.02}Ce$ (at%) in molten glass. Intermetallics, 2007;**15**(3):439–441

[15] Hotar V, Novotný F. Surface profile evaluation by fractal dimension and statistical tools, In: Proceedings of 11th International Conference on Fracture; Turin, Italy; 2005. p. 588

[16] Hotar A, Kratochvíl P, Hotar V. The corrosion resistance of Fe_3Al based iron aluminides in molten glasses. Kovovematerialy - Metallic Materials. 2009;**47**(4);247–252

[17] Hotar A, Hotar V. Fractal geometry used for evaluation of corrosion resistance of Fe-14Al-6Cr Wt. % against molten glass. Manufacturing Technology. 2015;**15**(2):543–541, Institute of Technology and Production Management University of J. E. Purkyne, Czech Republic; 2015

[18] Hotar A, Kratochvil P, Hotar V. The corrosion resistance of Fe_3Al-based iron aluminides in molten glasses. Kovove Materialy-Metallic Materials. 2009;**47**;247–252

[19] Hotar A, Hotar V, Novotný F. Corrosion behaviour of Fe-40Al-Zr (at.%) alloy in molten soda-lime glass. Kovove Materialy-Metallic Materials. 2014;**52**:149–155

[20] Hotar V, Novotný F, Reinischová H. Objective evaluation of corrugation test. Glass Technology—European Journal of Glass Science and Technology Part A. 2011;52(6):197–202

Chapter 8

Factors Affecting Accuracy and Precision in Measuring Material Surfaces

Jason A. Griggs

Additional information is available at the end of the chapter

http://dx.doi.org/10.5772/intechopen.68189

Abstract

The fractal dimensions of material surfaces are of interest because they can be related to material performance. Such surfaces include the fracture surfaces of broken specimens, surfaces abraded by airborne particles, and surfaces upon which coatings of another material have been applied. Scientists who study the fracture surfaces of failed medical implants stand to benefit greatly from fractal analysis. The origin of failure is often damaged or lost during retrieval of a failed implant, and evaluation of the undamaged portions of the fracture surface by relying on the self-similarity property of fractals may allow one to deduce the conditions that were present at the failure origin at the moment of failure. If the analysis of material surfaces will be used as an engineering tool, then it is important to identify the analysis methods that yield the most precise and accurate estimates of surface dimension. Eleven algorithms for calculating the surface dimension are compared. A method for correcting the bias of dimension estimates is presented. The sources of error involved in atomic force microscopy, optical microscopy, mechanical sectioning, and fabrication of specimen replicas are discussed.

Keywords: fractal surfaces, surface analysis, accuracy, precision, technique sensitivity, noise sensitivity, atomic force microscopy, confocal microscopy

1. Introduction

The fractal dimensions of material surfaces are of interest because they can be related to material performance. Such surfaces include the fracture surfaces of broken specimens, surfaces abraded by airborne particles, and surfaces upon which coatings of another material have been applied. In the case of fracture surfaces, the surface dimension is related to the fracture resistance of materials from which those specimens were made [1], and it can be a useful failure analysis tool. On abraded surfaces, such as a sandblasted surface, the surface dimension is inversely correlated to the surface roughness [2], and hence it is related to the strength with which an adhesive can be bonded to the surface [3]. In the case of a surface coating, the surface dimension is related to wettability of the surface [4], which may influence the way that the coating interacts with biological systems. There are many methods of profiling surfaces and many algorithms for determining the fractal dimension of a surface profile. If the analysis of fractal surfaces will be used as an engineering tool, then it is important to identify the analysis methods that yield the most precise and accurate measurements and to be aware of the possible sources of error.

2. Accuracy versus precision

The accuracy of a method of estimating the surface dimension refers to how small a difference there is between the true dimension and the average estimate provided by that method. If a perfectly accurate method is used many times to estimate the dimension of a fractal surface, then there will be no difference between the surface dimension and mean of the many estimates. The method would be unbiased. If the mean of the estimates is greater than the surface dimension, then the method has a positive bias. A systematic error in the other direction would indicate a negative bias. A method with larger bias (less accuracy) would provide a mean estimate that lies farther from the true surface dimension.

The precision of a method of estimating the surface dimension describes how close repeated measurements are to each other. A perfectly precise method would provide the same estimate every time it is used on the same surface, regardless of whether the estimate is accurate or not. Some measures that are used to rank the relative precision of different methods are standard deviation (SD), coefficient of variation (CV), and confidence interval. Most readers are already familiar with SD. The CV is the SD of a group of estimates divided by their mean. CV is considered by many to be a better descriptor of precision than SD when the methods under comparison have different levels of bias. The confidence interval spans a range of values, and researchers can be confident to a certain degree (expressed as a percent chance of being correct) that the true surface dimension lies somewhere between the lower and upper bounds of the interval. In situations where either a large number of replicates (generally 30 or more) are being averaged or where the estimates are assumed to follow a normal distribution (Gaussian distribution), the half-width of the confidence interval can be calculated by dividing the SD of the estimates by the square-root of the number of estimates (n) and multiplying by a factor from a statistical table (1.96 for 95% confidence). Lower and upper bounds of the confidence interval are then, respectively, calculated by subtracting and adding the interval half-width from the mean [Eq. (1)]. A precise method of surface analysis would exhibit a small SD, small CV, and narrow confidence interval.

$$95\% \text{ confidence interval} = \text{mean} \pm 1.96 \frac{\text{SD}}{\sqrt{n}}. \tag{1}$$

When comparing multiple methods of surface analysis, it is possible that the method with the best accuracy will not be the same as the method with the best precision. In this case, one must decide which is more important—accuracy or precision. When a method is accurate but not precise, the size of the confidence interval can be made smaller (improved) by increasing the number of replicates, n [Eq. (1)]. When a method is precise but not accurate, it can be used to test a series of calibration standards (surfaces with known fractal dimension) in order to determine the bias of the method and to determine how that bias changes over a range of surface dimensions. If a statistical model can be fit to describe the relationship between the biased estimates and the calibration dimensions, then the analysis method can be corrected to provide unbiased estimates. Then, it will be both accurate and precise. Therefore, precision is more important than accuracy when calibration standards are available. Consider the case of a biathlete firing a rifle at a target (**Figure 1**). The holes in the target on the lower left have good accuracy but poor precision. Although the group is centered on the bullseye, none of the shots were through the bullseye, and there is no calibration that can be performed to make the next shot more likely to hit the bullseye. However, the holes in the target on the upper right have poor accuracy and good precision. None of the shots hit the bullseye on the practice target, but the biathlete can adjust the screws on his or her aperture sight to increase the likelihood of hitting the target during the next practice session or during the upcoming competition.

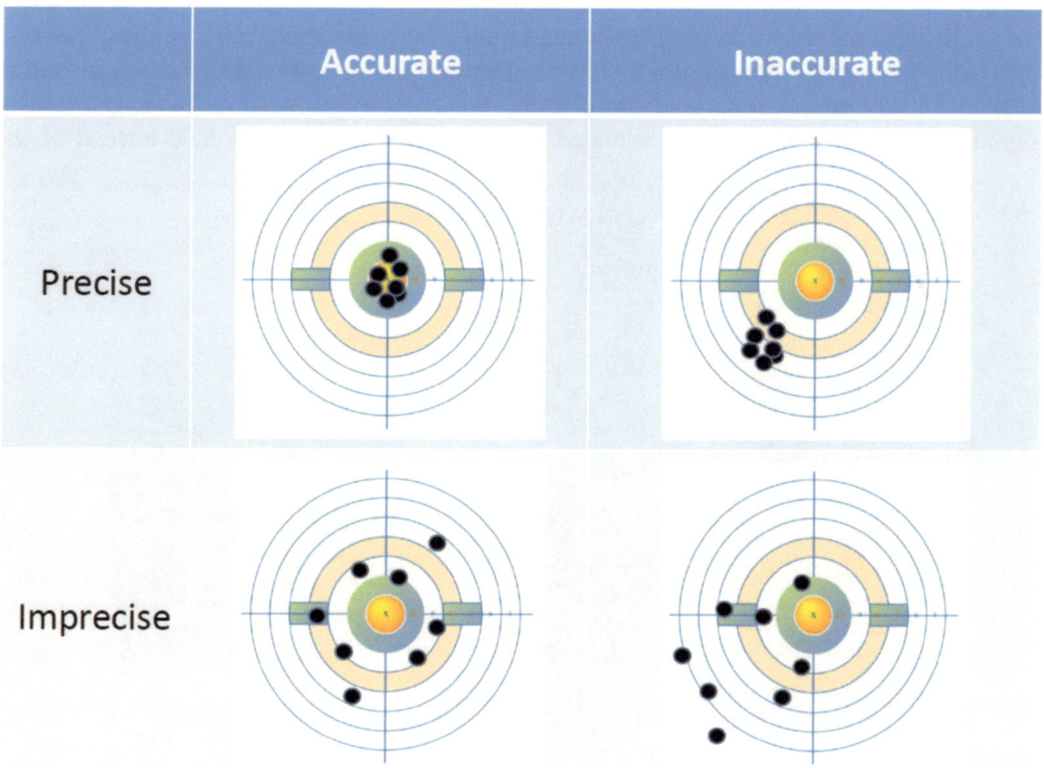

Figure 1. Illustration of the comparative importance of accuracy and precision.

3. Methods for profiling surfaces

Several types of methods (atomic force microscopy, optical scanning, and mechanical sectioning) have been used to capture profiles of material surfaces. The most common method is atomic force microscopy (AFM), which can be used either with (1) contact or tapping mode at the microscopic scale to feel the topography of the surface with a stylus attached to the end of a tiny cantilever beam, the deflections of which are magnified by observing the movements of a laser reflecting off of the back of the beam or (2) with scanning tunneling mode at the nanoscale to 'feel' the surface topography by recording fluctuations in the rate with which electrons tunnel through space and onto the surface from an electrically charged stylus as it rasters. Both of these mechanisms allow the construction of a virtual model of the surface, so unleveled surfaces can be virtually leveled to some degree after capture, which helps to maintain both accuracy and precision. AFM methods are moderately time consuming, but they offer fairly precise results with noise becoming a significant source of error only at the smallest scales and slow scan rates [5]. However, AFM is limited in the maximum amount of surface roughness (z-range) that it can accommodate.

Optical methods involve using a laser scanning confocal microscope or interferometer to build a virtual model of the surface. One advantage of the optical methods is that they have unlimited z-range. Another advantage is that some of the systems are portable and relatively inexpensive. The profiling can also be accomplished quite rapidly; however, there is a low signal-to-noise ratio compared with AFM. **Figure 2** shows a confocal microscope scan of a silicon nitride ceramic fracture surface that was captured by the author's research assistant (previously unpublished). This scan yielded estimated surface dimensions of over 2.50 instead of the

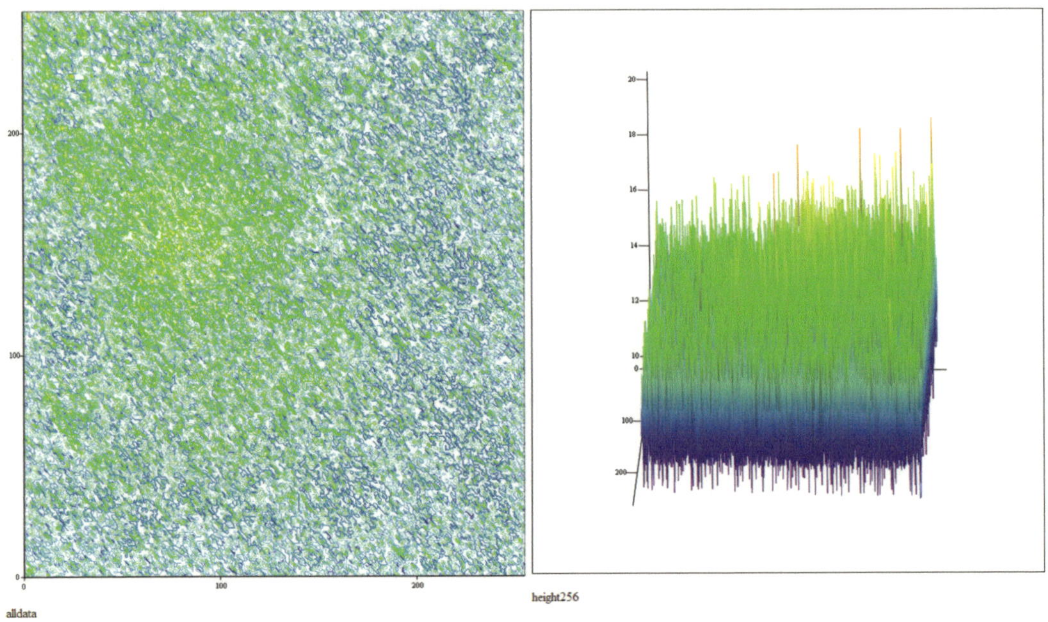

Figure 2. Confocal microscope scan of a silicon nitride fracture surface.

values produced by relatively noiseless AFM scans for the same material (2.12–2.14) (**Figure 3**). Applying noise filters in ImageJ eliminated the fractal nature of the data. Multiple attempts were made to improve signal-to-noise ratio by using different materials, different color lasers, two different confocal microscopes, and one optical interferometer, but all of those trials yielded similar results.

Mechanical sectioning requires coating either the specimen or an epoxy cast (replica) of the specimen with a reflective material such as gold, casting additional epoxy on top of the gold film, and cross-sectioning the resulting 'sandwich specimen' (usually by lapping/polishing with abrasive particles) to reveal a zero set that can be magnified and profiled using a metallographic optical microscope [6]. This method has three disadvantages: (1) It is labor intensive; (2) it relies on either the specimen being large enough or the technician being skilled enough to cross-section the surface in a level manner; (3) it limits the researcher to only algorithms that can be used on the zero sets of surfaces. Della Bona et al. studied the sensitivity of mechanical

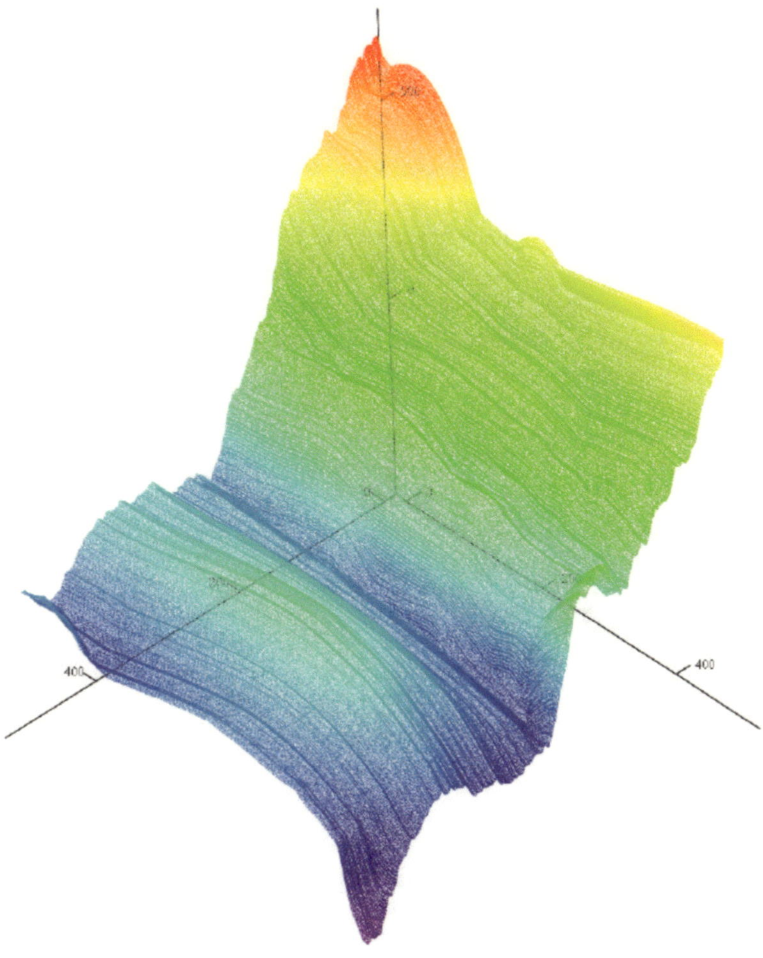

Figure 3. Atomic force microscope scan of a silicon nitride fracture surface.

sectioning to errors in leveling [7]. They analyzed the fracture surfaces of specimens made from two types of ceramic materials (baria silicate glass-ceramic and zinc selenide). After abrasive polishing of the epoxy replicas, Della Bona et al. estimated surface dimensions using the slit island Richardson algorithm. The estimated surface dimensions decreased with increasing error in leveling, and they decreased by 0.10–0.11 for only a seven-degree angle departure from the level (**Figure 4**).

Drummond et al. compared confocal microscopy with mechanical sectioning in analyzing the fractal dimensions of the fracture surfaces of three injection-molded dental ceramics [8]. They used the slit island Richardson algorithm to estimate the surface dimensions. Slit islands produced by mechanical sectioning yielded dimensions from 2.15 to 2.25, whereas confocal microscopy of the same specimens yielded dimensions from 2.71 to 2.91. It is possible that error in leveling the surfaces during sectioning was responsible for some of the discrepancy. However, it seems likely that noise in the confocal scans was responsible for much of the discrepancy. Artificial virtual surfaces that have been generated by fractal algorithms can have dimensions as high as 2.91, but real-world surfaces that have been created by fracture are limited by the physics of that process. They cannot have undercuts and have been reported to have dimensions ranging from 2.00 to 2.40, so measurements that yield higher dimensions should be viewed sceptically, and potential sources of the artifact should be identified.

A potential artifact in AFM and mechanical sectioning is the loss of microscopic details during creation of a specimen replica. Replicas are used in mechanical sectioning so that the original specimen need not be destroyed. Replicas are used for AFM because the original specimen is often too large to fit under the microscope. In the case of fracture surfaces, the surface may also be curved due to compression curl [9]. Even in cases when the entire fracture surface can be fit under the microscope, it can be difficult to perform coarse leveling on a large and curved

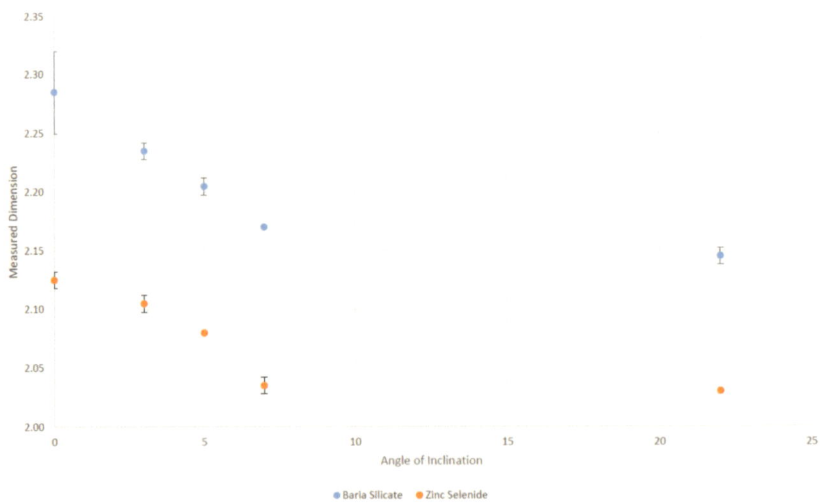

Figure 4. Effect of angle of cross-section on measured fractal dimension of baria silicate and zinc selenide fracture surfaces [7].

http://dx.doi.org/10.5772/intechopen.68189

specimen, and the resulting AFM scan will be too far from level for subsequent virtual leveling to be of any use. In these cases, it is necessary to replicate only a small portion of the curved surface and then to place the replica under the microscope. Joshi et al. used AFM to profile yttria-stabilized zirconia fracture surfaces and epoxy replicas of those surfaces [10]. They created a negative copy of each fracture surface using polyvinyl siloxane dental impression material, and then they cast a low viscosity, low shrinkage, slow curing epoxy into the impression to make a positive replica. The mean dimension of the epoxy replicas (2.245 ± 0.003) was not significantly different from that of the original surfaces (2.246 ± 0.007). However, a subsequent pilot study on using multiple generations of replicas has provided preliminary results that suggest the surface dimension continues to decrease with each successive iteration [previously unpublished].

4. Algorithms for calculating dimension

McMurphy et al. compared the accuracy, precision, and sensitivity-to-leveling of six different algorithms on virtual surfaces that were constructed by Brownian interpolation [11]. The surfaces were generated with known fractal dimensions so that the degree of bias could be assessed for each algorithm. Surfaces had dimensions of 2.10–2.40 to mimic ceramic fracture surfaces. The surfaces were analyzed using Minkowski cover (MC), root mean square (RMS) roughness versus area, Kolmogorov box (KB), Hurst exponent (HE), slit island box (SIB), and slit island Richardson (SIR) algorithms. KB was the most accurate with only a slight positive bias for surfaces having low dimension and a slight negative bias for surfaces having high dimension. The other algorithms exhibited a large negative bias (**Figure 5**). Fortunately, the bias of every algorithm was linearly related to the surface dimension, so all of these algorithms could be corrected to produce accurate (unbiased) estimates [**Table 1**, Eq. (2)].

$$D_{\text{unbiased}} = 2 + m(D_{\text{biased}} - 2) + b \tag{2}$$

MC exhibited the best precision (lowest CV) following bias correction. McMurphy et al. observed a wide range in precision with the least precise algorithms having three times the standard deviation compared to MC (**Table 2**). The surfaces were also analyzed at varying angles of inclination (3-, 5-, and 7-degree angles). KB exhibited great sensitivity to the angle of inclination, and SIB exhibited moderate sensitivity. The other algorithms were mostly insensitive to angulation on the Brownian interpolation surfaces. However, Brownian interpolation produces self-similar surfaces, and fracture surfaces are generally accepted to be self-affine. Therefore, these algorithms might exhibit a greater degree of sensitivity to angulation when used on fracture surfaces. **Figures 6** and **7** show the results of using RMS and SIR to analyze an AFM scan of a silica glass fracture surface (previously unpublished). Both algorithms show some sensitivity to angulation. RMS is not sensitive to rotation in the x-y plane (parallel versus perpendicular to the direction of crack propagation), but SIR is sensitive to rotation. RMS analyses the original surface, but SIR only analyses a zero set of the surface, and the orientation of cross-sectioning plane that produces the zero set has an influence on the result.

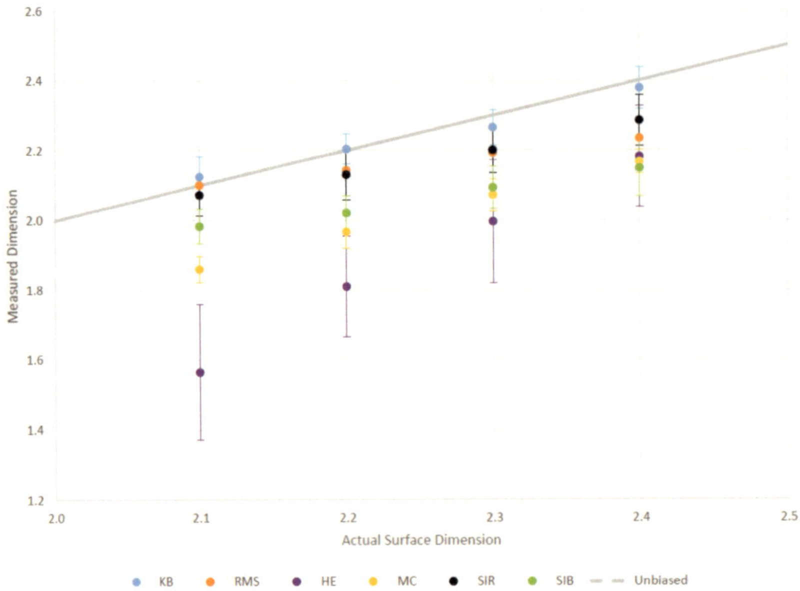

Figure 5. Discrepancy between measured and actual fractal dimensions of calibration surfaces that were generated using Brownian interpolation [11].

Algorithm	*R*	*m*	*b*
KB	0.9923	1.1826	−0.0406
RMS	0.9986	2.1890	−0.1193
HE	0.9973	0.4858	0.3043
MC	0.9994	0.9756	0.2318
SIR	0.9971	1.3917	0.0051
SIB	0.9949	1.7361	0.1397

Table 1. The slope (m), intercept (b), and correlation coefficient (R) for linear models to correct bias according to Eq. (1).

Algorithm	$D = 2.1$	$D = 2.2$	$D = 2.3$	$D = 2.4$
KB	2.1081 (0.0667)	2.2035 (0.0508)	2.2765 (0.0564)	2.4119 (0.0696)
RMS	2.1010 (0.0845)	2.1946 (0.0911)	2.3093 (0.1111)	2.3952 (0.0811)
HE	2.0918 (0.0951)	2.2123 (0.0702)	2.3026 (0.0874)	2.3933 (0.0716)
MC	2.0970 (0.0359)	2.2023 (0.0488)	2.3048 (0.0415)	2.3958 (0.0355)
SIR	2.1090 (0.0837)	2.1927 (0.1044)	2.2905 (0.0950)	2.4078 (0.0968)
SIB	2.1118 (0.0834)	2.1817 (0.0809)	2.3063 (0.1613)	2.4002 (0.1389)

Table 2. Mean (standard deviation) fractal dimension values after bias correction of six algorithms that were used for analyzing calibration surfaces (n = 10) with known dimensions (D = 2.1–2.4).

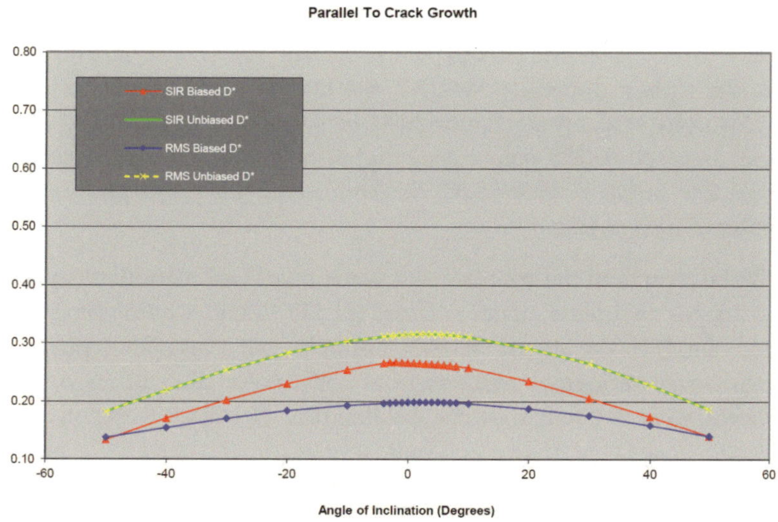

Figure 6. Comparison of sensitivity to angle of inclination for two algorithms used to estimate the fractal dimension of a silica glass fracture surface when the direction of inclination is parallel to the direction of crack propagation.

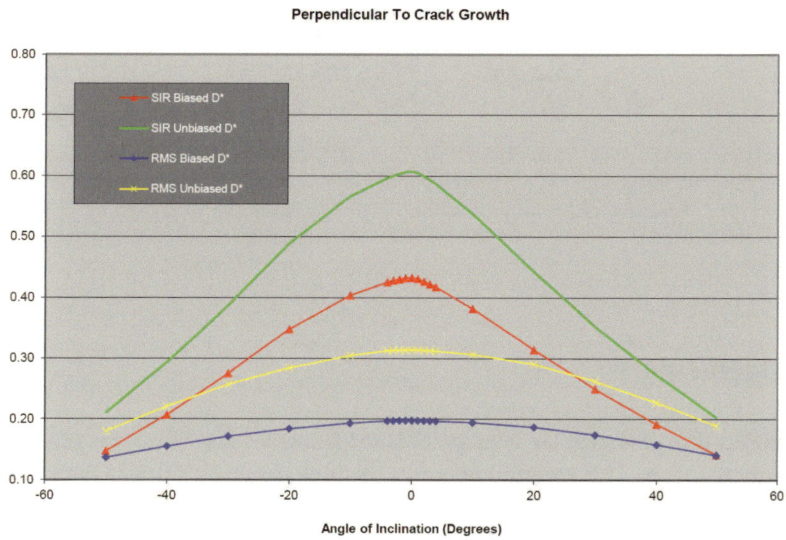

Figure 7. Comparison of sensitivity to angle of inclination for two algorithms used to estimate the fractal dimension of a silica glass fracture surface when the direction of inclination is perpendicular to the direction of crack propagation.

Mitchell & Bonnell investigated the accuracy and precision of a Fourier power spectrum algorithm on AFM (scanning tunneling mode) scans of sputtered gold films that were deposited on sodium silicate glass surfaces [5], and they generated model surfaces having dimensions of 2.4 and 2.7 to judge the degree of bias. The model surfaces were generated using both a Weierstrass-Mandelbrot function and Brownian motion. Only a linear trace along the surface was analyzed, so sensitivity to angulation could not be studied. The Fourier algorithm exhibited a large negative bias when the number of points sampled along the line was less than 500, and it exhibited no bias or a small positive bias for large numbers of points (in cases when the surface dimension was 2.4). In cases where the surface dimension was 2.7, the bias was consistently negative. This may explain why McMurphy et al. observed a negative bias for most algorithms as discussed above. McMurphy et al. analyzed 65,536 points per area but only 256 points on each line. Mitchell & Bonnell observed low precision (SD = 0.35) for samples of 100 points and high precision (SD = 0.04) for samples of 1000 points.

Williams and Beebe compared the accuracy and precision of four algorithms by analyzing scanning tunneling images of three material surfaces (highly oriented pyrolytic graphite, polished copper, and gold film) [12]. The algorithms were multiple-image variogram, power spectrum, slit island, and single-image variogram. Since surfaces with known dimensions were not analyzed, Williams and Beebe were not able to assess absolute bias. However, they noted that the power spectrum and single-image variogram algorithms consistently estimated much higher dimensions than the multiple-image variogram and slit island algorithms (**Table 3**). The slit island algorithm was not able to analyze the smoothest surface (graphite), and it exhibited a low precision for the other two surfaces. Multiple-image variography was deemed to be most reliable algorithm.

Algorithm	Graphite	Polished copper	Gold film
Multiple-image variogram	2.00–2.10	2.09–2.15	2.34–2.44
Power spectrum	2.45–2.62	2.39–2.52	2.34–2.50
Slit island	N/A	2.13–2.79	2.23–2.96
Single-image variogram	2.52–2.54	2.70–2.72	2.68–2.69

Table 3. Range of fractal dimension estimates produced by four algorithms that were used by Williams and Beebe to analyze three material surfaces.

Acknowledgements

The author would like to thank Dr. Susana Salazar Marocho for her attempts to troubleshoot the confocal microscope. The previously unpublished portions of data were supported by NIH/NIDCR grant 1R01 DE024333.

Author details

Jason A. Griggs

Address all correspondence to: jgriggs@umc.edu

Biomedical Materials Science, School of Dentistry, University of Mississippi Medical Center, Jackson, Mississippi, USA

References

[1] Mecholsky JJ, Passoja DE, Feinberg-Ringel KS. Quantitative analysis of brittle fracture surfaces using fractal geometry. Journal of the American Ceramic Society. 1989;**72**(1):60–65

[2] Persson BNJ. On the fractal dimension of rough surfaces. Tribology Letters. 2014;**54**:99–106

[3] Salazar Marocho SM, Manarão DS, Cesar PF, Griggs JA. Silica-coating protocols on Y-TZP: Roughness and fractal dimension. In: Watts DC, editor. Abstracts of the Academy of Dental Materials Annual Meeting; October 12-15; Chicago, IL: Elsevier; 2016. pp. e19–e20. Available from: www.sciencedirect.com

[4] Jain S, Williamson RS, Griggs JA, Roach MD. Fractal analysis of anodized titanium surfaces. In: Watts DC, editor. Abstracts of the Academy of Dental Materials Annual Meeting; October 12–15; Chicago, IL: Elsevier; 2016. pp. e81. Available from: www.sciencedirect.com

[5] Mitchell MW, Bonnell DA. Quantitative topographic analysis of fractal surfaces by scanning tunneling microscopy. Journal of Materials Research. 1990;**5**(10):2244–2254

[6] Hill TJ, Della Bona A, Mecholsky JJ. Establishing a protocol for measurements of fractal dimensions in brittle materials. Journal of Materials Science. 2001;**36**(11):2651–2657

[7] Della Bona A, Hill TJ, Mecholsky JJ. The effect of contour angle on fractal dimension measurements for brittle materials. Journal of Materials Science. 2001;**36**(11):2645–2650

[8] Drummond JL, Thompson M, Super BJ. Fracture surface examination of dental ceramics using fractal analysis. Dental Materials. 2005;**21**:586–589

[9] Quinn GD. Fractography of Glasses and Ceramics. 2nd ed. Gaithersburg, MD: National Institute of Standards and Technology; 2016. p. 640. DOI: http://dx.doi.org/10.6028/NIST.SP.960–16e2

[10] Joshi G, Roach M, Parsell D, Della Bona A, Griggs J. Fractal analysis of a dental Y-TZP using epoxy replicas. In: Giannobile WV, editor. AADR Annual Meeting & Exhibition; March 19–22; Charlotte, NC, London: SAGE Publishing; 2014. p. 798

[11] McMurphy TB, Harris CA, Griggs JA. Accuracy and precision of fractal dimension measured on model surfaces. Dental Materials. 2014;**30**(3):302–307. DOI: http://dx.doi.org/10.1016/j.dental.2013.11.015

[12] Williams JM, Beebe Jr TP. Analysis of fractal surfaces using scanning probe microscopy and multiple-image variography. 2. Results on fractal and nonfractal surfaces, observation of fractal crossovers, and comparison with other fractal analysis techniques. Journal of Physical Chemistry. 1993;**97**(23):6255–6260

Section 3

Applications, Scattering, Porosity, Turbulence

Chapter 9

On the Indicatrixes of Waves Scattering from the Random Fractal Anisotropic Surface

Alexander A. Potapov

Additional information is available at the end of the chapter

http://dx.doi.org/10.5772/intechopen.68187

Abstract

Millimeter and centimeter wave scattering from the random fractal anisotropic surface has been theoretically investigated. Designing of such surfaces is based on the modifications of non-differentiable two-dimensional Weierstrass function. Wave scattering on a random surface is interesting for many sections of physics, mathematics, biology, and so on. Questions of a radar location and radio physics take the predominating position here. There are many real surfaces and volumes in the nature that can be carried to fractal objects. At the same time, the description of processes of waves scattering of fractal objects differs from classical approaches markedly. There are many monographs in the world on the topic of classical methods of wave scattering but the number of books devoted to waves scattering on fractal stochastic surfaces is not enough. These results of estimation of three-dimensional scattering functions are a priority in the world and are important in radar of low-contrast targets near the surface of the earth and the sea.

Keywords: fractal, fractal surfaces, Kirchhoff approach, radio waves scattering, Weierstrass function, radar, low-contrast targets

1. Introduction

There are a lot of scientific and engineering problems, which can be successfully solved only with deep understanding of wave-scattering characteristics for statistically rough surface (see, e.g., [1–3] and references). In this section, we consider the main issues of theory of fractal wave scattering on the statistically rough surface as applied to problems of image creation by radar methods (RMs). These issues are crucial for radio location of low-contrast targets on the background of earth and sea surface.

In the general case, RM can be interpreted as a scattering specific effective squares (SESs), as a σ_* card (matrix) or as a signature (portrait) of object being sounded for the high angular resolution. SES card with fuzzy bounds corresponds to real RM for the wide-probing beam. RM resolution increase necessitates the use of complicated probing signals. Subject detail digital radar maps (DDRM or etalons) are often results of current image processing [4–8].

Currently, there are two general approaches of scattering on the statistically rough surface: method of small perturbation (SP) and Kirchhoff approach (tangent plane method (TPM)). These methods relate to two extreme cases of very small flat irregularities or smooth and large irregularities, respectively. Two-scale scattering model becomes a generalization of these methods. The model is a combination of small ripple (computations using SP) and large irregularities (computations using TPM). Review of these methods evolution is represented in Refs. [1–3].

Thus, before the present diffraction problems for the statistically rough surfaces took into account irregularities of only a single scale. Soon, it had been realized that multiscale surfaces lead to better fitting. As we have found out [6, 7] fractality accounting makes theoretical and experimental scattering patterns for earth cover in microwaves range closer. This fact is always interpreted (and has been interpreted now) as results of pure instrumental errors.

The aim of this work is to report systematically and consistently about theoretical solution of scattering problem for the random fractal anisotropic surface using Kirchhoff approach for the first time, to calculate scattering indicatrixes for radio microwaves, and to analyze the ensemble of indicatrixes obtained.

2. Formulation of the problem

Idea of fractal radio systems in the framework of fractal radio physics and radio electronics that was proposed and now is being consistently developed in the Institute of Radio Engineering and Electronics of the RAS (see, e.g., [5–48] and references) allows us to look at conventional radio physics methods in a new fashion. Currently, fractal radio physics and fractal radio location are the very active investigation areas, where significant applications have been obtained.

New problems that arise and being formulated are very important for every branch of science in the sense of its evolution. During the last 35 years, we succeeded in developing a number of important sections of fractal radio physics and fractal radio electronics that almost completes its main structure [6–8]. At once, these results reveal perspective of its modern applications and new relations between fractal physics and classical radio physics and electronics. It is necessary to note that for this course several monographs and more than 800 studies and 23 monographs were published (e.g., look at Refs. [5–48] and references).

Figure 1 shows us the main courses of works that are being carried out in the Institute of Radio Engineering and Electronics of the RAS and also information about the moment of its intensive growth beginning is demonstrated (for details, see Refs. [6, 7]). For such a "fractal" approach, it

http://dx.doi.org/10.5772/intechopen.68187

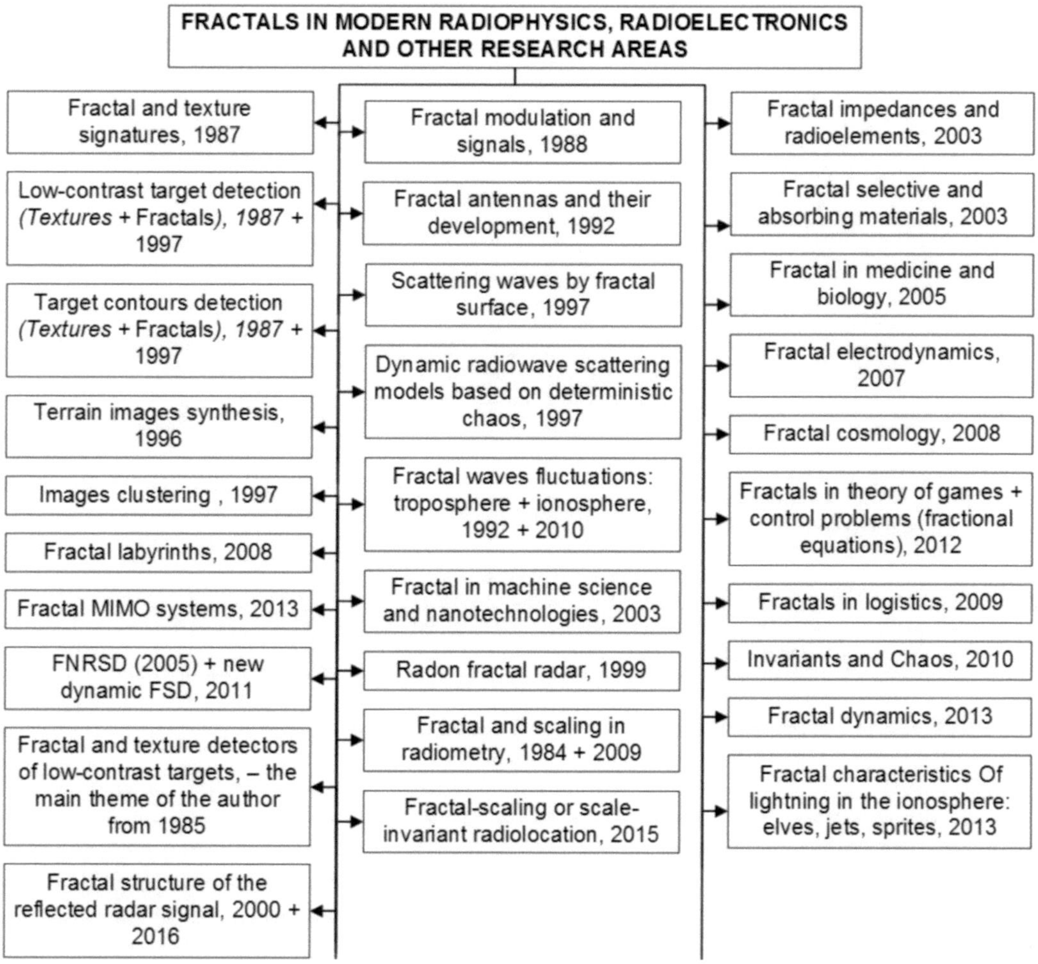

Figure 1. Sketch of development of a new information technology.

is natural to focus on analysis and also on the processing of radio physical signals (fields) only in space of fractional measuring using hypothesis of scaling and distributions with "heavy tales" or stable distributions. Note that scale transformations using scaling effects are widespread in up-to-date physics when different relations between thermodynamical values in renormgroup theory of phase changes are setting up [49].

Fractals belong to sets, which have extremely branched and irregular structure. In December 2005 in the USA, Mandelbrot approved [34] fractal classification that was developed by the author and is presented in **Figure 2**, where fractal features are characterized so long as there is a fractal structure with fractal dimension D in the space with topological dimension. Physical mathematical problems of the fractals theory and fractional measuring are represented in monographs [6–8] in detail.

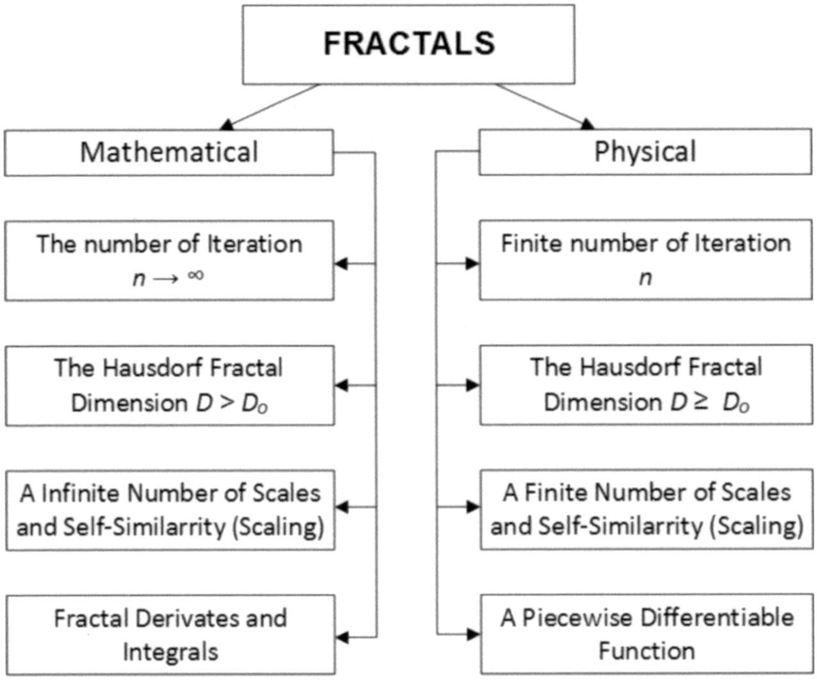

Figure 2. Classification and morphology of fractal sets and fractal set signatures.

In case of RM formation, the structure and parameters of wave field, which is generated by remote random surface at the field analysis area, depend on receiving point location and surface parameters. By taking into account these facts, we have to analyze the scattered field in a time-spatial continuum [5]. Therefore in the late 1970s of the XX century, the author formulated the problem of creating a theoretical modeling the band of millimeter and centimeter waves (MMW and CMW, respectively) for radar time-spatial signal by taking into account radio channel "antenna's aperture–atmosphere–targets–chaotic covering without vegetation" and the problem of creating of new features classes for radar targets recognition or radar signatures [5].

3. "Diffraction by fractals" ≠ "classical diffraction"

Effectiveness of radio physical investigations can be significantly improved by taking into account fractality of wave phenomena that are progressing at every stage of wave radiation, scattering, and propagation in different medium. In spite of pure scientific interest, there are practical applications to the radar and telecommunications problems solution and also to problems of mediums monitoring at different time-spatial scales.

Recently, interest to investigate wave scattering by rough surfaces that have non-Gaussian statistics has also grown. They often argue that correlation spatial coefficient of dispersive surface $\rho(\Delta x = x_2 - x_1, \Delta y = y_2 - y_1)$ cannot be exponential due to non-differentiability of respective random process. Sometimes in this case they use regularizing function about a zero point. Fundamental physical foundation of non-differentiable functions application for wave-scattering analysis was developed only after taking into account fractal theory, fractional-measuring theory, operators of integro-differentiation, and scaling relations in radio physical problems [6, 7, 20].

It is significant to note that Gaussian model is parabolic near the angle of incidence $\theta \approx 0$, while the exponential model is linear near the same point. Below, we consider in detail the approach to scattering of MMW and CMW by fractal random surface [5–7, 20, 39–44, 47].

At the present time, many works of foreign authors are related with wave interaction with fractal structures (see, e.g., respective chapters in monographs [6, 7]). Fractal surface implies the presence of irregularities of all scales with respect to scattered wavelength. Therefore, fractal wave front being non-differentiable does not have normal. In that way, conceptions of "ray trajectory" and "ray optics effects" are excluded. However, chords, which connect values of typical irregularity heights at the certain horizontal distances, still have finite root-mean-square slope. For this case, "topoteza" of fractal random surface is introduced; it is equal to the length of surface slope closeness to the unity [6, 7, 20].

Subject to all features, there are scattering models in the west of author works: (1) model of fractal heights and (2) model of fractal irregularities slopes. Thus, model No. 2 is once differentiable and has a slope that is changing continuously from point to point. This model leads to ray optics or to effects that are described using the conception of "ray." Such a kind of scattering was investigated together with radio waves propagation in the ionosphere [6, 7].

Electromagnetic waves scattering by fractal surfaces was investigated in detail in Refs. [50–58]. In Ref. [50], it was shown that diffraction by fractal surfaces fundamentally differs from diffraction by conventional random surfaces and some of classical statistical parameters like correlation length and root-mean-square deviation go to infinity. It is due to self-similarity of fractal surface. In Ref. [52], band-limited Weierstrass function was used. Less restrictions were imposed than the ones in Ref. [50]. The proposed function possesses both self-similarity property and still finite number of derivatives over a certain range under consideration. This relaxation of conditions of Weierstrass function allows performing analytical and numerical calculations.

Though there are many works on the creation and analysis of chaotic surfaces with the fractal structure [6, 7, 55–58], only few of them consider two-dimensional (2D) fractal surfaces. Corrugated surfaces that possess fractal properties only for one dimension (1D) were characterized in some works [52, 53, 59, 60]. In Refs. [39–44, 47, 61–63], modified Weierstrass function was used for designing 2D fractal chaotic surface. This function was derived from band-limited Weierstrass function. General solution for scattered field was obtained using Kirchhoff theory [1–3, 5–7, 61–65]. On this basis, we will carry out further calculations.

4. Fractal model of 2D chaotic surface

Modified 2D band-limited Weierstrass function has the view [6, 7, 20, 39–44, 47, 61–63]

$$W(x,y) = c_w \sum_{n=0}^{N-1} q^{(D-3)n} \sum_{m=1}^{M} \sin\left\{ Kq^n \left[x \cdot \cos\left(\frac{2\pi m}{M}\right) + y \cdot \sin\left(\frac{2\pi m}{M}\right) \right] + \phi_{nm} \right\}, \tag{1}$$

where c_w is the constant that provides unit normalization; $q > 1$- is the fundamental spatial frequency; D is the fractal dimension ($2<D<3$); K is the fundamental wave number; N and M are the number of tones; ϕ_{nm}- is an arbitrary phase that has a uniform distribution over the interval $[-\pi, \pi]$.

Eq. (1) is a combination of random structure and determined period. Function $W(x, y)$ is anisotropic in two directions if M and N are not very large. It has derivatives, and at the same time, it is self-similar. Respective surface is multiscale and roughness can vary depending on the scale being considered. Since the natural surfaces are neither purely random nor periodical and are often anisotropic [5, 40], the function that was proposed above is a good candidate for characterizing natural surfaces.

5. Relationships between statistical parameters of roughness measurements and fractal surface parameters

Such parameters as correlation length Γ, mean-root-square deviation σ, and spatial autocorrelation coefficient $\rho(\tau)$ are conventionally used for numerical characterization of rough surface. In this section of our work, these statistical parameters are introduced for the estimation of fractal dimension D influence and other fractal parameters influence on the surface roughness. Similar relationships are presented in Refs. [6, 7, 20] for 1D fractal surfaces. Derivations of σ and $\rho(\tau)$ for 2D fractal surfaces are cumbersome and tedious [61], and so we present here only some final results.

5.1. Mean square deviation

The mean-root-square deviation $\boldsymbol{\sigma}$ is determined as

$$\sigma = \left(\langle W^2(\vec{r})\rangle_s\right)^{1/2} \tag{2}$$

where $W(\vec{r}) = W(x,y)$; $\vec{r} = x\vec{\mathbf{I}} + y\vec{\mathbf{J}}$. Angle bracket implies ensemble averaging.

From Eqs. (1) and (2), we have

$$\sigma = c_w \left[\frac{M\left(1 - q^{2(D-3)N}\right)}{2\left(1 - q^{2(D-3)}\right)} \right]^{\frac{1}{2}}. \tag{3}$$

If $\sigma = 1$, then Eq. (3) is as follows:

$$c_w = \left[\frac{2\left(1-q^{2(D-3)}\right)}{M\left(1-q^{2(D-3)N}\right)}\right]^{\frac{1}{2}}. \tag{4}$$

Thus, Eqs. (1) and (4) are as follows:

$$W_{\text{H}}(x,y) = \left[\frac{2(1-q^{2(D-3)})}{M(1-q^{2(D-3)N})}\right]^{1/2} \sum_{n=0}^{N-1} q^{(D-3)n} \sum_{m=1}^{M} \sin\left\{Kq^n\left[x\cdot\cos\left(\frac{2\pi m}{M}\right)+y\cdot\sin\left(\frac{2\pi m}{M}\right)\right]+\phi_{nm}\right\}. \tag{5}$$

Eq. (5) is normalized with $\sigma = 1$. A normalized function will be used in the following sections for the analysis and modeling of wave field scattered by fractal surfaces. Surface becomes more isotropic with the increase of N and M. It is important to notice that $W_u(x, y)$ characterizes mathematical fractals only if $N \to \infty$ и $M \to \infty$.

5.2. Coefficient of spatial autocorrelation and of correlation length

Now, let us turn to the consideration of spatial autocorrelation coefficient ρ(τ) and correlation length Γ. By definition

$$\rho(\tau) = \frac{\langle W_{\text{H}}(\vec{r}+\vec{\tau})W_{\text{H}}(\vec{r})\rangle_s}{\sigma^2} \tag{6}$$

$$\text{where } \tau = (\Delta x^2 + \Delta y^2)^{\frac{1}{2}}. \tag{7}$$

From Eqs. (5) and (6), we have

$$\rho(\tau) = \left[\frac{\left(1-q^{2(D-3)}\right)}{M\left(1-q^{2(D-3)N}\right)}\right] \sum_{n=0}^{N-1} q^{2(D-3)n} \sum_{m=1}^{M} \cos\left[Kq^n\tau\cdot\cos\left(\theta-\frac{2\pi\cdot m}{M}\right)\right], \tag{8}$$

$$\text{where } \sin\theta = \frac{\Delta y}{\tau},\ \cos\theta = \frac{\Delta x}{\tau}. \tag{9}$$

The average spatial autocorrelation coefficient

$$\tilde{\rho}(\tau) = \langle\rho(\tau)\rangle_s = \left[\frac{\left(1-q^{2(D-3)}\right)}{\left(1-q^{2(D-3)N}\right)}\right] \sum_{n=0}^{N-1} q^{2(D-3)n} J_0(Kq^n\tau), \tag{10}$$

where $J_0(Kq^n\tau)$ is the zero-order Bessel function of the first kind.

Correlation length Γ is defined as the first root of ρ(τ) = 1/e when τ increases from zero. From relationship (8)

$$\left[\frac{(1-q^{2(D-3)})}{M(1-q^{2(D-3)N})}\right]\sum_{n=0}^{N-1} q^{2(D-3)n}\sum_{m=1}^{M}\cos\left[Kq^n\Gamma\cdot\cos\left(\theta-\frac{2\pi\cdot m}{M}\right)\right]=1/e. \tag{11}$$

Similarly from Eq. (10), the average correlation length is defined $\breve{\Gamma}$:

$$\frac{(1-q^{2(D-3)})}{M(1-q^{2(D-3)N})}\sum_{n=0}^{N-1} q^{2(D-3)n}J_0(Kq^n\tilde{\Gamma})=1/e. \tag{12}$$

From Eqs. (10)–(12), one can find relationships between average correlation length $\breve{\Gamma}$, fractal dimension D, and also q. There are dependences $\breve{\Gamma}$ on q and D shown in **Figures 3** and **4**, respectively. It is shown that with an increased value of D, $\breve{\Gamma}$ decreases more rapidly for the same variation of q. It is shown in **Figure 4** that the value of $\breve{\Gamma}$ reduces steadily with the increase of D value. However, $\breve{\Gamma}$ does not change when q = 1.01.

Consequently, the mean correlation length $\breve{\Gamma}$ is sensitive to fractal dimension D with the exception of cases when q is close to unity. These results imply that the value of fractal surface irregularities is mainly determined by fractal parameter D.

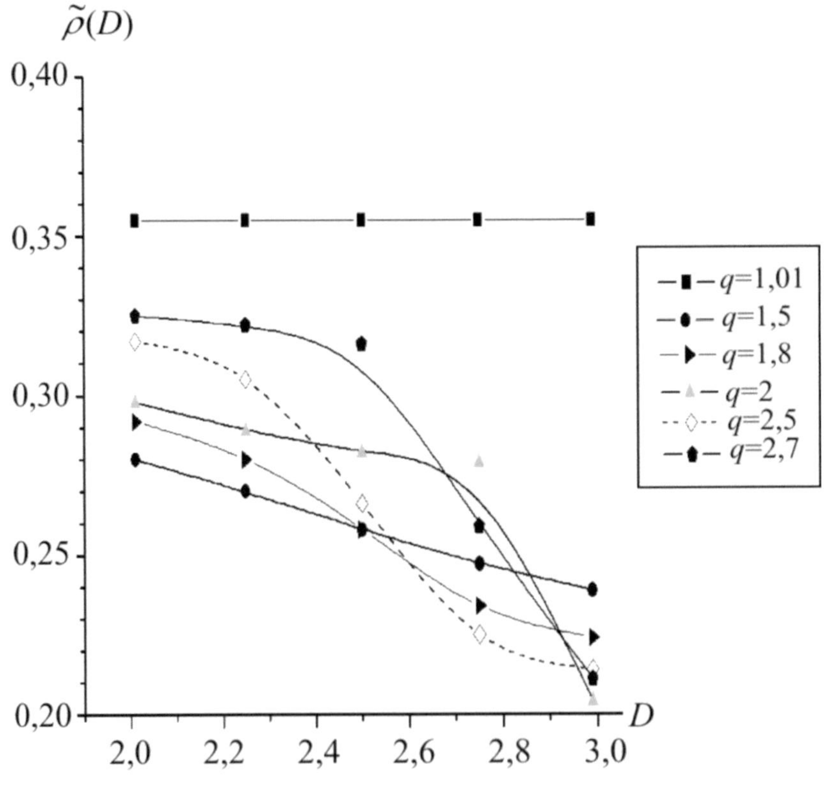

Figure 3. Average correlation length $\tilde{\Gamma}$ as function of q.

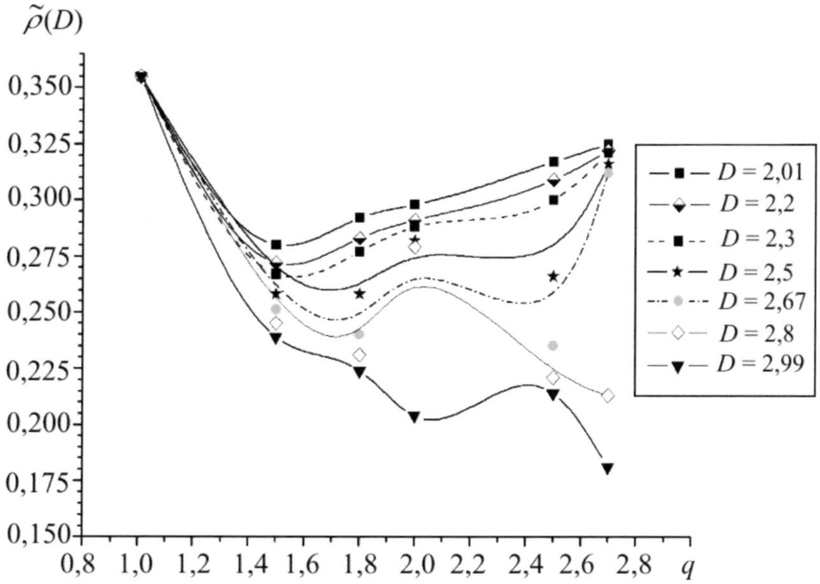

Figure 4. Average correlation length $\tilde{\Gamma}$ as function of D.

6. Memoir about the basic foundation of wave-scattering theory by fractal surfaces

As mentioned above, Kirchhoff approach has been already used for the analysis of wave scattering by fractal surfaces [6, 7, 20, 39–44, 47, 50–63]. This theory will be used in our work for numerical analysis of a field scattered by fractal chaotic surfaces. Conventional conditions of Kirchhoff approach are the following: irregularities are large scale, irregularities are smooth and flat. In the following calculations, we assume that observation is carried out from Fraunhofer zone, incident wave is plane and monochromatic, there are no points with infinite gradient on the surface, Fresnel coefficient V_0 is constant for this surface, and surface scales are much greater than incident wavelength.

6.1. Scattered field

Scattering geometry is presented in **Figure 5**. Then, *scattered field* $\psi_p(\vec{r})$ that interacts with surface square S of $\mathbf{2L_x} \times \mathbf{2L_y}$ when $-\boldsymbol{L_x} \leq \boldsymbol{x} \leq \boldsymbol{L_x}$ and $-\boldsymbol{L_y} \leq \boldsymbol{y} \leq \boldsymbol{L_y}$ are equal to [1–3, 5–7, 20, 61]:

$$\psi_p(\vec{r}) = -\frac{ik \cdot \exp(ikr)}{4\pi \cdot r} 2F(\theta_1, \theta_2, \theta_3) \int_S \exp[ik\phi(x_0, y_0)]dx_0 dy_0 + \psi_\kappa. \quad (13)$$

In Eq. (13), we used the following notations:

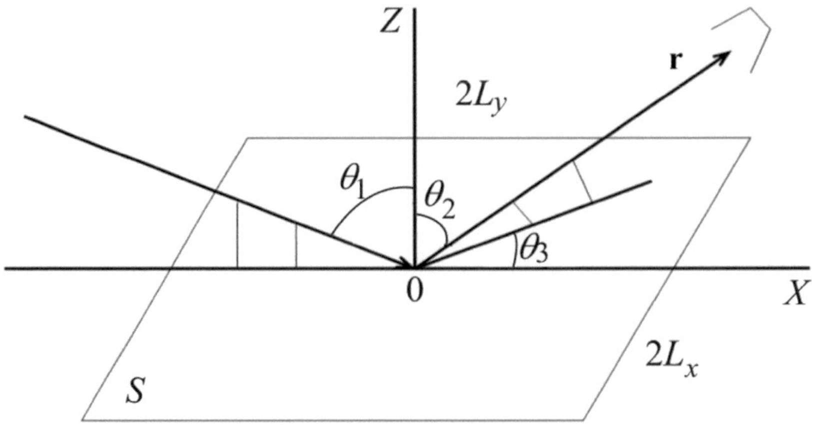

Figure 5. Scattering geometry: θ_1, incident angle; θ_2– scattering angle; and θ_3 azimuth angle.

$$F(\theta_1, \theta_2, \theta_3) = \frac{1}{2}\left(\frac{Aa}{C} + \frac{Bb}{C} + c\right), \tag{14}$$

$$\phi(x_0, y_0) = Ax_0 + By_0 + Ch(x_0, y_0), \tag{15}$$

$$h(x_0, y) = \sigma \cdot W_{\text{H}}(x_0, y_0), \tag{16}$$

$$a = V_0(\sin\theta_1 - \sin\theta_2\cos\theta_3), \tag{17}$$

$$b = V_0(\sin\theta_2 \cdot \sin\theta_3), \tag{18}$$

$$c = V_0(\cos\theta_1 + \cos\theta_2), \tag{19}$$

$$A = \sin\theta_1 - \sin\theta_2\cos\theta_3, \tag{20}$$

$$B = -\sin\theta_2\sin\theta_3, \tag{21}$$

$$C = -(\cos\theta_1 + \cos\theta_2), \tag{22}$$

$$\begin{aligned}\psi_{\text{к}} = -\frac{ik \cdot \exp(ikr)}{4\pi \cdot r}\Bigg\{\frac{ia}{kc}\int [\exp(ik\phi(X, y_0)) - \exp(ik\phi(-X, y_0))] \cdot dy_0 + \\ + \frac{ib}{kc}\int \left[\exp\left(ik\phi(x_0, Y)\right) - \exp\left(ik\phi(x_0, -Y)\right)\right] \cdot dx_0\Bigg\}.\end{aligned} \tag{23}$$

Component $\psi_{\text{к}}$ relates to edge effect. From Eqs. (15) and (16), we have

$$\exp[ik\phi(x_0, y_0)] = \exp\{ik[Ax_0 + By_0 + C\sigma \cdot W_u(x_0, y_0)]\}. \tag{24}$$

In Eq. (24), the third exponent is expressed as

$$\begin{aligned}\exp[ikC\sigma\cdot W_u(x_0,y_0)] &= \exp\left\{ikC\sigma\left[\frac{2\left(1-q^{2(D-3)}\right)}{M\left(1-q^{2(D-3)N}\right)}\right]^{1/2}\sum_{n=0}^{N-1}q^{(D-3)n}\sum_{m=1}^{M}\times\right.\\ &\left.\times\cos\left[Kq^n\cdot\left(x_0\cdot\cos\left(\frac{2\pi\cdot m}{M}\right)+y_0\sin\left(\frac{2\pi\cdot m}{M}\right)+\phi_{nm}-\frac{\pi}{2}\right)\right]\right\}=\\ &=\prod_{n=1}^{N-1}\prod_{m=1}^{M}\exp\left\{ikC\sigma\left[\frac{2\left(1-q^{2(D-3)}\right)}{M\left(1-q^{2(D-3)N_0}\right)}\right]^{1/2}q^{(D-3)n}\times\right.\\ &\left.\times\cos\left[x_0\cdot\cos\left(\frac{2\pi\cdot m}{M}\right)+y_0\sin\left(\frac{2\pi\cdot m}{M}\right)+\phi_{nm}-\frac{\pi}{2}\right]\right\}.\end{aligned}\tag{25}$$

From expression (Eq. (25)) and by taking into account expansion (Eq. (26)) and relationship (Eq. (27)):

$$\exp(iz\cos\varphi)=\sum_{u=-\infty}^{+\infty}iuJ_u(z)\exp(iu\varphi),\tag{26}$$

$$c_f=kC\sigma\left[\frac{2\left(1-q^{2(D-3)}\right)}{M\left(1-q^{2(D-3)N}\right)}\right]^{1/2}.\tag{27}$$

we obtain

$$\begin{aligned}\exp[ikC\sigma\cdot W_u(x_0,y_0)] &= \prod_{n=0}^{N-1}\prod_{m=1}^{M}\sum_{u=-\infty}^{+\infty}J_{u_{nm}}\left(c_f q^{(D-3)n}\right)\times\\ &\times\exp\left\{iu\left[Kq^n\cdot\left(x_0\cdot\cos\left(\frac{2\pi\cdot m}{M}\right)+y_0\sin\left(\frac{2\pi\cdot m}{M}\right)+\phi_{nm}-\frac{\pi}{2}\right)\right]\right\}.\end{aligned}\tag{28}$$

Eq. (28) can be written as

$$\begin{aligned}\exp[ikC\sigma\cdot W_u(x_0,y_0)] &= \sum_{u_{1,0}=-\infty}^{+\infty}\cdots\sum_{u_{1,0}=-\infty}^{+\infty}\cdots\sum_{u_{1,0}=-\infty}^{+\infty}\cdots\sum_{u_{1,0}=-\infty}^{+\infty}\cdots\sum_{u_{1,0}=-\infty}^{+\infty}\times\\ &\times\left[\prod_{n=0}^{N-1}\prod_{m=1}^{M}J_{u_{nm}}\left(c_f q^{(D-3)n}\right)\right]\cdot\exp\left\{iK\left[\sum_{n=0}^{N-1}q^{(D-3)n}\sum_{m=1}^{M}u_{nm}\cos\left(\frac{2\pi\cdot m}{M}\right)\right]x_0\right\}\times\\ &\times\exp\left\{iK\left[\sum_{n=0}^{N-1}q^{(D-3)n}\sum_{m=1}^{M}u_{nm}\sin\left(\frac{2\pi\cdot m}{M}\right)\right]y_0\right\}\times\\ &\times\exp\left(i\sum_{n=0}^{N-1}\sum_{m=1}^{M}u_{nm}\phi_{nm}\right).\end{aligned}\tag{29}$$

As result from Eqs. (13)–(23) and (29) ***field*** $\psi_p(\vec{r})$ ***scattered*** from finite site S is

$$\begin{aligned}\psi_{\mathrm{p}}(\vec{r}) &= -\frac{iL_xL_yk\cdot\exp(ikr)}{\pi r}2F(\theta_1,\theta_2,\theta_3)\\ &\times\sum_{u_{1,0}=-\infty}^{+\infty}\cdots\sum_{u_{1,N-1}=-\infty}^{+\infty}\cdots\sum_{u_{2,0}=-\infty}^{+\infty}\cdots\sum_{u_{2,N-1}=-\infty}^{+\infty}\cdots\sum_{u_{M,N-1}=-\infty}^{+\infty}\times\\ &\times\left[\prod_{n=0}^{N-1}\prod_{m=1}^{M}J_{u_{mn}}(c_jq^{(D-3)n})\right]\exp\left(i\sum_{n=0}^{N-1}\sum_{m=1}^{M}u_{mn}\phi_{mn}\right)\times\\ &\times\mathrm{sinc}(\varphi_cL_x)\cdot\mathrm{sinc}(\varphi_sL_y)+\psi_{\kappa},\end{aligned} \tag{30}$$

$$\mathrm{sinc}(x)\equiv\frac{\sin(x)}{x}, \tag{31}$$

$$\varphi_c=kA+K\sum_{n=0}^{N-1}q^n\sum_{m=1}^{M}u_{nm}\cos\left(\frac{2\pi\cdot m}{M}\right),\ \varphi_s=kB+K\sum_{n=0}^{N-1}q^n\sum_{m=1}^{M}u_{nm}\sin\left(\frac{2\pi\cdot m}{M}\right). \tag{32}$$

6.2. Average-scattered field

A more convenient parameter for the characterization of scattered field properties is *average-scattered field* $\tilde{\psi}_{\mathrm{p}}(\vec{\boldsymbol{r}})$:

$$\tilde{\psi}_{\mathrm{p}}(\vec{r})=\langle\psi_{\mathrm{p}}(\vec{r})\rangle_s \tag{33}$$

Eqs. (32) and (33) are defined as follows:

$$\tilde{\psi}_{\mathrm{p}}(\vec{r})=-\frac{iL_xL_yk\cdot\exp(ikr)}{\pi r}2F(\theta_1,\theta_2,\theta_3)\left[\prod_{n=0}^{N-1}J_0^M(c_fq^{(D-3)n})\right]\mathrm{sinc}(kAL_x)\mathrm{sinc}(kBL_y)+\psi_{\kappa} \tag{34}$$

Assume that the outside area $-L_x\le x_0\le L_x и -L_y\le y_0\le L_y$ surface S is smooth, that is,

$$h(\pm X,\pm Y)\equiv 0, \tag{35}$$

$$\text{where } X>L_x,\ Y>L_y. \tag{36}$$

Then, Eq. (23) can be written as

$$\psi_{\kappa}=-\frac{ik\cdot\exp(ikr)}{\pi\cdot r}\left(\frac{Aa}{C}+\frac{Bb}{C}\right)\lim_{X\to L_x^+}X\cdot\mathrm{sinc}(kAX)\cdot\lim_{Y\to L_y^+}Y\cdot\mathrm{sinc}(kBY). \tag{37}$$

6.3. Scattering indicatrixes for field

Scattering indicatrixes for field $\boldsymbol{\rho}_{\psi}$ is defined as

$$\rho_{\psi}=\frac{\psi_{\mathrm{p}}(\vec{r})}{\psi_{\mathrm{p}0}(\vec{r})}, \tag{38}$$

where field scattered from perfectly smooth surface $\psi_{\mathrm{p}0}(\vec{r})$ in a specular direction is expressed as

$$\psi_{p0}(\vec{r}) = -\frac{2L_xL_yik \cdot \exp(ikr)\cos\theta_1}{\pi \cdot r}. \tag{39}$$

Average-scattering indicatrix $\tilde{\rho}_\psi$ can be obtained after normalization:

$$\tilde{\rho}_\psi = \frac{\tilde{\psi}_{sc}(\mathbf{r})}{\psi_{sc0}(\mathbf{r})}. \tag{40}$$

Assume that surface gradients much less than incident angle is θ_1, then from Eqs. (30), (37)–(39) we have

$$\begin{aligned}\tilde{\rho}_\psi &= \frac{F(\theta_1, \theta_2, \theta_3)}{\cos\theta_1}\left[\prod_{n=0}^{N-1} J_0^M\left(c_f q^{(D-3)n}\right)\right]\mathrm{sinc}(kAL_x)\mathrm{sinc}(kBL_y)+\\ &+\frac{1}{2L_xL_y\cos\theta_1}\left(\frac{Aa}{C}+\frac{Bb}{C}\right)\lim_{X\to L_x^+} X\cdot\mathrm{sinc}(kAX)\cdot\lim_{Y\to L_y^+} Y\cdot\mathrm{sinc}(kBY).\end{aligned} \tag{41}$$

In specular direction $\theta_1 = \theta_2, \theta_3 = 0$ and coefficients are the $A = 0$, $B = 0$, $a = 0$, $b = 0$. Using Eqs. (17)–(22), we can write average-scattering indicatrixes $\tilde{\rho}_\psi$, which was defined in Eq. (40), as

$$\tilde{\rho}_\psi = \left[\prod_{n=0}^{N-1} J_0^M(c_f q^{(D-3)n})\right], \tag{42}$$

$$\text{where } c_f = -2k\sigma\cdot\cos\theta_1\left[\frac{2\left(1-q^{2(D-3)}\right)}{M\left(1-q^{2(D-3)N}\right)}\right]^{\frac{1}{2}}. \tag{43}$$

Thus, $\tilde{\rho}_\psi$ relates to parameters k, σ, θ_1, q, D, N, M. If $c_f q^{(D-3)n} < 1$, then in second approximation $\tilde{\rho}_\psi$ we have

$$\tilde{\rho}_\psi = 1 - 2(k\sigma\cdot\cos\theta_1)^2. \tag{44}$$

Eq. (44) shows that in specular direction $\tilde{\rho}_\psi$ depends on the wavelength of incident radiation, σ of rough surface, and incident angle θ_1. This result coincides with conventional results for Gaussian random surfaces [1]. Thus, fractal surfaces have diffraction properties that are similar to the ones of Gaussian random surfaces in a specular direction. This result involves a previous one [26], which was used as main assumption for mean-root-square scattering cross section measurement on this surface with specular ray measurement.

6.4. Average field intensity

Now, let us find *scattering indicatrixes for average field intensity* $\tilde{\rho}_I$. The intensity of scattered field is defined as

$$I(\vec{r}) = \psi_p(\vec{r})\psi_p^*(\vec{r}). \tag{45}$$

The average intensity of scattered field is obtained by Eq. (45) averaging:

$$\tilde{I}(\vec{r}) = \langle I(\vec{r}) \rangle_S. \tag{46}$$

From Eqs. (30), (45), and (46), we have

$$\begin{aligned}\tilde{I}(\vec{r}) = &\left[\tfrac{L_x L_y k}{\pi \cdot r} 2F(\theta_1, \theta_2, \theta_3)\right]^2 \times \\ &\times \sum_{u_{1,0}=-\infty}^{+\infty} \cdots \sum_{u_{1,N-1}=-\infty}^{+\infty} \cdots \sum_{u_{2,0}=-\infty}^{+\infty} \cdots \sum_{u_{2,N-1}=-\infty}^{+\infty} \cdots \sum_{u_{M,N-1}=-\infty}^{+\infty} \times \\ &\times \left[\prod_{n=0}^{N-1}\prod_{m=1}^{M} J_{u_{mn}}\left(c_j q^{(D-3)n}\right)\right]^2 \times \\ &\mathrm{sinc}^2(\varphi_c L_x) \cdot \mathrm{sinc}^2(\varphi_s L_y).\end{aligned} \tag{47}$$

6.5. Scattering indicatrix for average field intensity

In a similar manner as stated above, here we define ***scattering indicatrix for average field intensity*** $\tilde{\rho}_I \equiv g$:

$$g(\vec{r}) \equiv \tilde{\rho}_I = \frac{\tilde{I}(\vec{r})}{I_0}, \tag{48}$$

$$\text{where } I_0 = \psi_{p0}(\vec{r}) \cdot \psi_{p0}^*(\vec{r}). \tag{49}$$

Based on the assumptions that were proposed in the beginning of this section, we can write Eq. (48) as

$$\begin{aligned}g \approx \frac{F^2(\theta_1, \theta_2, \theta_3)}{\cos^2\theta_1}&\left\{\left[1 - \frac{1}{2}(kC\sigma)^2\right] \cdot \mathrm{sinc}^2(kAL_x)\mathrm{sinc}^2(kBL_y)\right. \\ &+ \frac{1}{4}C_f^2 \sum_{n=0}^{N-1}\sum_{m=1}^{M} q^{2(D-3)n} \mathrm{sinc}^2\left[\left(kA + Kq^n \cos\frac{2\pi \cdot m}{M}\right)L_x\right] + \\ &\left. + \mathrm{sinc}^2\left[\left(kB + Kq^n \sin\frac{2\pi \cdot m}{M}\right)L_y\right]\right\},\end{aligned} \tag{50}$$

where values with the order higher than $c_f^2 q^{2(D-3)n}$ in Eqs. (48) and (49) are negligible.

Statistical parameter of scattered field σ_I is defined as

$$\frac{\sigma_I = \tilde{I}(\vec{r}) - \tilde{\psi}_p^2(\vec{r})}{I_0,} \tag{51}$$

that here corresponds to the mean-root-square value of average-scattered field.

Let us compare the view of Eq. (34) with the first term in Eq. (50). It is obvious that the first term in Eq. (50) is equal to the expression for $\tilde{\psi}_p^2(\vec{r})$ that represents specular ray and side lobes. Thus, δ_I is determined only by the second term in Eq. (50) that relates to scattering by surface roughness. The second moment of scattered field σ_I can be useful for diffraction studying away from specular direction and also for the determination of the influence of fractal parameters on inverse-scattering pattern. The advantage of such a presentation is that in the consideration it is sufficient to discount only average coefficients. Thus, it is necessary to measure phase components that relate with scattered wave front.

6.6. Results clarification

In Ref. [52], *approximate formula of average field intensity for the problem of scattering by fractal phase screen* was presented. As it is explained in Ref. [61], this formula includes some errors. Below, details are explained and presented in Ref. [61]. Surface model in Ref. [52] is specified by *Weierstrass function* (see also expression (6.77) in monograph [7]:

$$\phi(x) = \frac{\sqrt{2\sigma_\phi[1 - b^{(2D-4)}]^{1/2}}}{[b^{(2D-4)N_1} - b^{(2D-4)(N_2+1)}]^{1/2}} \sum_{n=N_1}^{N_2} b^{(D-2)n} \cos(2\pi s b^n x + \phi_n), \tag{52}$$

where b is the fundamental spatial frequency; D is the fractal dimension, which varies over interval from 1 to 2; s is the scaling factor; ϕ_n is the phase that is distributed uniformly over the $[0, 2\pi]$. Number of harmonics in function (Eq. (52)) is determined by $N = N_2 - N_1 + 1$.

Average-scattered field intensity is determined by Eq. (22) in Ref. [52] (or by Eq. (6.96) in monograph [7] in the form of weighted array of Bessel functions):

$$\begin{aligned} \langle I(x) \rangle = & \frac{L^4}{\lambda^2 z^2} \sum_{q_1=-\infty}^{\infty} \sum_{q_2=-\infty}^{\infty} \dots \sum_{q_N=-\infty}^{\infty} J_{q_1}^2(C_{N_1}) J_{q_2}^2(C_{N_1+1}) \dots J_{q_N}^2(C_{N_2}) \times \\ & \times \operatorname{sinc}^2\left[L\left(\frac{x}{\lambda_z} - sq_1 b - sq_2 b^{N_1+1} - \dots sq^{N_2}\right)\right] \operatorname{sinc}^2\left(\frac{Lx}{\lambda z}\right), \end{aligned} \tag{53}$$

$$\text{where } C_n = \frac{\sqrt{2\sigma_\phi[1 - b^{(2D-4)} b^{(D-2)n}]^{1/2}}}{[b^{(2D-4)N_1} - b^{(2D-4)(N_2+1)}]^{1/2}} \tag{54}$$

L is the phase screen size, x, y are the coordinate values in intensity observation plane at a distance of z from the phase screen, and λ is the incident wavelength.

In Eqs. (B2) and (B3) in Ref. [52], typographical errors were made. In Eq. (53), term sq_1b must be: $sq_1 b^{N_1}$, which is clear from expression (6.95) in monograph [7]. In line with Eq. (B2), Eq. (B3) must be

$$C_n = \frac{\sqrt{2\sigma_\phi[1 - b^{(2D-4)}]^{1/2} b^{(D-2)n}}}{[b^{(2D-4)N_1} - b^{(2D-4)(N_2+1)}]^{1/2}}. \tag{55}$$

Approximate expression for the average intensity is derived from Eq. (22) in Ref. [12] (see also expression (6.97) in monograph [6]):

$$\langle I(x)\rangle = \frac{L^4}{\lambda^2 z^2}\left\{(1-\sigma_\phi^2)\mathrm{sinc}^2(Lx/\lambda z) + \sum_{n=-\infty}^{+\infty}(C_n^2/4)\mathrm{sinc}^2[L(x/\lambda z - sb^n)]\right\}\mathrm{sinc}^2(Ly/\lambda z). \quad (56)$$

After the correction, we have

$$\langle I(x)\rangle = \frac{L^4}{\lambda^2 z^2}\left\{(1-\sigma_\phi^2)\mathrm{sinc}^2(Lx/\lambda z) + \sum_{n=N_1}^{N_2}(C_n^2/4)\mathrm{sinc}^2[L(x/\lambda z - sb^n)]\right\}\mathrm{sinc}^2(Ly/\lambda z). \quad (57)$$

Accurate derivation of Eq. (57) looks like this [61]:

$$\left.\begin{aligned} J_{u_i}(C_n) &= \left(\tfrac{C_n}{2}\right)^u \sum_{j=-\infty}^{\infty}\frac{(-1)^j (C_n/2)^{2j}}{j!(u+j)!} \\ J_{u_i}^2(C_n) &= \left(\tfrac{C_n}{2}\right)^{2u}\left[\sum_{j=-\infty}^{\infty}\tfrac{(-1)^j (C_n/2)^{2j}}{j!(u+j)!}\right]^2 \end{aligned}\right\}, \quad (58)$$

where u_i is the integer and $u_i \in (q_1, \ldots, q_N)$.

Since terms with the order higher than C_n^2 are negligible, then $u_i \in \{0, 1\}$, $\sum u_i = 0$ or 1.

Thus

$$\left.\begin{aligned} J_0(C_n) &= \left(\tfrac{C_n}{2}\right)^u \sum_{j=-\infty}^{\infty}\frac{(-1)^j (C_n/2)^{2j}}{j!(u+j)!} \approx 1 - \frac{1}{4}C_n^2 \\ J_0^2(C_n) &= 1 - \frac{1}{2}C_n^2 \end{aligned}\right\}, \quad (59)$$

$$J_1^2(C_n) \approx \frac{1}{4}C_n^2. \quad (60)$$

From Eqs. (53), (59), and (60), we have

$$\begin{aligned} \langle I(x)\rangle = \frac{L^4}{\lambda^2 z^2}\Bigg[& J_0^2(C_{N_1})\ldots J_0^2(C_{N_2})\mathrm{sinc}^2\left(\frac{Lx}{\lambda z}\right) + \\ & + J_1^2(C_n)J_0^2(C_{N_1+1})\ldots J_0^2(C_{N_2})\mathrm{sinc}^2\left(\frac{Lx}{\lambda z} - sb^{N_1}\right) + \\ & + J_0^2(C_{N_1})J_1^2(C_{N_1+1})\ldots J_0^2(C_{N_2})\mathrm{sinc}^2\left(\frac{Lx}{\lambda z} - sb^{N_1+1}\right) + \\ & + \ldots + J_0^2(C_{N_1})\ldots J_1^2(C_{N_2})\mathrm{sinc}^2\left(\frac{Lx}{\lambda z} - sb^{N_2}\right)\Bigg]\mathrm{sinc}^2\left(\frac{Ly}{\lambda z}\right), \end{aligned} \quad (61)$$

where

$$J_0^2(C_{N_1})\ldots J_0^2(C_{N_2}) \approx \left(1 - \frac{1}{2}C_{N_1}^2\right)\ldots\left(1 - \frac{1}{2}C_{N_2}^2\right) \approx 1 - \frac{1}{2}\sum_{n=N_1}^{N_2} C_n^2 = 1 - \sigma_\phi^2, \quad (62)$$

http://dx.doi.org/10.5772/intechopen.68187

$$\left.\begin{array}{l} J_1^2(C_{N_1})\ldots J_0^2(C_{N_2}) \approx \frac{1}{4}C_{N_1}^2 \\ \cdot \\ \cdot \\ \cdot \\ J_0^2(C_{N_1})\ldots J_1^2(C_{N_2}) \approx \frac{1}{4}C_{N_2}^2 \end{array}\right\}. \tag{63}$$

So, from Eqs. (61)–(63) finally we obtain the expression for average intensity in Fraunhofer zone:

$$\langle I(x)\rangle = \frac{L^4}{\lambda^2 z^2}\left\{\left(1-\sigma_\phi^2\right)\mathrm{sinc}^2(Lx/\lambda z) + + \sum_{n=N_1}^{N_2}(C_n^2/4)\mathrm{sinc}^2[(L(x/\lambda z - sb^n)]\right\}\mathrm{sinc}^2(Ly/\lambda z). \tag{64}$$

7. Results of the theoretical investigations of scattering indicatrixes in MW range

In **Figures 6–80**, we present a thorough array of typical kinds of dispersing fractal surfaces with the basis of Weierstrass function, and also 3D-scattering indicatrixes and their cross sections that

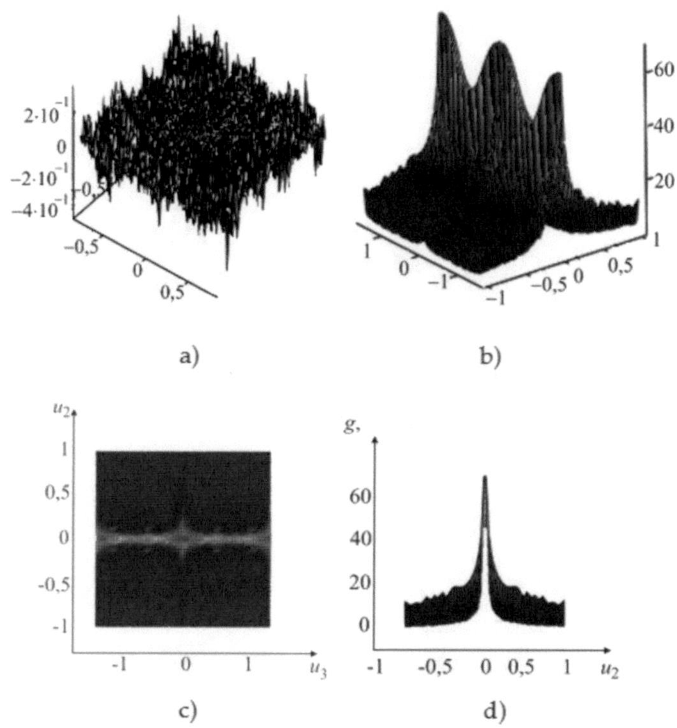

Figure 6. The fractal surface and the scattering indicatrix $g(\theta_2, \theta_3)$ when λ = 2.2 mm and θ_1 = 0°: (a) fractal surface for D = 2.2; N = M = 10; q = 2.7; (b) $g(\theta_2, \theta_3)$; (c) $g(\theta_2, \theta_3)$, top view; (d) $g(\theta_2, \theta_3)$, side view.

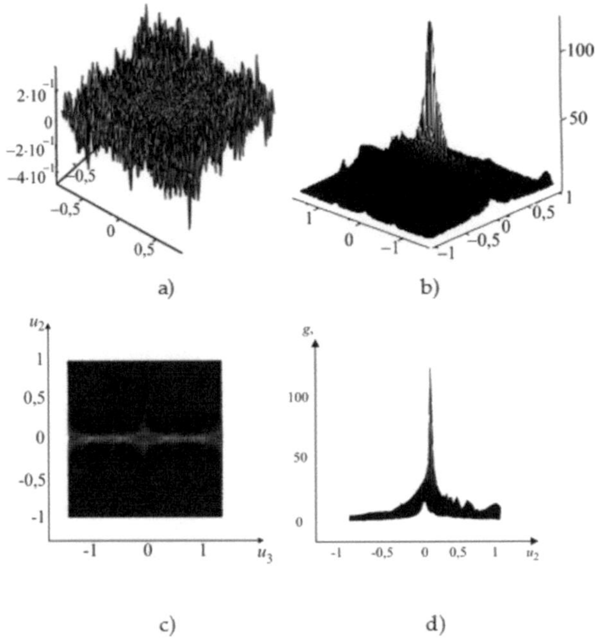

Figure 7. The fractal surface and the scattering indicatrix $g(\theta_2, \theta_3)$ when $\lambda = 2.2$ mm and $\theta_1 = 5°$: (a) fractal surface for $D = 2.2$; $N = M = 10$; $q = 2.7$; (b) $g(\theta_2, \theta_3)$; (c) $g(\theta_2, \theta_3)$, top view; (d) $g(\theta_2, \theta_3)$, side view.

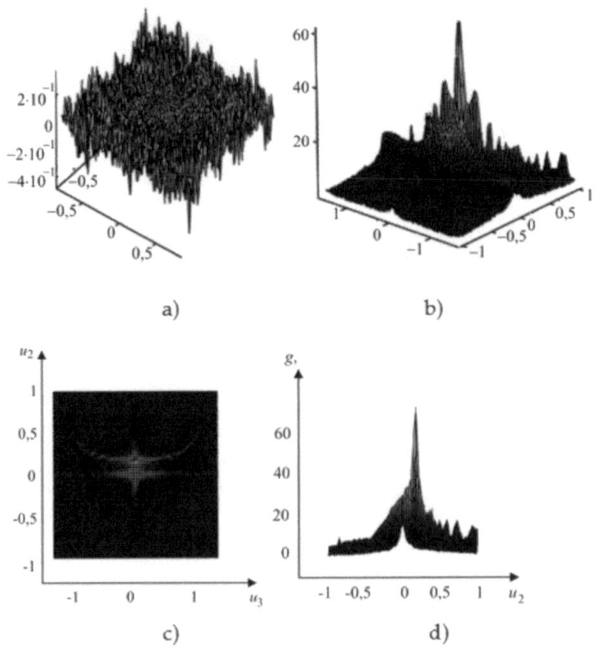

Figure 8. The fractal surface and the scattering indicatrix $g(\theta_2, \theta_3)$ when $\lambda = 2.2$ mm and $\theta_1 = 10°$: (a) fractal surface for $D = 2.2$; $N = M = 10$; $q = 2.7$; (b) $g(\theta_2, \theta_3)$; (c) $g(\theta_2, \theta_3)$, top view; (d) $g(\theta_2, \theta_3)$, side view.

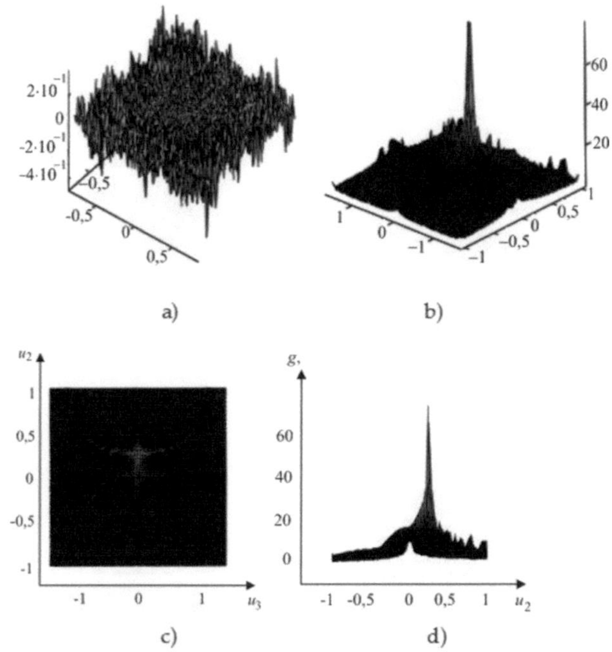

Figure 9. The fractal surface and the scattering indicatrix g(θ2, θ3) when $\lambda = 2.2$ mm and $\theta_1 = 15°$: (a) fractal surface for $D = 2.2$; $N = M = 10$; $q = 2.7$; (b) g(θ2, θ3); (c) g(θ2, θ3), top view; (d) g(θ2, θ3), side view.

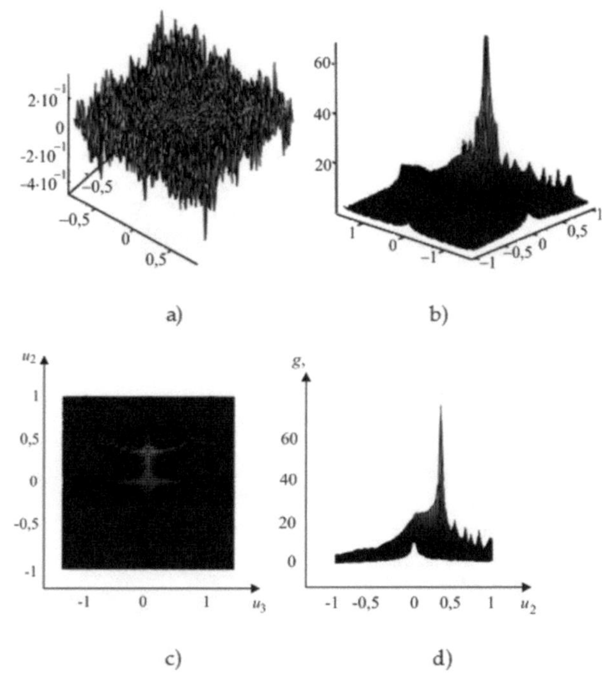

Figure 10. The fractal surface and the scattering indicatrix $g(\theta_2, \theta_3)$ when $\lambda = 2.2$ mm and $\theta_1 = 20°$: (a) fractal surface for $D = 2.2$; $N = M = 10$; $q = 2.7$; (b) $g(\theta_2, \theta_3)$; (c) $g(\theta_2, \theta_3)$, top view; (d) $g(\theta_2, \theta_3)$, side view.

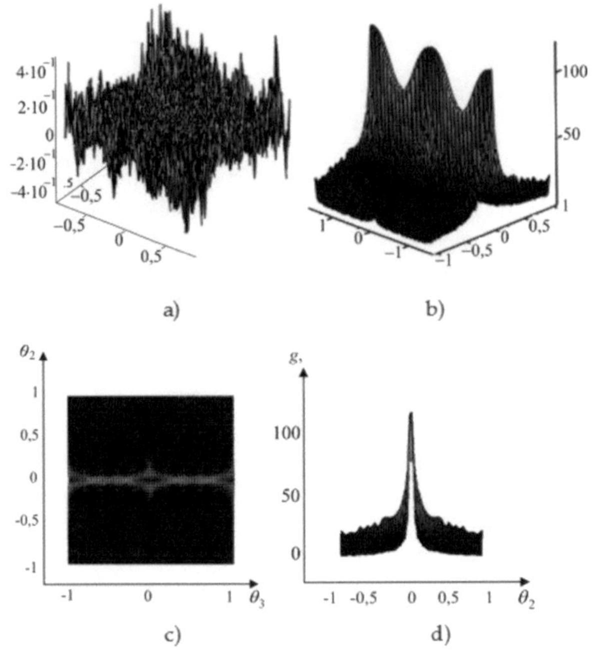

Figure 11. The fractal surface and the scattering indicatrix $g(\theta_2, \theta_3)$ when λ = 2.2 mm and $\theta_1 = 0°$: (a) fractal surface for $D = 2.2$; $N = M = 20$; $q = 2.7$; (b) $g(\theta_2, \theta_3)$; (c) $g(\theta_2, \theta_3)$, top view; (d) $g(\theta_2, \theta_3)$, side view.

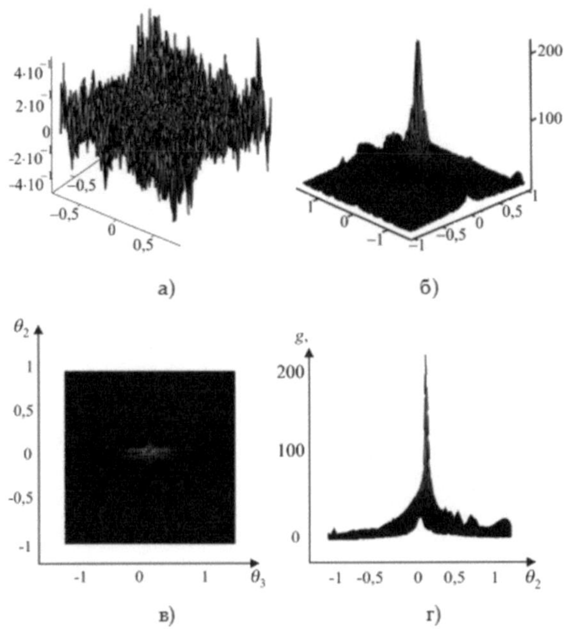

Figure 12. The fractal surface and the scattering indicatrix $g(\theta_2, \theta_3)$ when λ = 2.2 mm and $\theta_1 = 5°$: (a) fractal surface for $D = 2.2$; $N = M = 20$; $q = 2.7$; (b) $g(\theta_2, \theta_3)$; (c) $g(\theta_2, \theta_3)$, top view; (d) $g(\theta_2, \theta_3)$, side view.

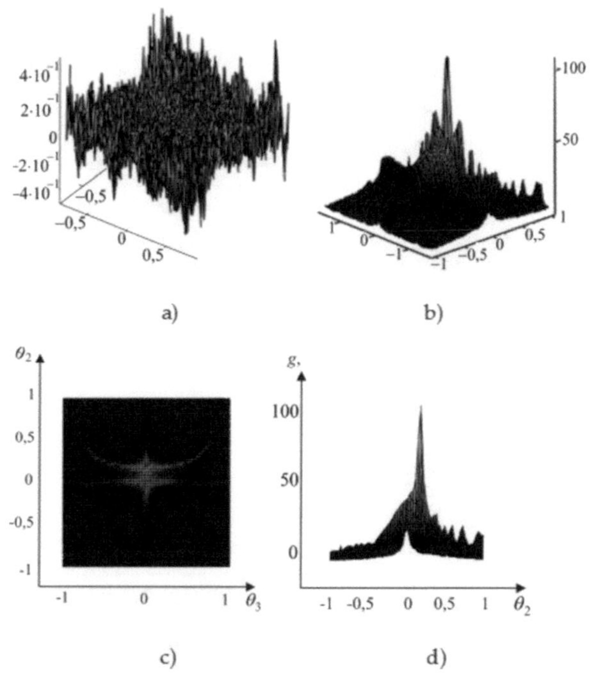

Figure 13. The fractal surface and the scattering indicatrix $g(\theta_2, \theta_3)$ when $\lambda = 2.2$ mm and $\theta_1 = 10°$: (a) fractal surface for $D = 2.2$; $N = M = 20$; $q = 2.7$; (b) $g(\theta_2, \theta_3)$; (c) $g(\theta_2, \theta_3)$, top view; (d) $g(\theta_2, \theta_3)$, side view.

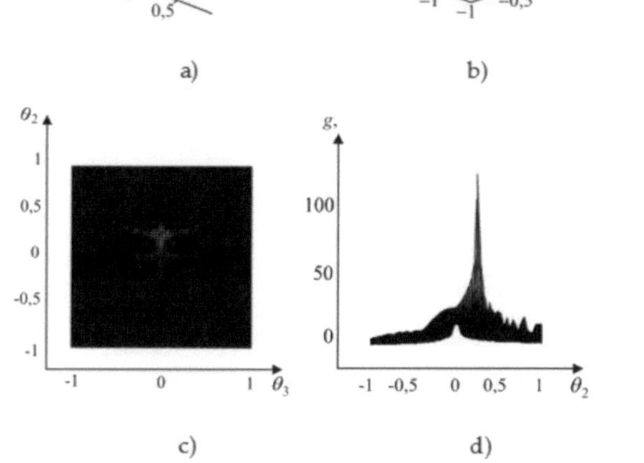

Figure 14. The fractal surface and the scattering indicatrix $g(\theta_2, \theta_3)$ when $\lambda = 2.2$ mm and $\theta_1 = 15°$: (a) fractal surface for $D = 2.2$; $N = M = 20$; $q = 2.7$; (b) $g(\theta_2, \theta_3)$; (c) $g(\theta_2, \theta_3)$, top view; (d) $g(\theta_2, \theta_3)$, side view.

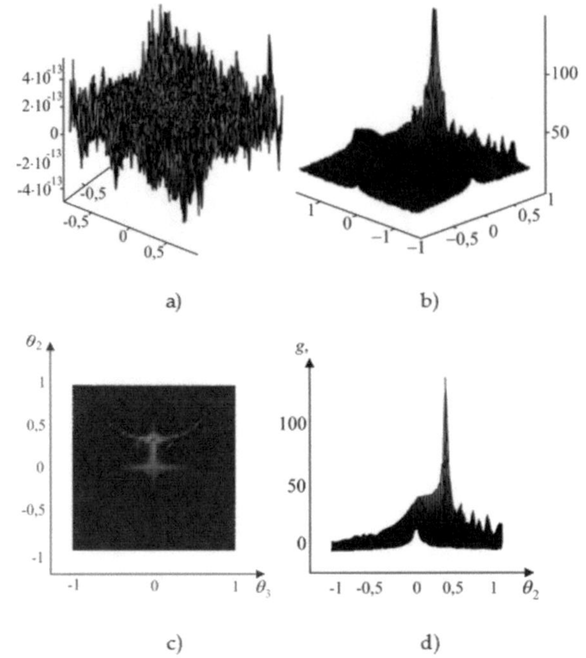

Figure 15. The fractal surface and the scattering indicatrix $g(\theta_2, \theta_3)$ when $\lambda = 2.2$ mm and $\theta_1 = 20°$: (a) fractal surface for $D = 2.2$; $N = M = 20$; $q = 2.7$; (b) $g(\theta_2, \theta_3)$; (c) $g(\theta_2, \theta_3)$, top view; (d) $g(\theta_2, \theta_3)$, side view.

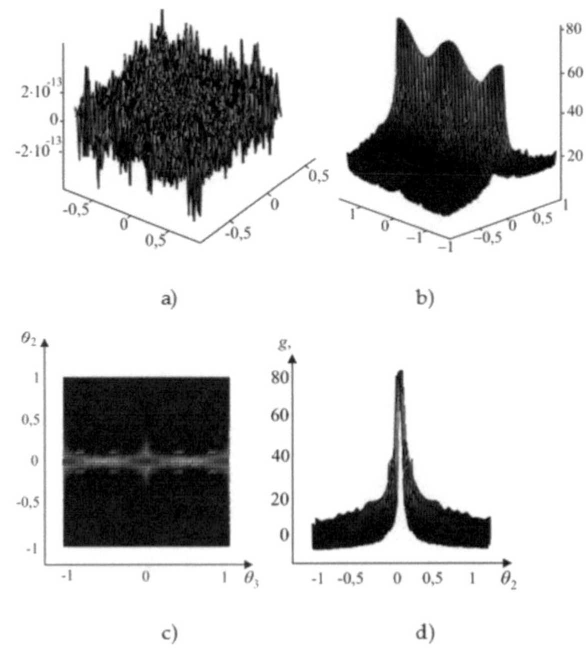

Figure 16. The fractal surface and the scattering indicatrix $g(\theta_2, \theta_3)$ when $\lambda = 2.2$ mm and $\theta_1 = 0°$: (a) fractal surface for $D = 2.5$; $N = M = 10$; $q = 2.7$; (b) $g(\theta_2, \theta_3)$; (c) $g(\theta_2, \theta_3)$, top view; (d) $g(\theta_2, \theta_3)$, side view.

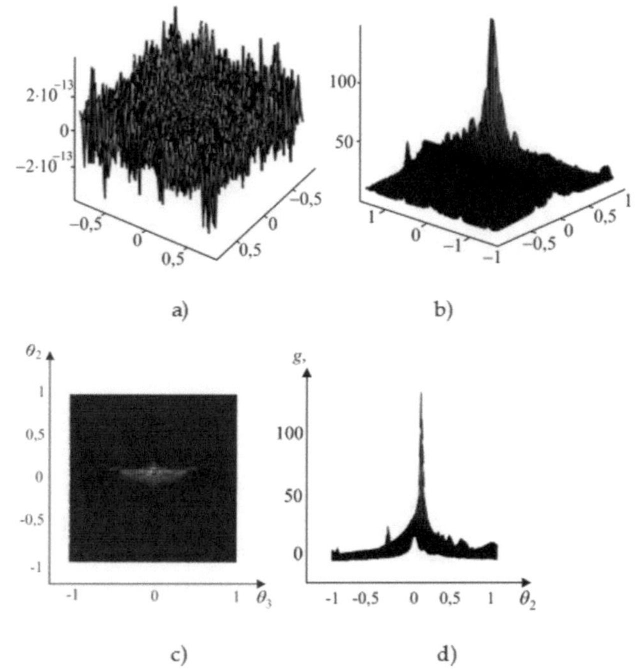

Figure 17. The fractal surface and the scattering indicatrix $g(\theta_2, \theta_3)$ when $\lambda = 2.2$ mm and $\theta_1 = 5°$: (a) fractal surface for $D = 2.5$; $N = M = 10$; $q = 2.7$; (b) $g(\theta_2, \theta_3)$; (c) $g(\theta_2, \theta_3)$, top view; (d) $g(\theta_2, \theta_3)$, side view.

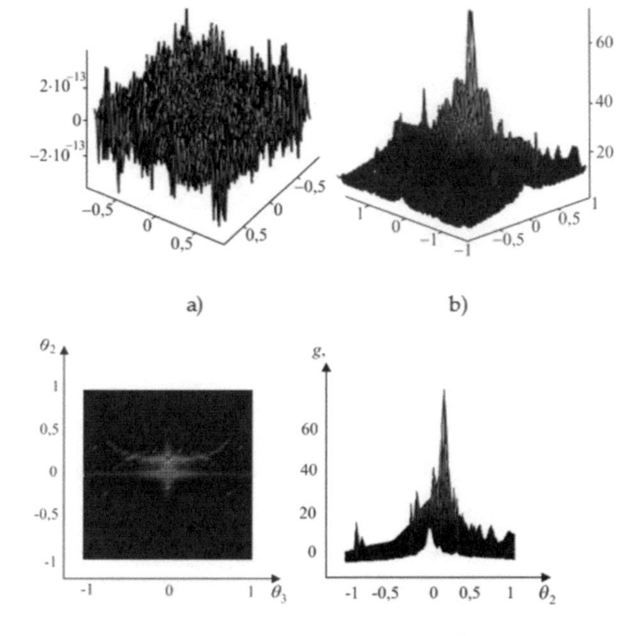

Figure 18. The fractal surface and the scattering indicatrix $g(\theta_2, \theta_3)$ when $\lambda = 2.2$ mm and $\theta_1 = 10°$: (a) fractal surface for $D = 2.5$; $N = M = 10$; $q = 2.7$; (b) $g(\theta_2, \theta_3)$; (c) $g(\theta_2, \theta_3)$, top view; (d) $g(\theta_2, \theta_3)$, side view.

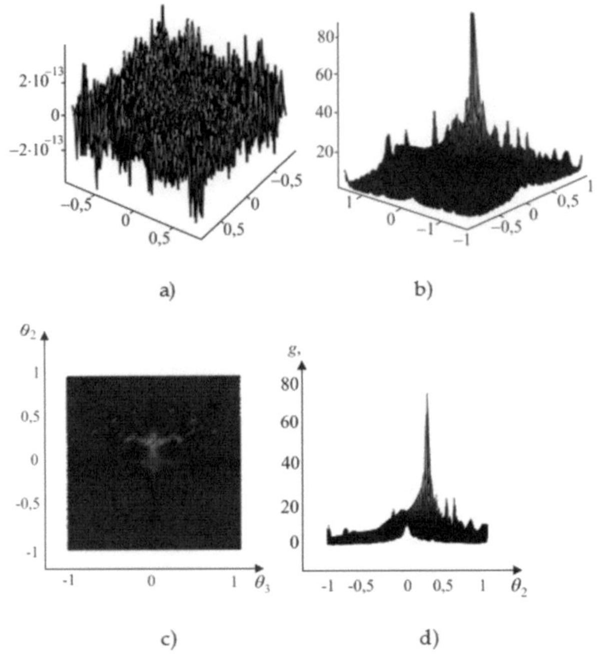

Figure 19. The fractal surface and the scattering indicatrix $g(\theta_2, \theta_3)$ when $\lambda = 2.2$ mm and $\theta_1 = 15°$: (a) fractal surface for $D = 2.5$; $N = M = 10$; $q = 2.7$; (b) $g(\theta_2, \theta_3)$; (c) $g(\theta_2, \theta_3)$, top view; (d) $g(\theta_2, \theta_3)$, side view.

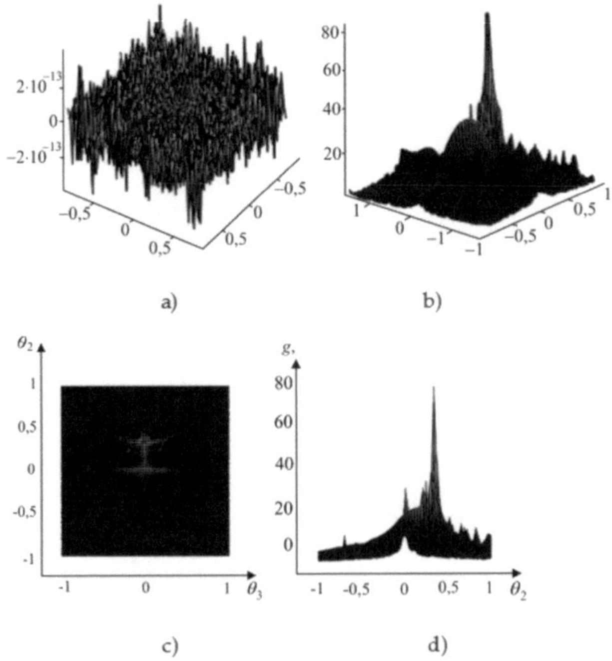

Figure 20. The fractal surface and the scattering indicatrix $g(\theta_2, \theta_3)$ when $\lambda = 2.2$ mm and $\theta_1 = 20°$: (a) fractal surface for $D = 2.5$; $N = M = 10$; $q = 2.7$; (b) $g(\theta_2, \theta_3)$; (c) $g(\theta_2, \theta_3)$, top view; (d) $g(\theta_2, \theta_3)$, side view.

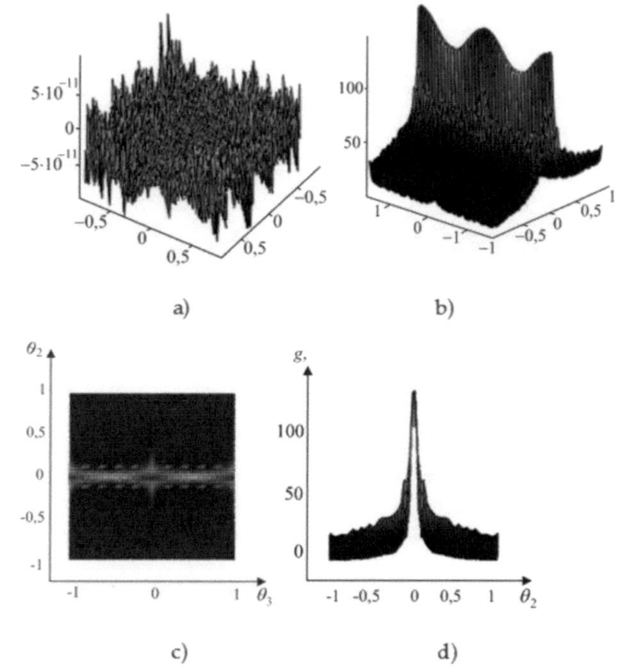

Figure 21. The fractal surface and the scattering indicatrix $g(\theta_2, \theta_3)$ when $\lambda = 2.2$ mm and $\theta_1 = 0°$: (a) fractal surface for $D = 2.5$; $N = M = 20$; $q = 2.7$; (b) $g(\theta_2, \theta_3)$; (c) $g(\theta_2, \theta_3)$, top view; (d) $g(\theta_2, \theta_3)$, side view.

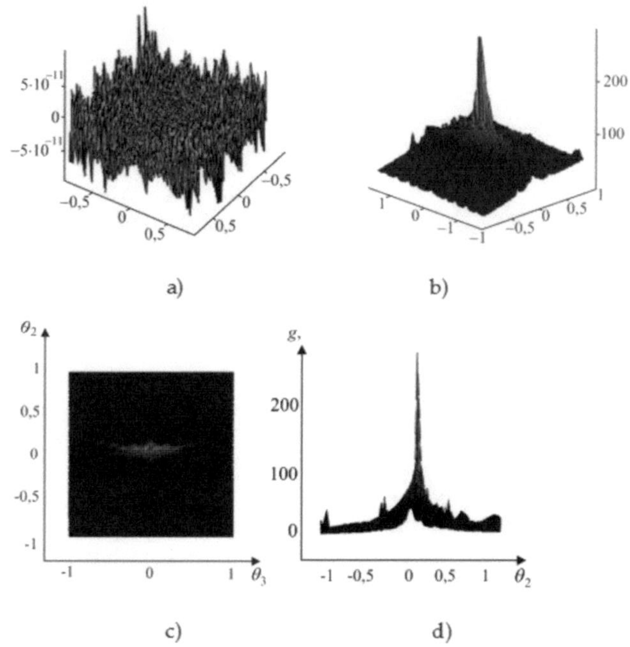

Figure 22. The fractal surface and the scattering indicatrix $g(\theta_2, \theta_3)$ when $\lambda = 2.2$ mm and $\theta_1 = 5°$: (a) fractal surface for $D = 2.5$; $N = M = 20$; $q = 2.7$; (b) $g(\theta_2, \theta_3)$; (c) $g(\theta_2, \theta_3)$, top view; (d) $g(\theta_2, \theta_3)$, side view.

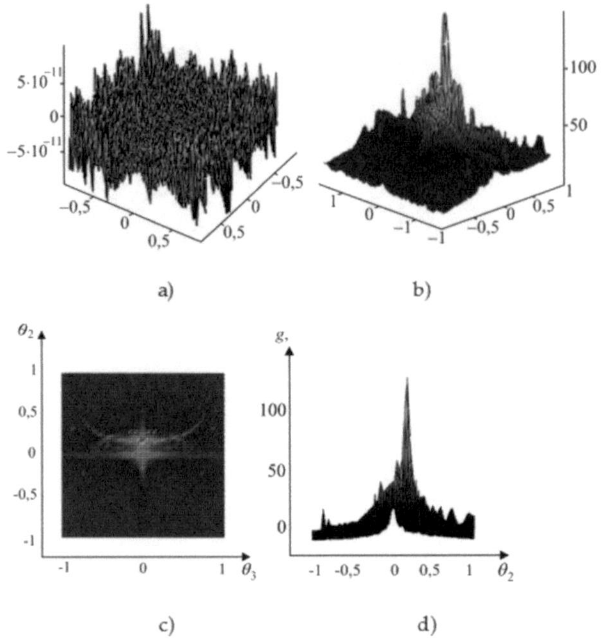

Figure 23. The fractal surface and the scattering indicatrix $g(\theta_2, \theta_3)$ when $\lambda = 2.2$ mm and $\theta_1 = 10°$: (a) fractal surface for $D = 2.5$; $N = M = 20$; $q = 2.7$; (b) $g(\theta_2, \theta_3)$; (c) $g(\theta_2, \theta_3)$, top view; (d) $g(\theta_2, \theta_3)$, side view.

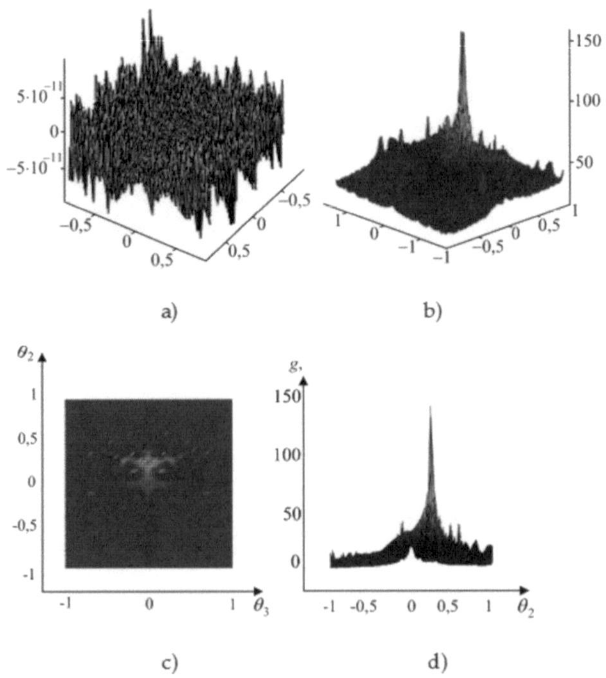

Figure 24. The fractal surface and the scattering indicatrix $g(\theta_2, \theta_3)$ when $\lambda = 2.2$ mm and $\theta_1 = 15°$: (a) fractal surface for $D = 2.5$; $N = M = 20$; $q = 2.7$; (b) $g(\theta_2, \theta_3)$; (c) $g(\theta_2, \theta_3)$, top view; (d) $g(\theta_2, \theta_3)$, side view.

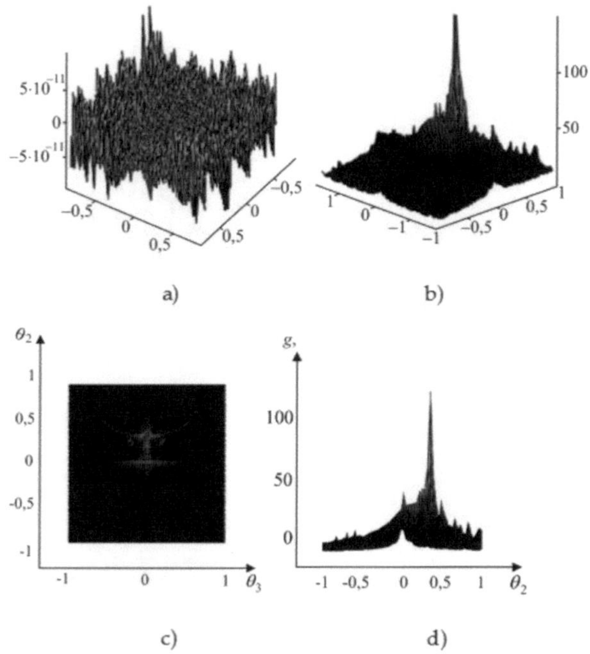

Figure 25. The fractal surface and the scattering indicatrix $g(\theta_2, \theta_3)$ when $\lambda = 2.2$ mm and $\theta_1 = 20°$: (a) fractal surface for $D = 2.5$; $N = M = 20$; $q = 2.7$; (b) $g(\theta_2, \theta_3)$; (c) $g(\theta_2, \theta_3)$, top view; (d) $g(\theta_2, \theta_3)$, side view.

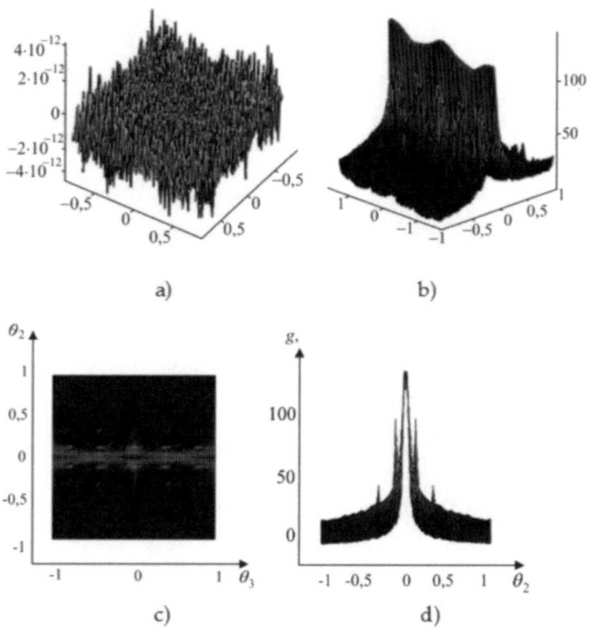

Figure 26. The fractal surface and the scattering indicatrix $g(\theta_2, \theta_3)$ when $\lambda = 2.2$ mm and $\theta_1 = 0°$: (a) fractal surface for $D = 2.8$; $N = M = 10$; $q = 2.7$; (b) $g(\theta_2, \theta_3)$; (c) $g(\theta_2, \theta_3)$, top view; (d) $g(\theta_2, \theta_3)$, side view.

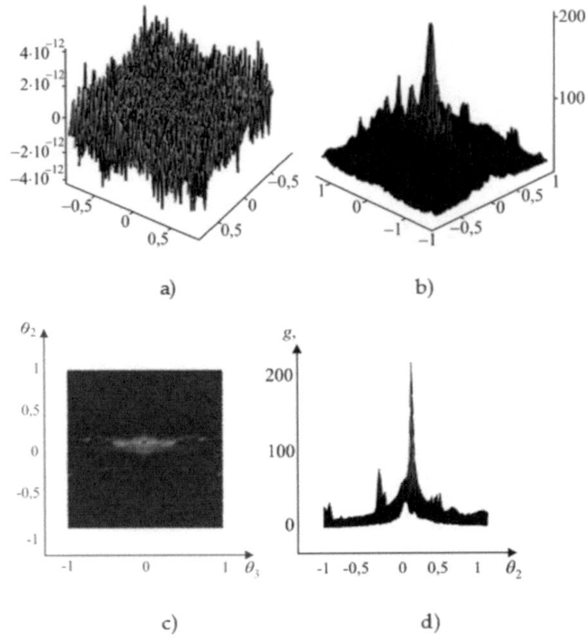

Figure 27. The fractal surface and the scattering indicatrix $g(\theta_2, \theta_3)$ when $\lambda = 2.2$ mm and $\theta_1 = 5°$: (a) fractal surface for $D = 2.8$; $N = M = 10$; $q = 2.7$; (b) $g(\theta_2, \theta_3)$; (c) $g(\theta_2, \theta_3)$, top view; (d) $g(\theta_2, \theta_3)$, side view.

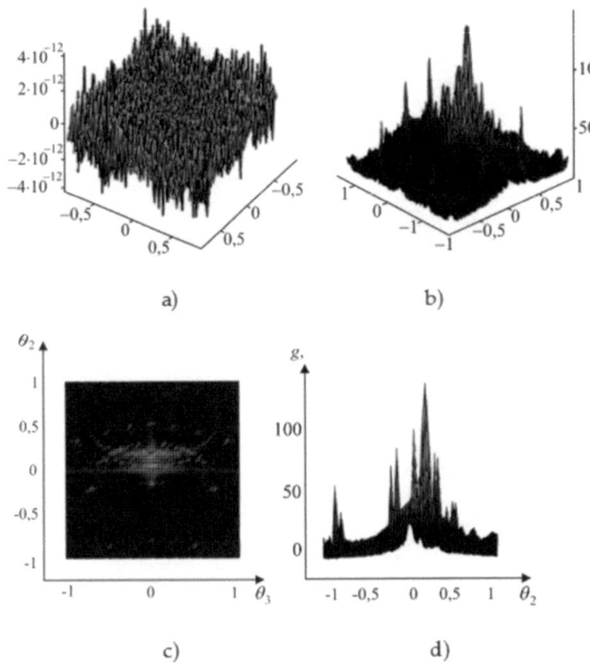

Figure 28. The fractal surface and the scattering indicatrix $g(\theta_2, \theta_3)$ when $\lambda = 2.2$ mm and $\theta_1 = 10°$: (a) fractal surface for $D = 2.8$; $N = M = 10$; $q = 2.7$; (b) $g(\theta_2, \theta_3)$; (c) $g(\theta_2, \theta_3)$, top view; (d) $g(\theta_2, \theta_3)$, side view.

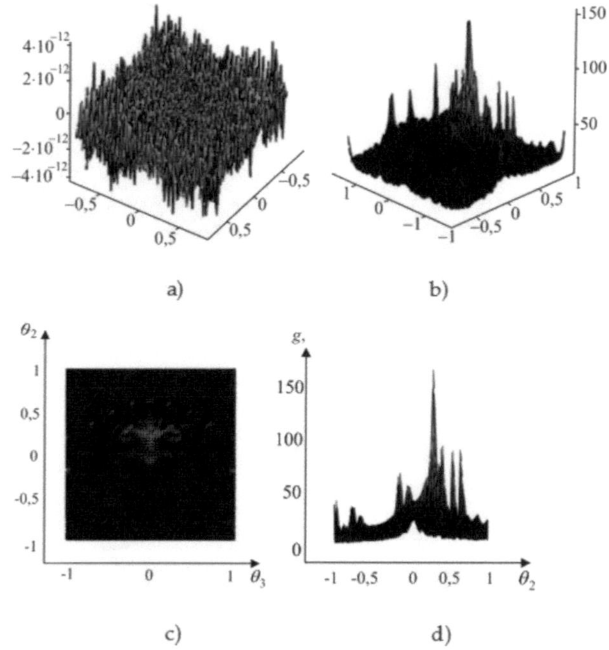

Figure 29. The fractal surface and the scattering indicatrix $g(\theta_2, \theta_3)$ when $\lambda = 2.2$ mm and $\theta_1 = 15°$: (a) fractal surface for $D = 2.8$; $N = M = 10$; $q = 2.7$; (b) $g(\theta_2, \theta_3)$; (c) $g(\theta_2, \theta_3)$, top view; (d) $g(\theta_2, \theta_3)$, side view.

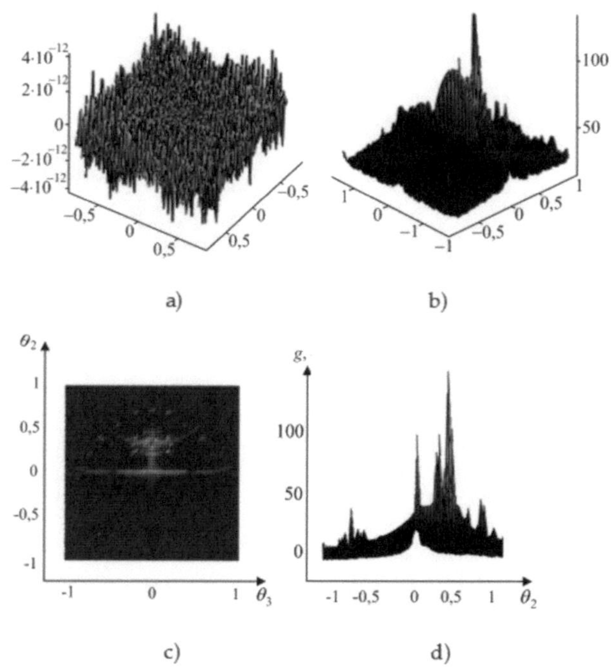

Figure 30. The fractal surface and the scattering indicatrix $g(\theta_2, \theta_3)$ when $\lambda = 2.2$ mm and $\theta_1 = 20°$: (a) fractal surface for $D = 2.8$; $N = M = 10$; $q = 2.7$; (b) $g(\theta_2, \theta_3)$; (c) $g(\theta_2, \theta_3)$, top view; (d) $g(\theta_2, \theta_3)$, side view.

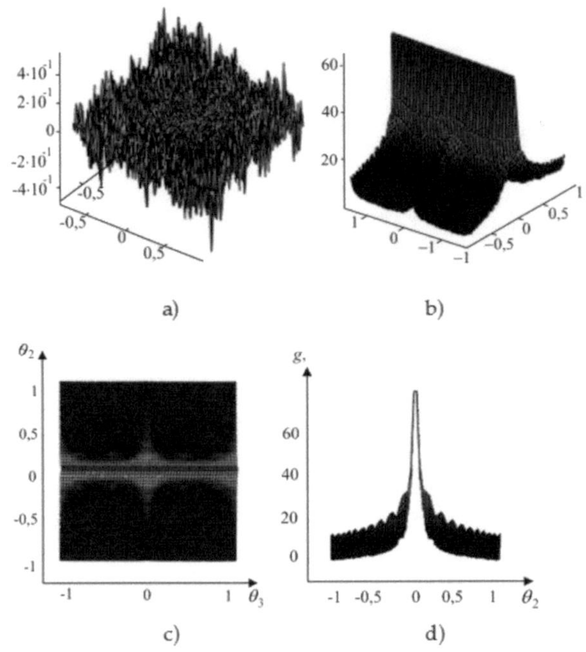

Figure 31. The fractal surface and the scattering indicatrix $g(\theta_2, \theta_3)$ when $\lambda = 8.6$ mm and $\theta_1 = 0°$: (a) fractal surface for $D = 2.2$; $N = M = 10$; $q = 2.7$; (b) $g(\theta_2, \theta_3)$; (c) $g(\theta_2, \theta_3)$, top view; (d) $g(\theta_2, \theta_3)$, side view.

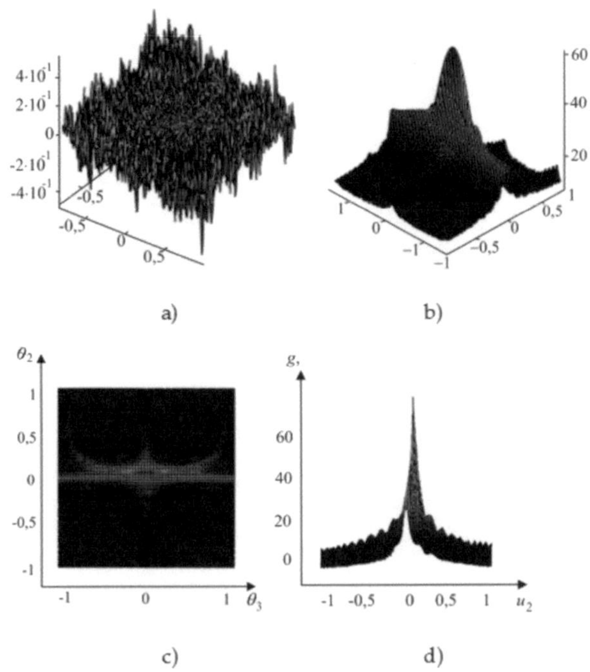

Figure 32. The fractal surface and the scattering indicatrix $g(\theta_2, \theta_3)$ when $\lambda = 8.6$ mm and $\theta_1 = 5°$: (a) fractal surface for $D = 2.2$; $N = M = 10$; $q = 2.7$; (b) $g(\theta_2, \theta_3)$; (c) $g(\theta_2, \theta_3)$, top view; (d) $g(\theta_2, \theta_3)$, side view.

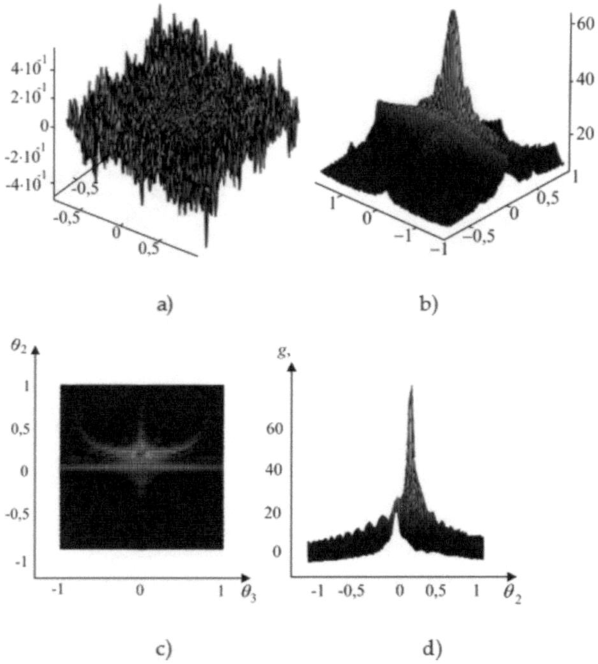

Figure 33. The fractal surface and the scattering indicatrix $g(\theta_2, \theta_3)$ when $\lambda = 8.6$ mm and $\theta_1 = 10°$: (a) fractal surface for $D = 2.2$; $N = M = 10$; $q = 2.7$; (b) $g(\theta_2, \theta_3)$; (c) $g(\theta_2, \theta_3)$, top view; (d) $g(\theta_2, \theta_3)$, side view.

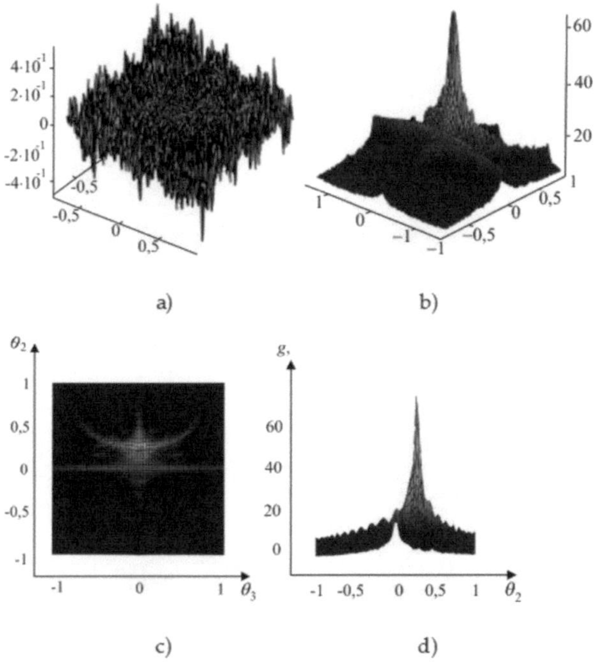

Figure 34. The fractal surface and the scattering indicatrix $g(\theta_2, \theta_3)$ when $\lambda = 8.6$ mm and $\theta_1 = 15°$: (a) fractal surface for $D = 2.2$; $N = M = 10$; $q = 2.7$; (b) $g(\theta_2, \theta_3)$; (c) $g(\theta_2, \theta_3)$, top view; (d) $g(\theta_2, \theta_3)$, side view.

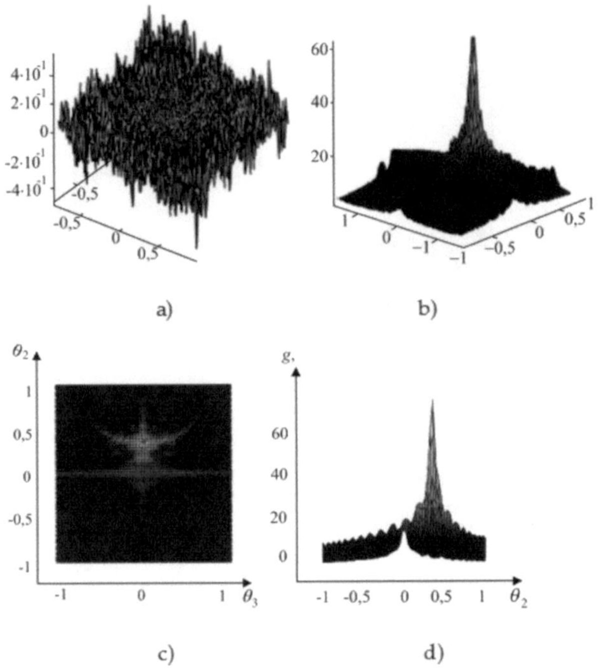

Figure 35. The fractal surface and the scattering indicatrix $g(\theta_2, \theta_3)$ when $\lambda = 8.6$ mm and $\theta_1 = 20°$: (a) fractal surface for $D = 2.2$; $N = M = 10$; $q = 2.7$; (b) $g(\theta_2, \theta_3)$; (c) $g(\theta_2, \theta_3)$, top view; (d) $g(\theta_2, \theta_3)$, side view.

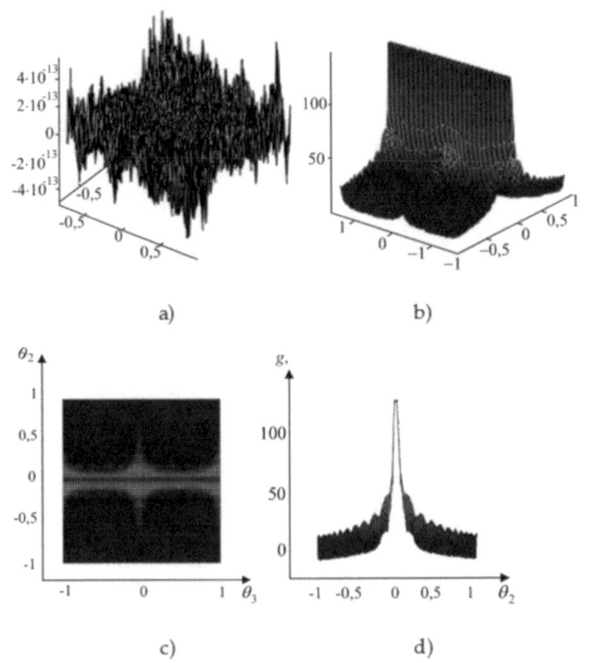

Figure 36. The fractal surface and the scattering indicatrix $g(\theta_2, \theta_3)$ when $\lambda = 8.6$ mm and $\theta_1 = 0°$: (a) fractal surface for $D = 2.2$; $N = M = 20$; $q = 2.7$; (b) $g(\theta_2, \theta_3)$; (c) $g(\theta_2, \theta_3)$, top view; (d) $g(\theta_2, \theta_3)$, side view.

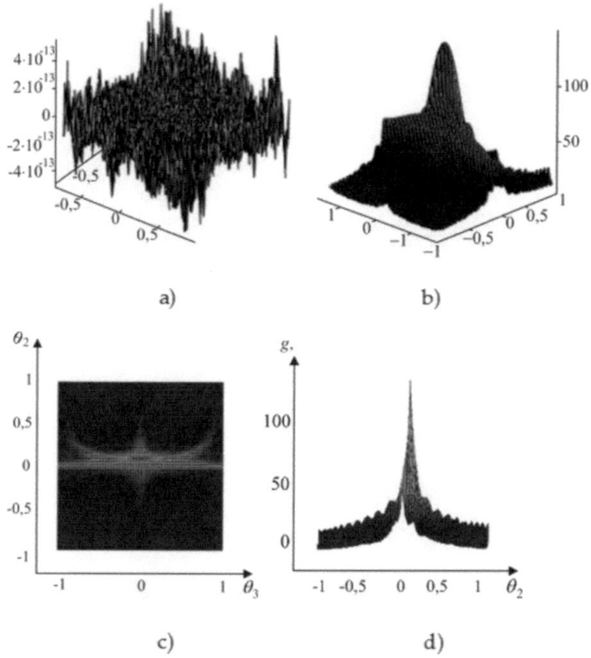

Figure 37. The fractal surface and the scattering indicatrix $g(\theta_2, \theta_3)$ when $\lambda = 8.6$ mm and $\theta_1 = 5°$: (a) fractal surface for $D = 2.2$; $N = M = 20$; $q = 2.7$; (b) $g(\theta_2, \theta_3)$; (c) $g(\theta_2, \theta_3)$, top view; (d) $g(\theta_2, \theta_3)$, side view.

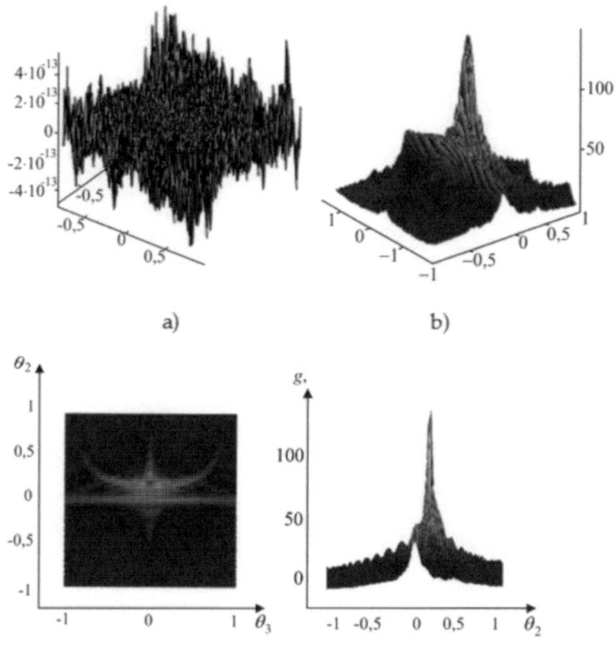

Figure 38. The fractal surface and the scattering indicatrix $g(\theta_2, \theta_3)$ when $\lambda = 8.6$ mm and $\theta_1 = 10°$: (a) fractal surface for $D = 2.2$; $N = M = 20$; $q = 2.7$; (b) $g(\theta_2, \theta_3)$; (c) $g(\theta_2, \theta_3)$, top view; (d) $g(\theta_2, \theta_3)$, side view.

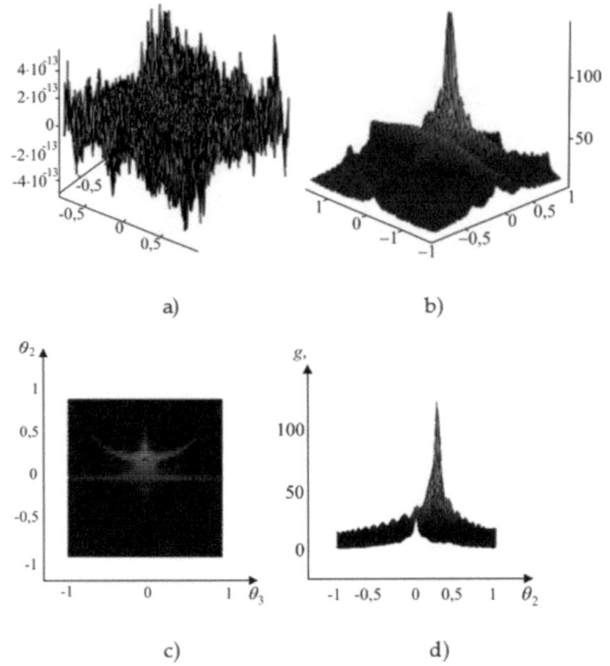

Figure 39. The fractal surface and the scattering indicatrix $g(\theta_2, \theta_3)$ when $\lambda = 8.6$ mm and $\theta_1 = 15°$: (a) fractal surface for $D = 2.2$; $N = M = 20$; $q = 2.7$; (b) $g(\theta_2, \theta_3)$; (c) $g(\theta_2, \theta_3)$, top view; (d) $g(\theta_2, \theta_3)$, side view.

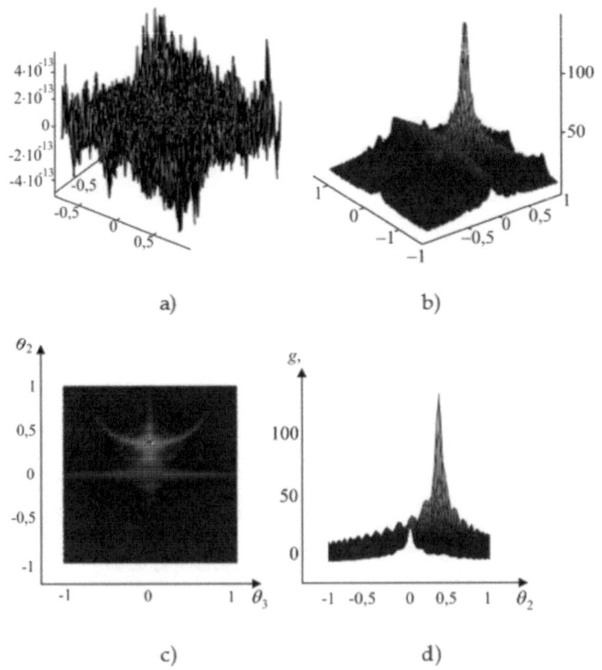

Figure 40. The fractal surface and the scattering indicatrix $g(\theta_2, \theta_3)$ when $\lambda = 8.6$ mm and $\theta_1 = 20°$: (a) fractal surface for $D = 2.2$; $N = M = 20$; $q = 2.7$; (b) $g(\theta_2, \theta_3)$; (c) $g(\theta_2, \theta_3)$, top view; (d) $g(\theta_2, \theta_3)$, side view.

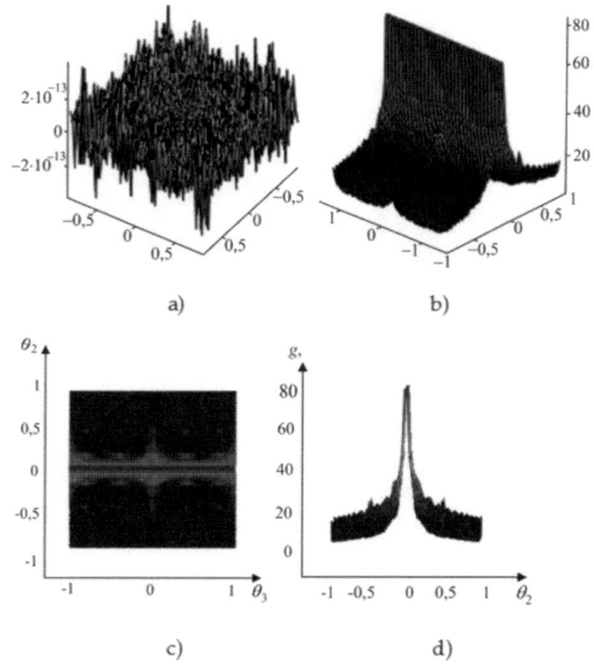

Figure 41. The fractal surface and the scattering indicatrix $g(\theta_2, \theta_3)$ when $\lambda = 8.6$ mm and $\theta_1 = 0°$: (a) fractal surface for $D = 2.5$; $N = M = 10$; $q = 2.7$; (b) $g(\theta_2, \theta_3)$; (c) $g(\theta_2, \theta_3)$, top view; (d) $g(\theta_2, \theta_3)$, side view.

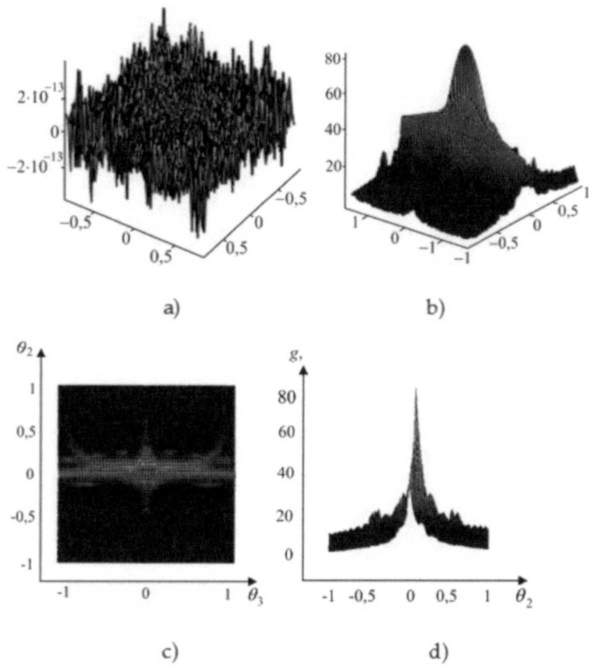

Figure 42. The fractal surface and the scattering indicatrix $g(\theta_2, \theta_3)$ when $\lambda = 8.6$ mm and $\theta_1 = 5°$: (a) fractal surface for $D = 2.5$; $N = M = 10$; $q = 2.7$; (b) $g(\theta_2, \theta_3)$; (c) $g(\theta_2, \theta_3)$, top view; (d) $g(\theta_2, \theta_3)$, side view.

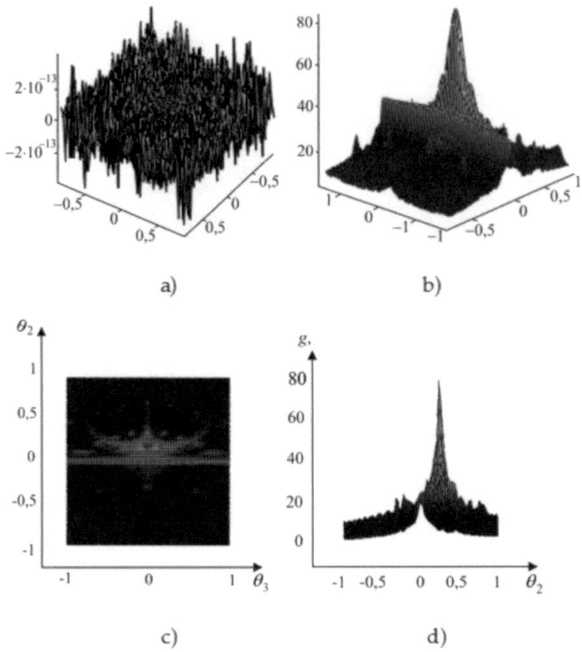

Figure 43. The fractal surface and the scattering indicatrix $g(\theta_2, \theta_3)$ when $\lambda = 8.6$ mm and $\theta_1 = 10°$: (a) fractal surface for $D = 2.5$; $N = M = 10$; $q = 2.7$; (b) $g(\theta_2, \theta_3)$; (c) $g(\theta_2, \theta_3)$, top view; (d) $g(\theta_2, \theta_3)$, side view.

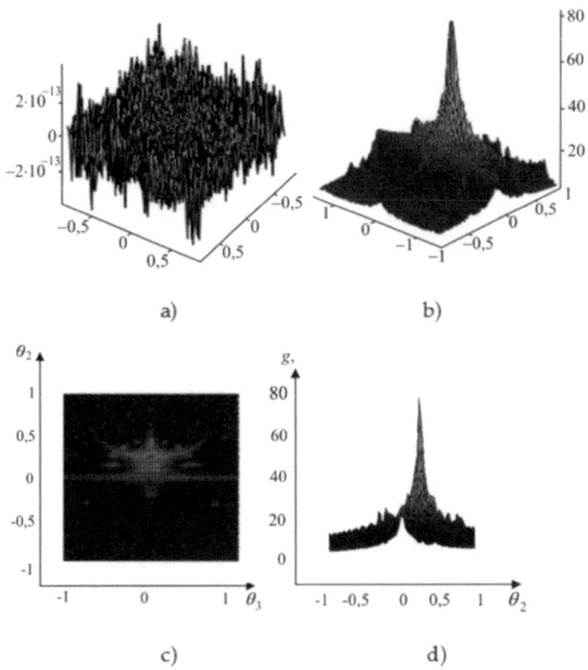

Figure 44. The fractal surface and the scattering indicatrix $g(\theta_2, \theta_3)$ when $\lambda = 8.6$ mm and $\theta_1 = 15°$: (a) fractal surface for $D = 2.5$; $N = M = 10$; $q = 2.7$; (b) $g(\theta_2, \theta_3)$; (c) $g(\theta_2, \theta_3)$, top view; (d) $g(\theta_2, \theta_3)$, side view.

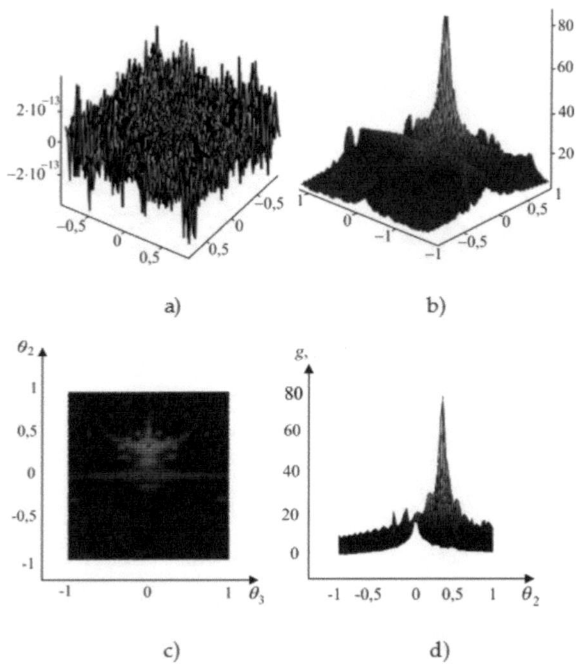

Figure 45. The fractal surface and the scattering indicatrix $g(\theta_2, \theta_3)$ when $\lambda = 8.6$ mm and $\theta_1 = 20°$: (a) fractal surface for $D = 2.5$; $N = M = 10$; $q = 2.7$; (b) $g(\theta_2, \theta_3)$; (c) $g(\theta_2, \theta_3)$, top view; (d) $g(\theta_2, \theta_3)$, side view.

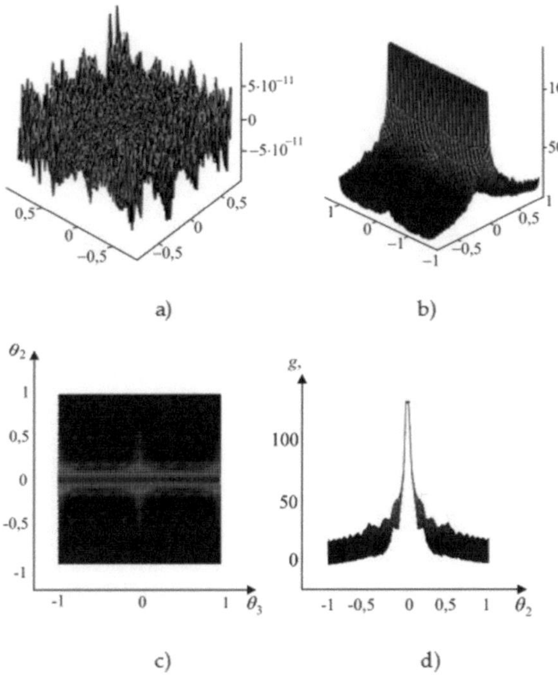

Figure 46. The fractal surface and the scattering indicatrix $g(\theta_2, \theta_3)$ when $\lambda = 8.6$ mm and $\theta_1 = 0°$: (a) fractal surface for $D = 2.5$; $N = M = 20$; $q = 2.7$; (b) $g(\theta_2, \theta_3)$; (c) $g(\theta_2, \theta_3)$, top view; (d) $g(\theta_2, \theta_3)$, side view.

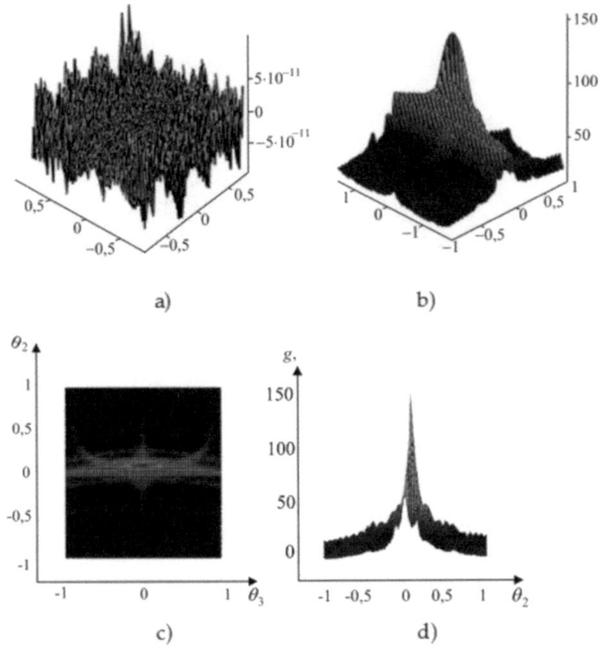

Figure 47. The fractal surface and the scattering indicatrix $g(\theta_2, \theta_3)$ when $\lambda = 8.6$ mm and $\theta_1 = 5°$: (a) fractal surface for $D = 2.5$; $N = M = 20$; $q = 2.7$; (b) $g(\theta_2, \theta_3)$; (c) $g(\theta_2, \theta_3)$, top view; (d) $g(\theta_2, \theta_3)$, side view.

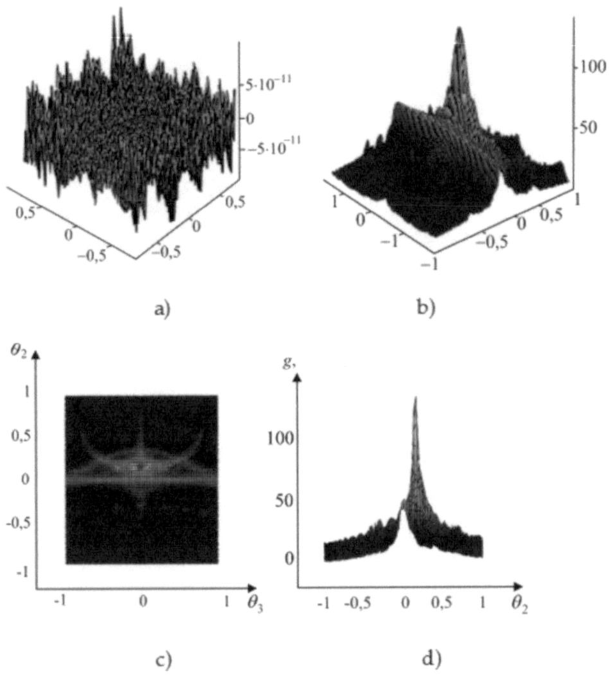

Figure 48. The fractal surface and the scattering indicatrix $g(\theta_2, \theta_3)$ when $\lambda = 8.6$ mm and $\theta_1 = 10°$: (a) fractal surface for $D = 2.5$; $N = M = 20$; $q = 2.7$; (b) $g(\theta_2, \theta_3)$; (c) $g(\theta_2, \theta_3)$, top view; (d) $g(\theta_2, \theta_3)$, side view.

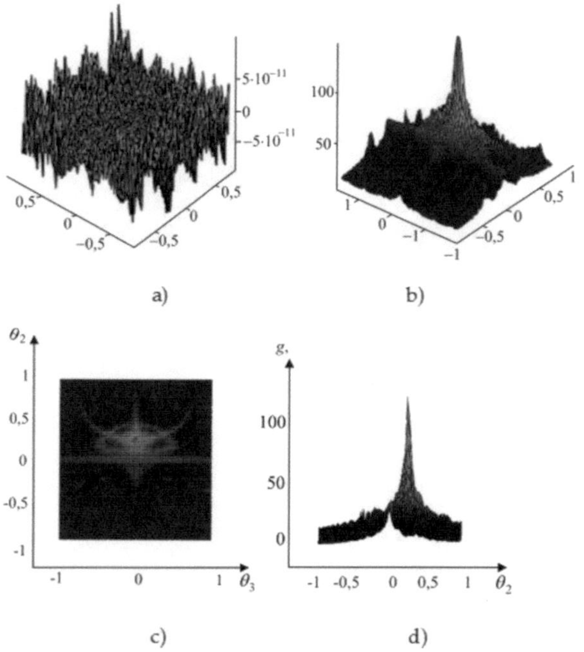

Figure 49. The fractal surface and the scattering indicatrix $g(\theta_2, \theta_3)$ when $\lambda = 8.6$ mm and $\theta_1 = 15°$: (a) fractal surface for $D = 2.5$; $N = M = 20$; $q = 2.7$; (b) $g(\theta_2, \theta_3)$; (c) $g(\theta_2, \theta_3)$, top view; (d) $g(\theta_2, \theta_3)$, side view.

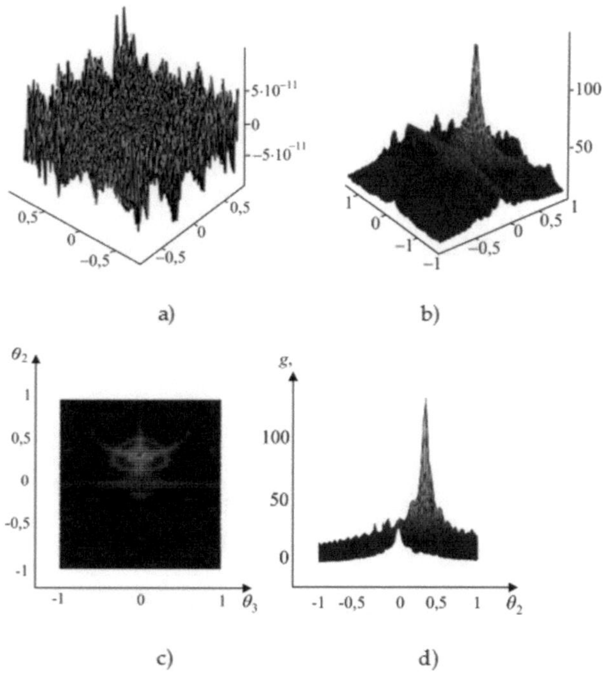

Figure 50. The fractal surface and the scattering indicatrix $g(\theta_2, \theta_3)$ when $\lambda = 8.6$ mm and $\theta_1 = 20°$: (a) fractal surface for $D = 2.5$; $N = M = 20$; $q = 2.7$; (b) $g(\theta_2, \theta_3)$; (c) $g(\theta_2, \theta_3)$, top view; (d) $g(\theta_2, \theta_3)$, side view.

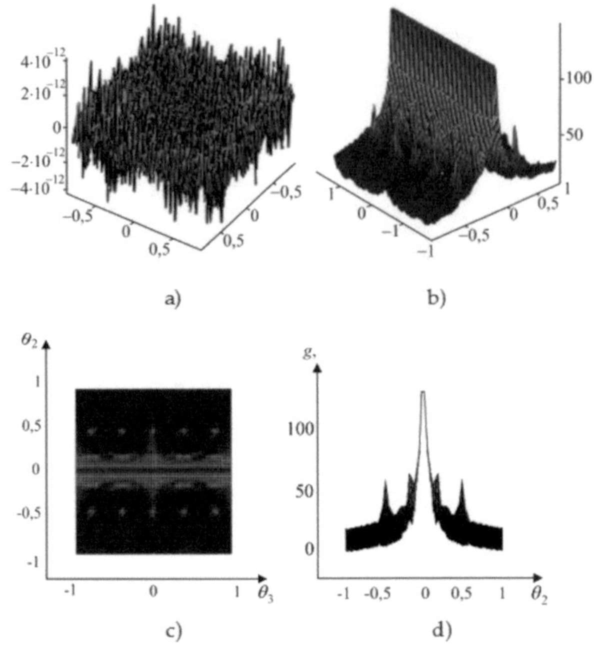

Figure 51. The fractal surface and the scattering indicatrix $g(\theta_2, \theta_3)$ when $\lambda = 8.6$ mm and $\theta_1 = 0°$: (a) fractal surface for $D = 2.8$; $N = M = 10$; $q = 2.7$; (b) $g(\theta_2, \theta_3)$; (c) $g(\theta_2, \theta_3)$, top view; (d) $g(\theta_2, \theta_3)$, side view.

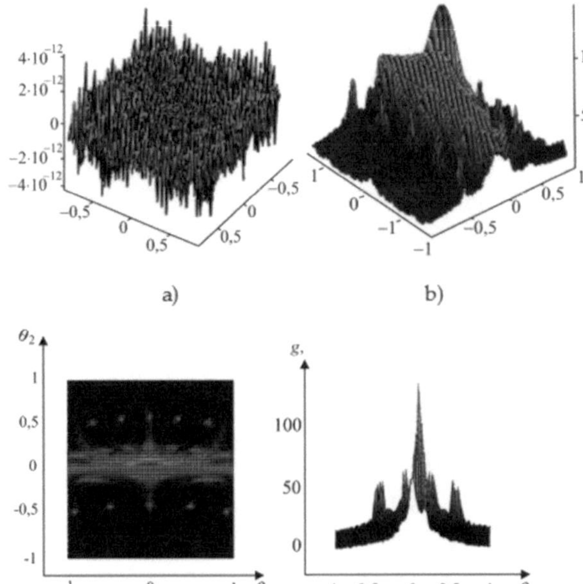

Figure 52. The fractal surface and the scattering indicatrix $g(\theta_2, \theta_3)$ when $\lambda = 8.6$ mm and $\theta_1 = 5°$: (a) fractal surface for $D = 2.8$; $N = M = 10$; $q = 2.7$; (b) $g(\theta_2, \theta_3)$; (c) $g(\theta_2, \theta_3)$, top view; (d) $g(\theta_2, \theta_3)$, side view.

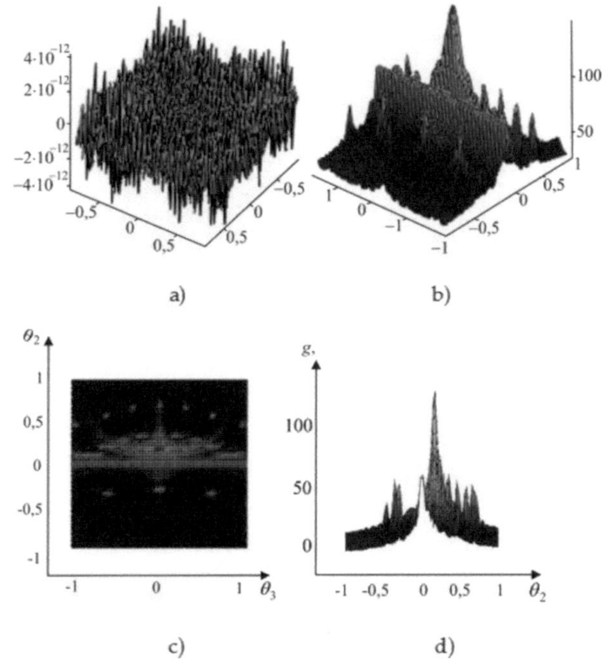

Figure 53. The fractal surface and the scattering indicatrix $g(\theta_2, \theta_3)$ when $\lambda = 8.6$ mm and $\theta_1 = 10°$: (a) fractal surface for $D = 2.8$; $N = M = 10$; $q = 2.7$; (b) $g(\theta_2, \theta_3)$; (c) $g(\theta_2, \theta_3)$, top view; (d) $g(\theta_2, \theta_3)$, side view.

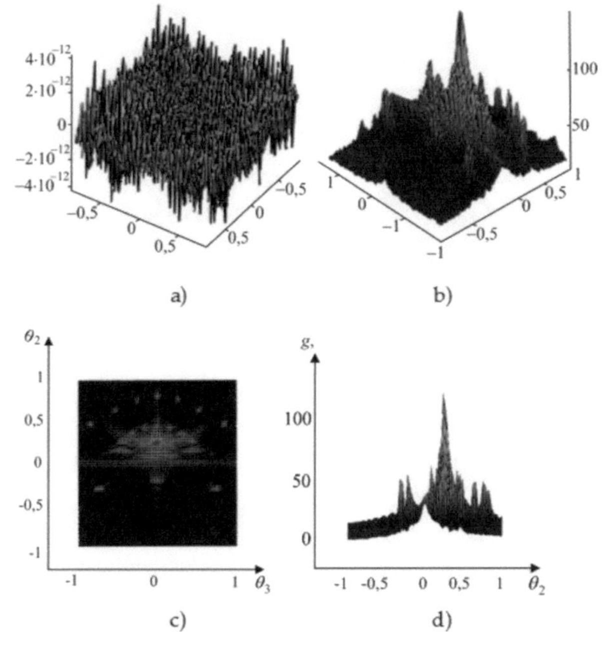

Figure 54. The fractal surface and the scattering indicatrix $g(\theta_2, \theta_3)$ when $\lambda = 8.6$ mm and $\theta_1 = 15°$: (a) fractal surface for $D = 2.8$; $N = M = 10$; $q = 2.7$; (b) $g(\theta_2, \theta_3)$; (c) $g(\theta_2, \theta_3)$, top view; (d) $g(\theta_2, \theta_3)$, side view.

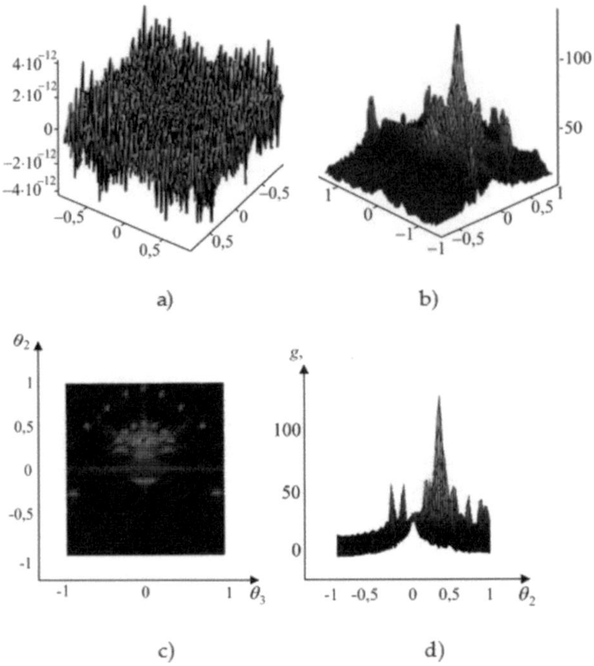

Figure 55. The fractal surface and the scattering indicatrix $g(\theta_2, \theta_3)$ when $\lambda = 8.6$ mm and $\theta_1 = 20°$: (a) fractal surface for $D = 2.8$; $N = M = 10$; $q = 2.7$; (b) $g(\theta_2, \theta_3)$; (c) $g(\theta_2, \theta_3)$, top view; (d) $g(\theta_2, \theta_3)$, side view.

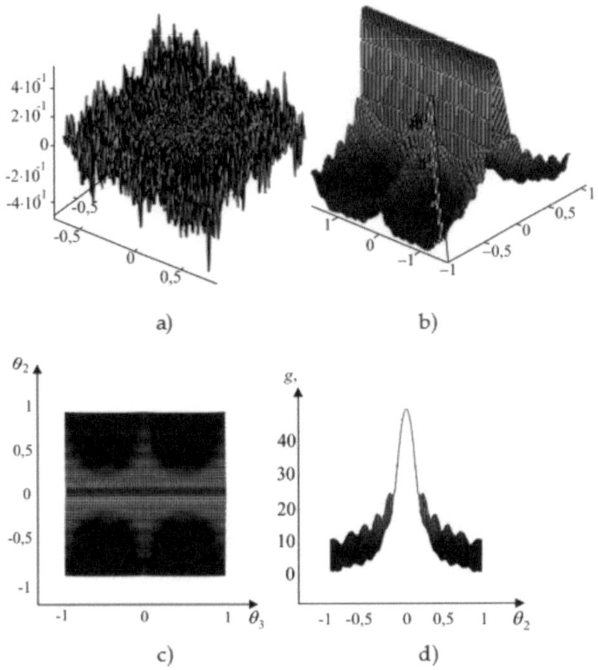

Figure 56. The fractal surface and the scattering indicatrix $g(\theta_2, \theta_3)$ when $\lambda = 30$ mm and $\theta_1 = 0°$: (a) fractal surface for $D = 2.2$; $N = M = 10$; $q = 2.7$; (b) $g(\theta_2, \theta_3)$; (c) $g(\theta_2, \theta_3)$, top view; (d) $g(\theta_2, \theta_3)$, side view.

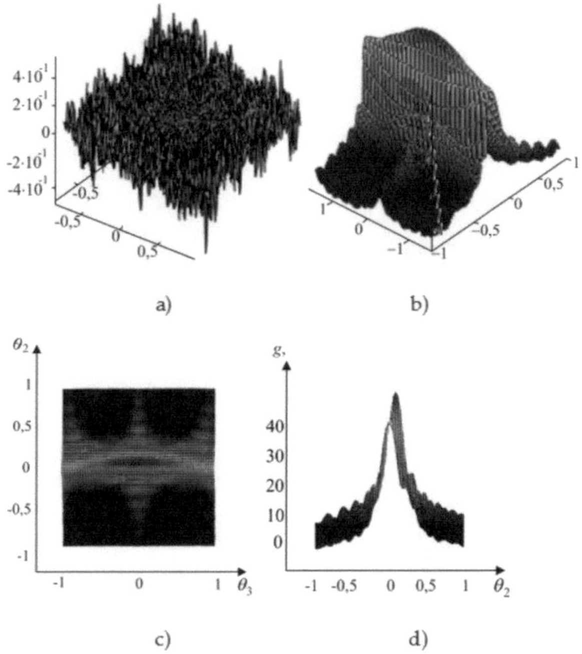

Figure 57. The fractal surface and the scattering indicatrix $g(\theta_2, \theta_3)$ when $\lambda = 30$ mm and $\theta_1 = 5°$: (a) fractal surface for $D = 2.2$; $N = M = 10$; $q = 2.7$; (b) $g(\theta_2, \theta_3)$; (c) $g(\theta_2, \theta_3)$, top view; (d) $g(\theta_2, \theta_3)$, side view.

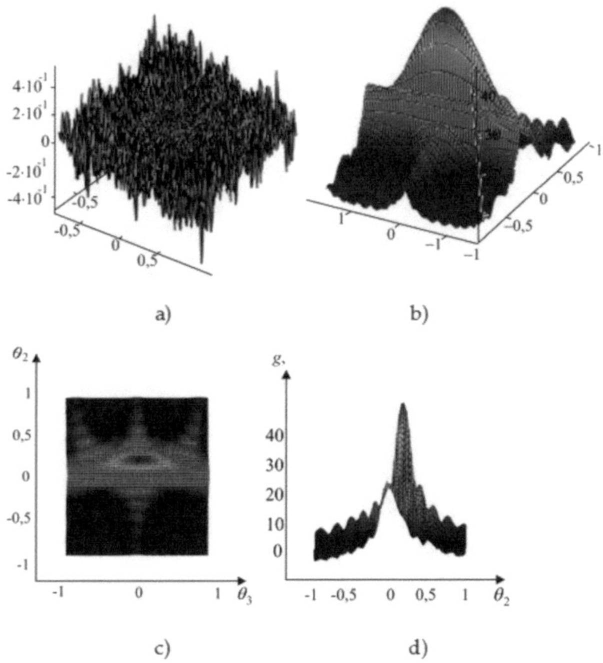

Figure 58. The fractal surface and the scattering indicatrix $g(\theta_2, \theta_3)$ when $\lambda = 30$ mm and $\theta_1 = 10°$: (a) fractal surface for $D = 2.2$; $N = M = 10$; $q = 2.7$; (b) $g(\theta_2, \theta_3)$; (c) $g(\theta_2, \theta_3)$, top view; (d) $g(\theta_2, \theta_3)$, side view.

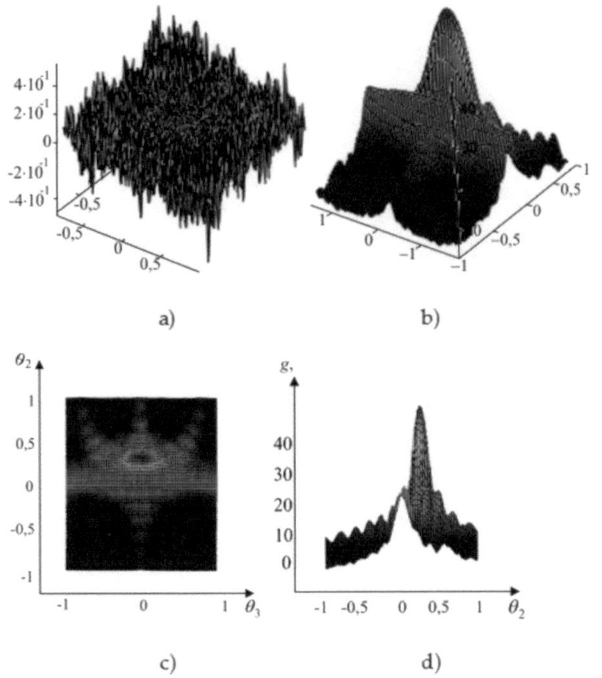

Figure 59. The fractal surface and the scattering indicatrix $g(\theta_2, \theta_3)$ when $\lambda = 30$ mm and $\theta_1 = 15°$: (a) fractal surface for $D = 2.2$; $N = M = 10$; $q = 2.7$; (b) $g(\theta_2, \theta_3)$; (c) $g(\theta_2, \theta_3)$, top view; (d) $g(\theta_2, \theta_3)$, side view.

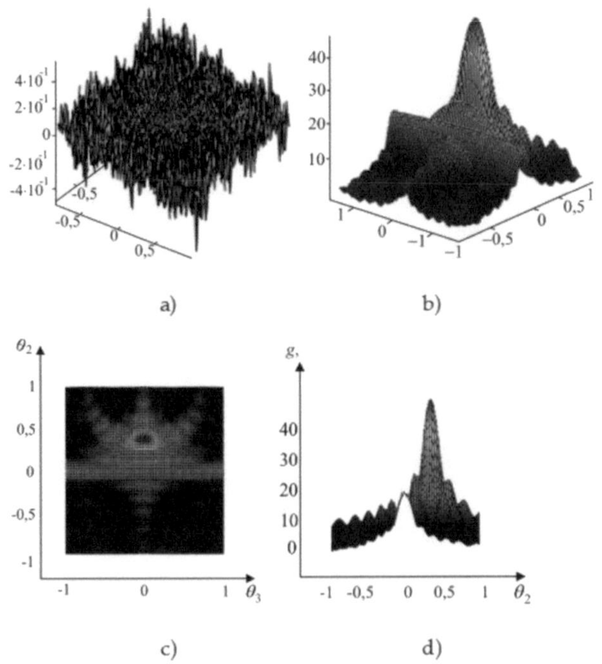

Figure 60. The fractal surface and the scattering indicatrix $g(\theta_2, \theta_3)$ when $\lambda = 30$ mm and $\theta_1 = 20°$: (a) fractal surface for $D = 2.2$; $N = M = 10$; $q = 2.7$; (b) $g(\theta_2, \theta_3)$; (c) $g(\theta_2, \theta_3)$, top view; (d) $g(\theta_2, \theta_3)$, side view.

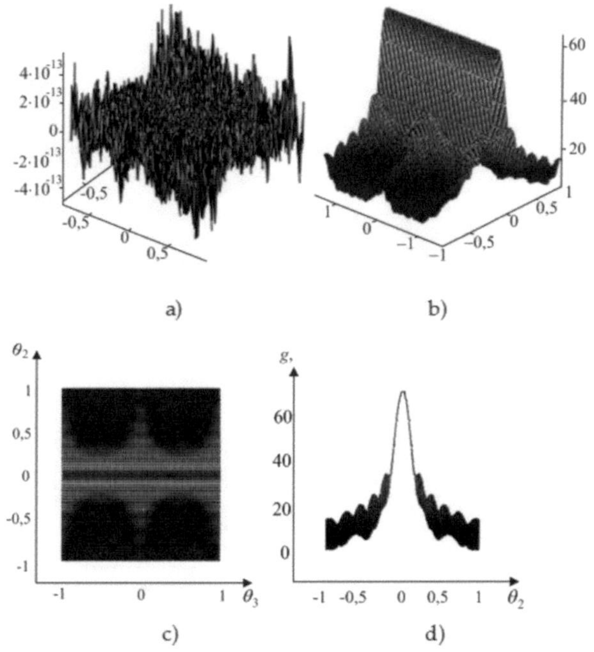

Figure 61. The fractal surface and the scattering indicatrix $g(\theta_2, \theta_3)$ when $\lambda = 30$ mm and $\theta_1 = 0°$: (a) fractal surface for $D = 2.2$; $N = M = 20$; $q = 2.7$; (b) $g(\theta_2, \theta_3)$; (c) $g(\theta_2, \theta_3)$, top view; (d) $g(\theta_2, \theta_3)$, side view.

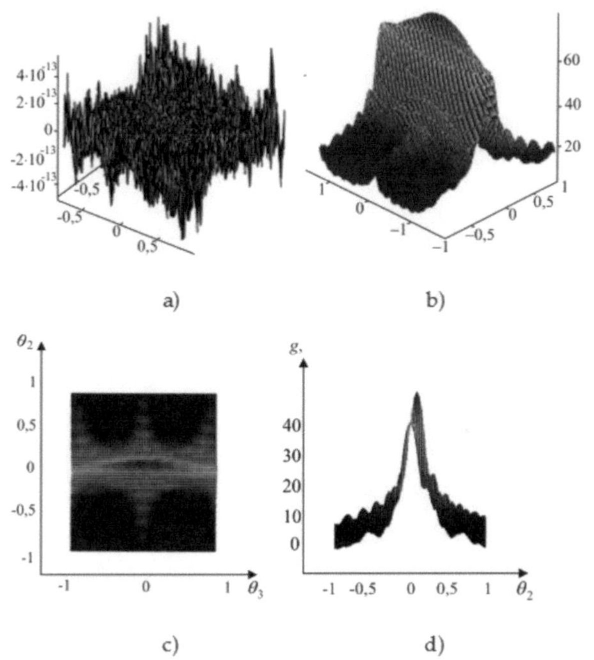

Figure 62. The fractal surface and the scattering indicatrix $g(\theta_2, \theta_3)$ when $\lambda = 30$ mm and $\theta_1 = 5°$: (a) fractal surface for $D = 2.2$; $N = M = 20$; $q = 2.7$; (b) $g(\theta_2, \theta_3)$; (c) $g(\theta_2, \theta_3)$, top view; (d) $g(\theta_2, \theta_3)$, side view.

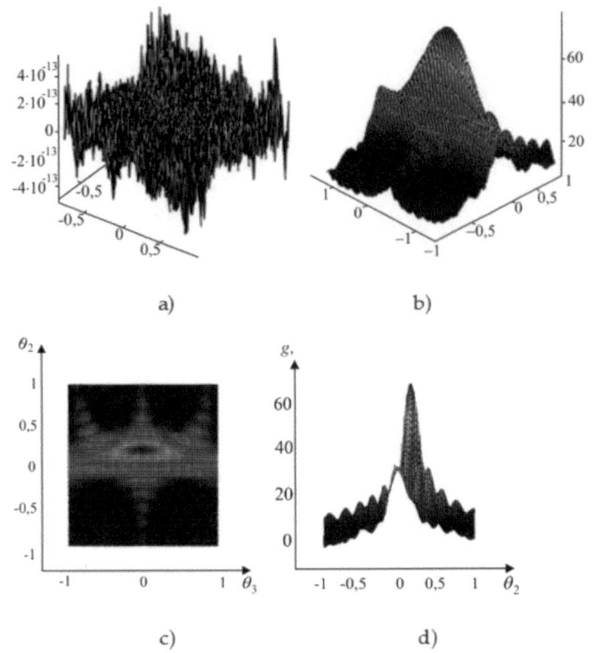

Figure 63. The fractal surface and the scattering indicatrix $g(\theta_2, \theta_3)$ when $\lambda = 30$ mm and $\theta_1 = 10°$: (a) fractal surface for $D = 2.2$; $N = M = 20$; $q = 2.7$; (b) $g(\theta_2, \theta_3)$; (c) $g(\theta_2, \theta_3)$, top view; (d) $g(\theta_2, \theta_3)$, side view.

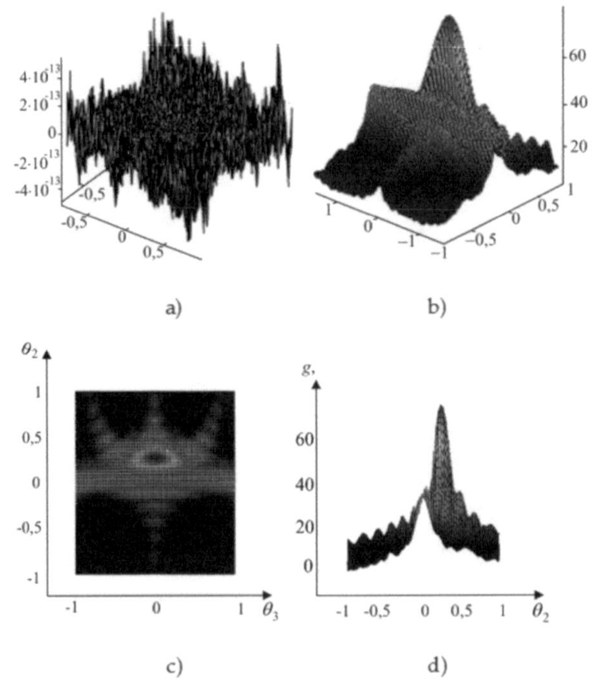

Figure 64. The fractal surface and the scattering indicatrix $g(\theta_2, \theta_3)$ when $\lambda = 30$ mm and $\theta_1 = 15°$: (a) fractal surface for $D = 2.2$; $N = M = 20$; $q = 2.7$; (b) $g(\theta_2, \theta_3)$; (c) $g(\theta_2, \theta_3)$, top view; (d) $g(\theta_2, \theta_3)$, side view.

http://dx.doi.org/10.5772/intechopen.68187

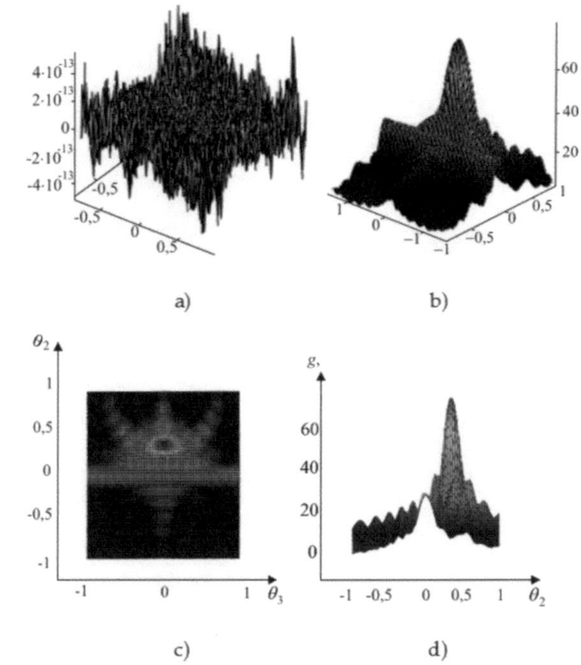

Figure 65. The fractal surface and the scattering indicatrix $g(\theta_2, \theta_3)$ when $\lambda = 30$ mm and $\theta_1 = 20°$: (a) fractal surface for $D = 2.2$; $N = M = 20$; $q = 2.7$; (b) $g(\theta_2, \theta_3)$; (c) $g(\theta_2, \theta_3)$, top view; (d) $g(\theta_2, \theta_3)$, side view.

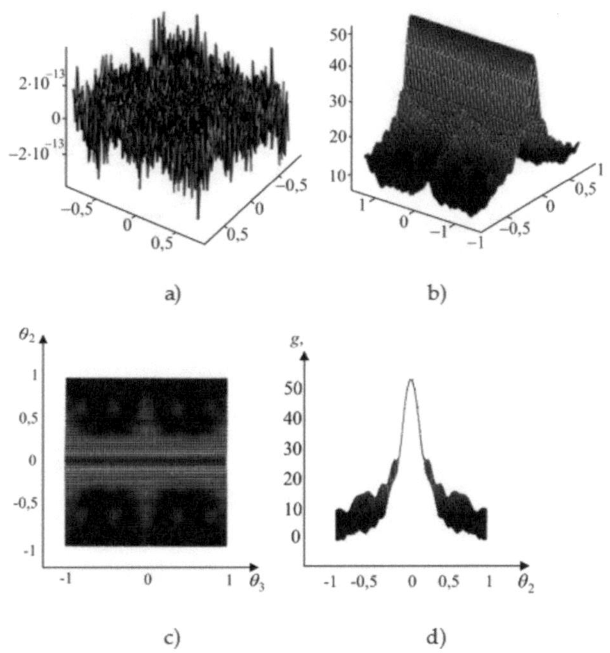

Figure 66. The fractal surface and the scattering indicatrix $g(\theta_2, \theta_3)$ when $\lambda = 30$ mm and $\theta_1 = 0°$: (a) fractal surface for $D = 2.5$; $N = M = 10$; $q = 2.7$; (b) $g(\theta_2, \theta_3)$; (c) $g(\theta_2, \theta_3)$, top view; (d) $g(\theta_2, \theta_3)$, side view.

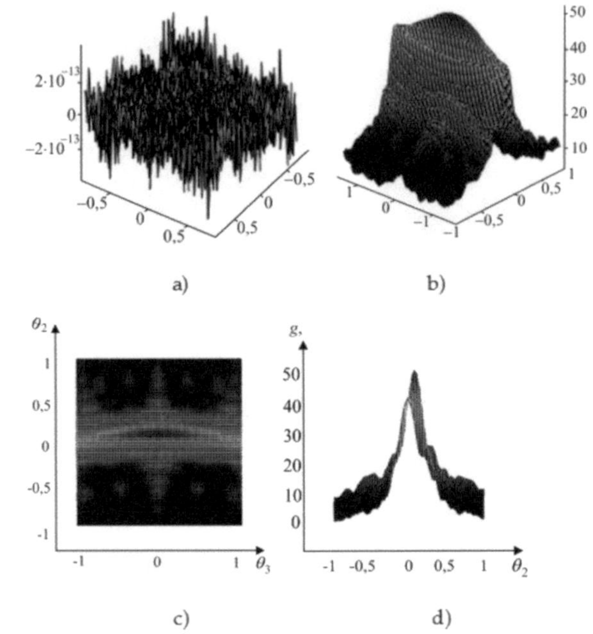

Figure 67. The fractal surface and the scattering indicatrix $g(\theta_2, \theta_3)$ when $\lambda = 30$ mm and $\theta_1 = 5°$: (a) fractal surface for $D = 2.5$; $N = M = 10$; $q = 2.7$; (b) $g(\theta_2, \theta_3)$; (c) $g(\theta_2, \theta_3)$, top view; (d) $g(\theta_2, \theta_3)$, side view.

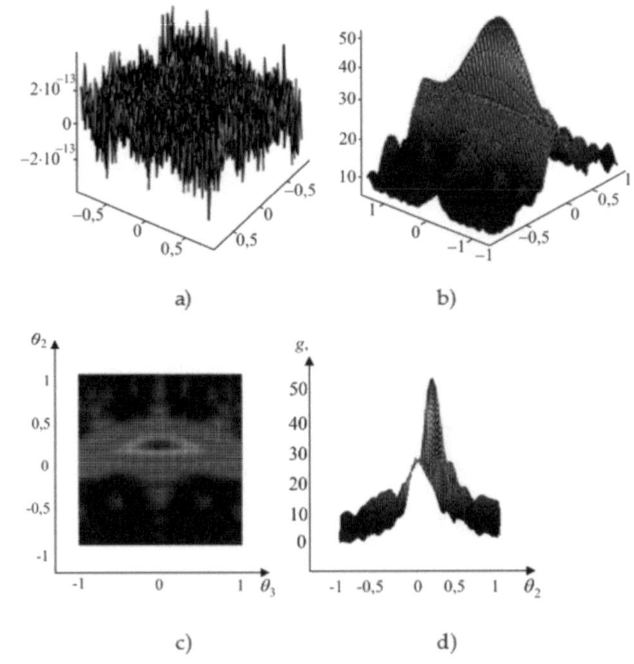

Figure 68. The fractal surface and the scattering indicatrix $g(\theta_2, \theta_3)$ when $\lambda = 30$ mm and $\theta_1 = 10°$: (a) fractal surface for $D = 2.5$; $N = M = 10$; $q = 2.7$; (b) $g(\theta_2, \theta_3)$; (c) $g(\theta_2, \theta_3)$, top view; (d) $g(\theta_2, \theta_3)$, side view.

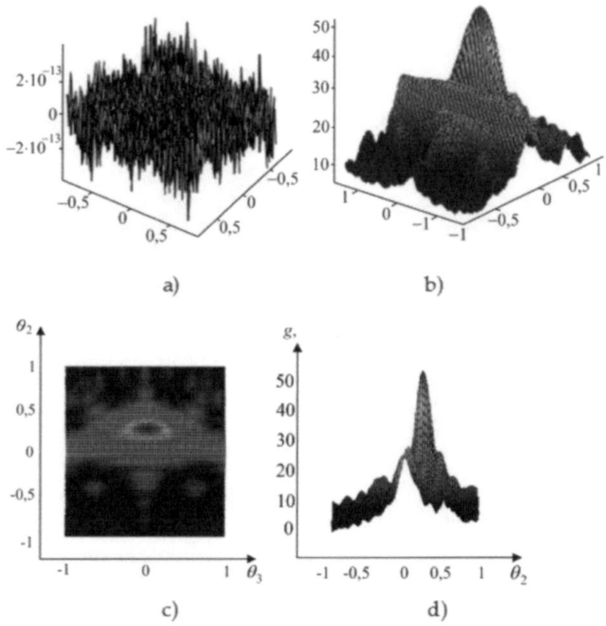

Figure 69. The fractal surface and the scattering indicatrix $g(\theta_2, \theta_3)$ when $\lambda = 30$ mm and $\theta_1 = 15°$: (a) fractal surface for $D = 2.5$; $N = M = 10$; $q = 2.7$; (b) $g(\theta_2, \theta_3)$; (c) $g(\theta_2, \theta_3)$, top view; (d) $g(\theta_2, \theta_3)$, side view.

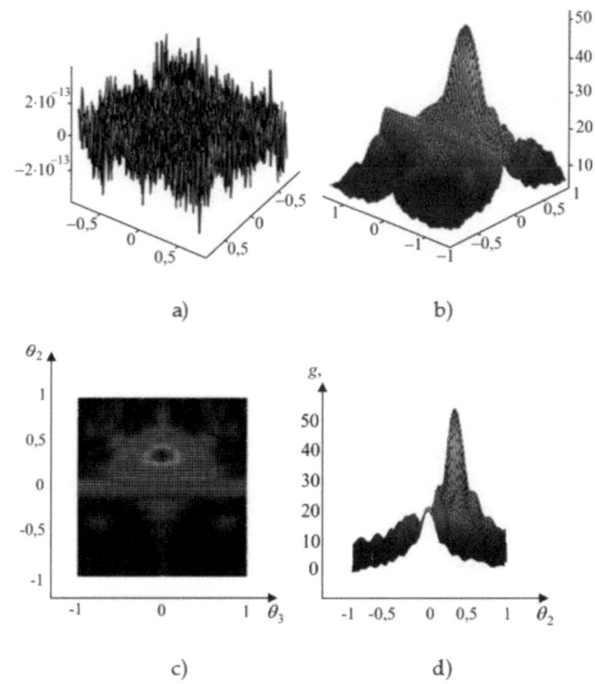

Figure 70. The fractal surface and the scattering indicatrix $g(\theta_2, \theta_3)$ when $\lambda = 30$ mm and $\theta_1 = 20°$: (a) fractal surface for $D = 2.5$; $N = M = 10$; $q = 2.7$; (b) $g(\theta_2, \theta_3)$; (c) $g(\theta_2, \theta_3)$, top view; (d) $g(\theta_2, \theta_3)$, side view.

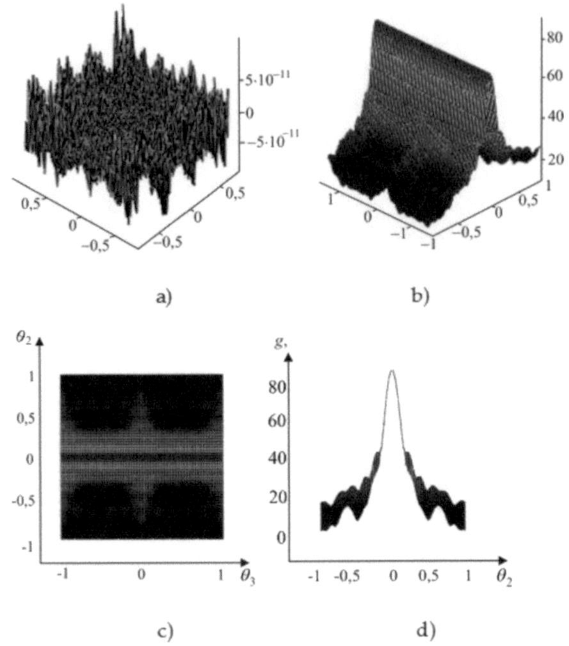

Figure 71. The fractal surface and the scattering indicatrix $g(\theta_2, \theta_3)$ when $\lambda = 30$ mm and $\theta_1 = 0°$: (a) fractal surface for $D = 2.5$; $N = M = 20$; $q = 2.7$; (b) $g(\theta_2, \theta_3)$; (c) $g(\theta_2, \theta_3)$, top view; (d) $g(\theta_2, \theta_3)$, side view.

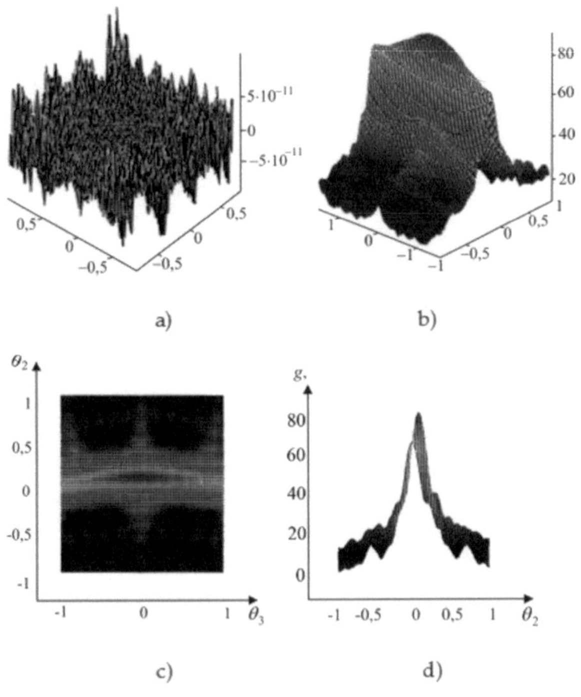

Figure 72. The fractal surface and the scattering indicatrix $g(\theta_2, \theta_3)$ when $\lambda = 30$ mm and $\theta_1 = 5°$: (a) fractal surface for $D = 2.5$; $N = M = 20$; $q = 2.7$; (b) $g(\theta_2, \theta_3)$; (c) $g(\theta_2, \theta_3)$, top view; (d) $g(\theta_2, \theta_3)$, side view.

http://dx.doi.org/10.5772/intechopen.68187

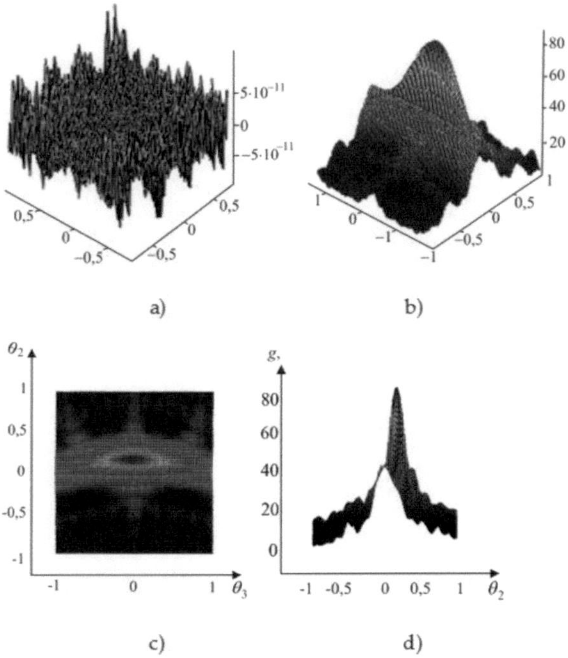

Figure 73. The fractal surface and the scattering indicatrix $g(\theta_2, \theta_3)$ when $\lambda = 30$ mm and $\theta_1 = 10°$: (a) fractal surface for $D = 2.5$; $N = M = 20$; $q = 2.7$; (b) $g(\theta_2, \theta_3)$; (c) $g(\theta_2, \theta_3)$, top view; (d) $g(\theta_2, \theta_3)$, side view.

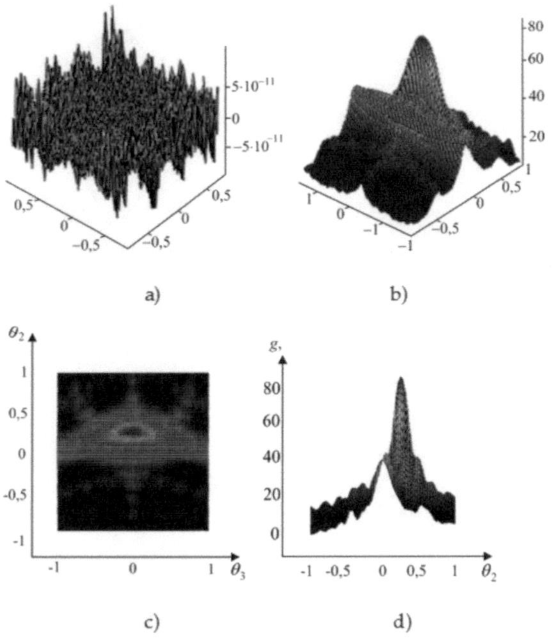

Figure 74. The fractal surface and the scattering indicatrix $g(\theta_2, \theta_3)$ when $\lambda = 30$ mm and $\theta_1 = 15°$: (a) fractal surface for $D = 2.5$; $N = M = 20$; $q = 2.7$; (b) $g(\theta_2, \theta_3)$; (c) $g(\theta_2, \theta_3)$, top view; (d) $g(\theta_2, \theta_3)$, side view.

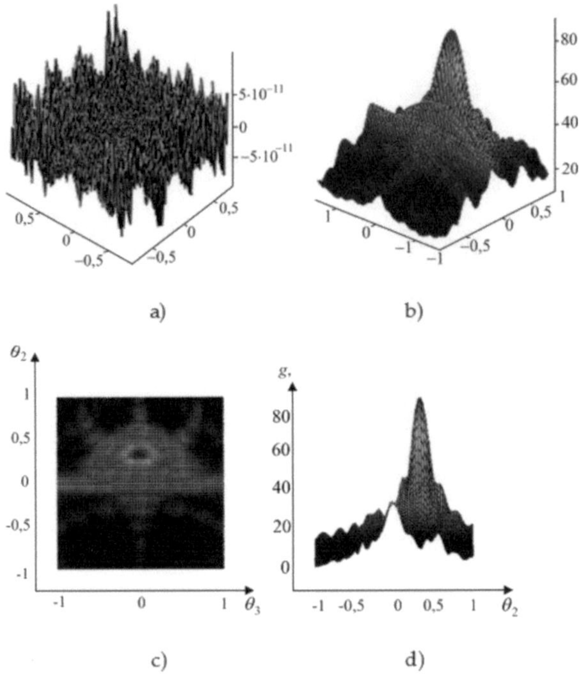

Figure 75. The fractal surface and the scattering indicatrix $g(\theta_2, \theta_3)$ when $\lambda = 30$ mm and $\theta_1 = 20°$: (a) fractal surface for $D = 2.5$; $N = M = 20$; $q = 2.7$; (b) $g(\theta_2, \theta_3)$; (c) $g(\theta_2, \theta_3)$, top view; (d) $g(\theta_2, \theta_3)$, side view.

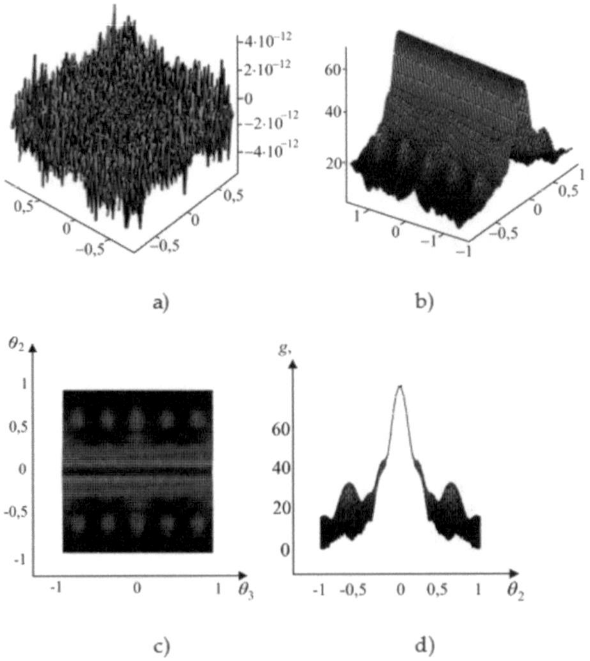

Figure 76. The fractal surface and the scattering indicatrix $g(\theta_2, \theta_3)$ when $\lambda = 30$ mm and $\theta_1 = 0°$: (a) fractal surface for $D = 2.8$; $N = M = 10$; $q = 2.7$; (b) $g(\theta_2, \theta_3)$; (c) $g(\theta_2, \theta_3)$, top view; (d) $g(\theta_2, \theta_3)$, side view.

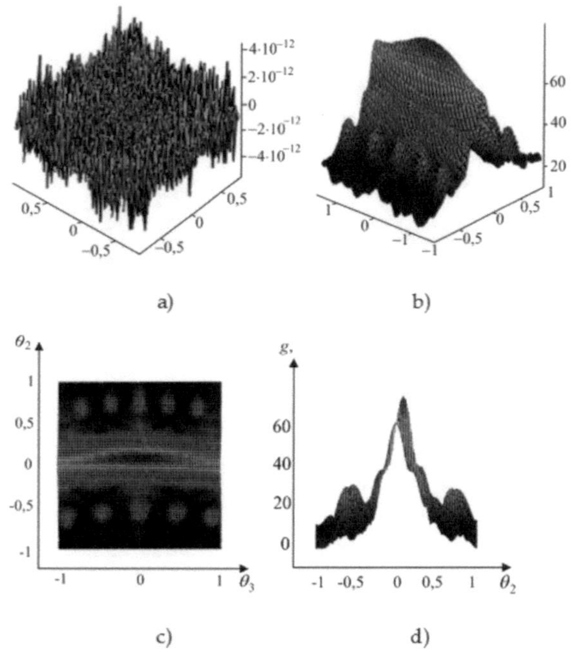

Figure 77. The fractal surface and the scattering indicatrix $g(\theta_2, \theta_3)$ when $\lambda = 30$ mm and $\theta_1 = 5°$: (a) fractal surface for $D = 2.8$; $N = M = 10$; $q = 2.7$; (b) $g(\theta_2, \theta_3)$; (c) $g(\theta_2, \theta_3)$, top view; (d) $g(\theta_2, \theta_3)$, side view.

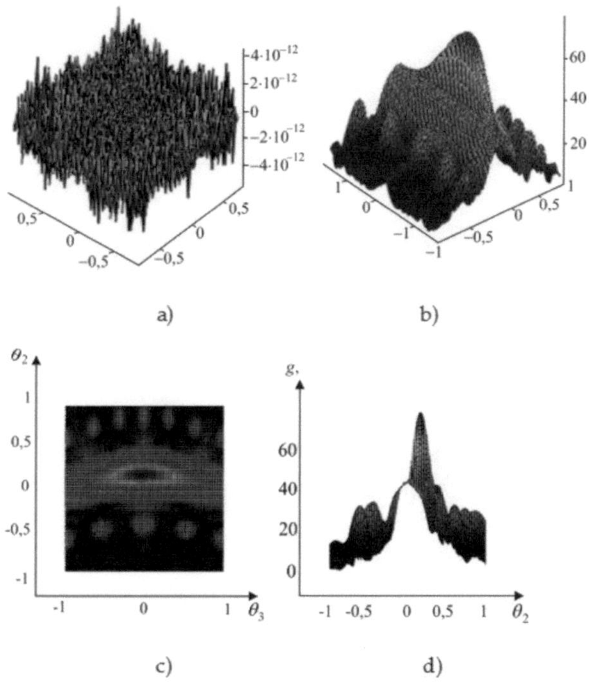

Figure 78. The fractal surface and the scattering indicatrix $g(\theta_2, \theta_3)$ when $\lambda = 30$ mm and $\theta_1 = 10°$: (a) fractal surface for $D = 2.8$; $N = M = 10$; $q = 2.7$; (b) $g(\theta_2, \theta_3)$; (c) $g(\theta_2, \theta_3)$, top view; (d) $g(\theta_2, \theta_3)$, side view.

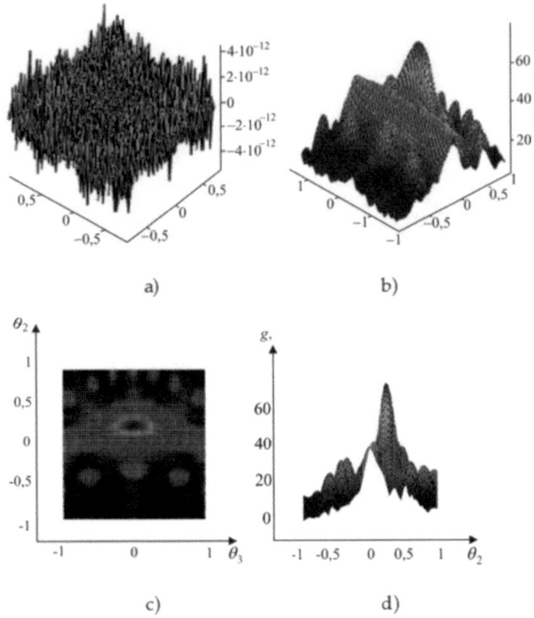

Figure 79. The fractal surface and the scattering indicatrix $g(\theta_2, \theta_3)$ when $\lambda = 30$ mm and $\theta_1 = 15°$: (a) fractal surface for $D = 2.8$; $N = M = 10$; $q = 2.7$; (b) $g(\theta_2, \theta_3)$; (c) $g(\theta_2, \theta_3)$, top view; (d) $g(\theta_2, \theta_3)$, side view.

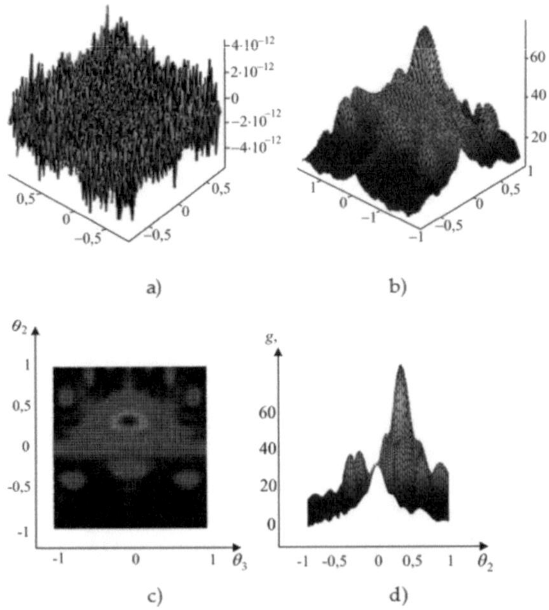

Figure 80. The fractal surface and the scattering indicatrix $g(\theta_2, \theta_3)$ when $\lambda = 30$ mm and $\theta_1 = 20°$: (a) fractal surface for $D = 2.8$; $N = M = 10$; $q = 2.7$; (b) $g(\theta_2, \theta_3)$; (c) $g(\theta_2, \theta_3)$, top view; (d) $g(\theta_2, \theta_3)$, side view.

were calculated in the summer of 2006 for the wavelengths $\lambda = 2.2$ mm, $\lambda = 8.6$ mm and $\lambda = 3.0$ cm for the different values of fractal dimension D and different scattering geometry, respectively. It is significant to note that in this work there is only part of all of our theoretical results obtained for these courses. Some of results for this course that relates to "Fractal Electrodynamics" (this conception appeared for the first time in the USA in the monographs [66, 67]; see also native monographs [6, 7]) were published by us earlier in works [8, 15, 40].

8. Conclusion

Now on the basis of large scattering characteristics data array, we can arrive at some significant conclusions. When D has small value, the main part of energy is scattered in the specular direction. Side lobes appear due to Bragg scattering. The number of side lobes and their intensity increases with an increased value of fractal dimension D of the dispersive surface. Angular range of the side lobes also increases with an increase of D when higher spatial frequencies become significant. Radio wave that interacts with a fractal can be viewed as a yardstick to probe rough surfaces by means of spatial frequencies selection on the basis of Bragg diffraction conditions [6, 7]. In the case of small D values, classical and fractal approaches for scattered field solution coincide with each other. In practice, sizes of illuminated area must be at least two times greater than the main period of a surface structure in order to obtain fractal parameters information from scattering patterns.

Undoubtedly, fractal describing of the wave-scattering process [5–10, 15, 63, 72, 73, 77] will result in establishing new physical laws in the wave theory. Author is sure that the use of *fractal theory* and *determined chaos* jointly with *formalism of the apparatus of fractional operators* in the just considered problems allows to generate more valid radio physical and radar models that decrease significantly discrepancies between theory and measurements.

This work reviews in detail a variety of modern wave-scattering problems that appear in theoretical and applied areas of radio physics and radiolocation when the theory of *integer and fractional measuring* is used in general case. In other words, the use of *dissipative system dynamics formalism* (fractality, fractional operators, non-Gaussian statistics, distributions with heavy tales, mode of determined chaos, existing of strange attractors in the phase space of reflected signals, their topology, etc.) allows us to expect that classical problem of wave scattering by random mediums will be area of productive investigations in the future as before.

All results presented here are the priority ones in the world, and it is a basis material for the further development and foundation of practical application of fractal approaches in radio location, electronics, and radio physics and also for generating fundamentally new fractal elements/devices and fractal radio systems [5–48, 62, 63, 68–83]. These results can be applied widely for fractal antennas modeling, fractal frequency-selective structures modeling, solid-state physics, physics of nanostructures, and for the synthesis of nano-materials.

Acknowledgements

This work was supported in part by the project of International Science and Technology Center No. 0847.2 (2000–2005, USA), Russian Foundation for Basic Research (projects No. 05-07-90349, 07-07-07005, 07-07-12054, 07-08-00637, 11-07-00203) and also was supported in part by the project "Leading Talents of Guangdong Province," No. 00201502 (2016–2020) in the JiNan University (China, Guangzhou).

Author details

Alexander A. Potapov

Address all correspondence to: potapov@cplire.ru

1 Kotel'nikov Institute of Radio Engineering and Electronics of Russian Academy of Sciences, Moscow, Russia

2 JNU-IRE RAS Joint Laboratory of Information Technology and Fractal Processing of Signals, JiNan University, Guangzhou, China

References

[1] Bass FG, Fuks IM. Wave Scattering from Statistically Rough Surfaces. Oxford: Pergamon Press; 1979. p. 532

[2] Rytov SM, Kravtsov YA, Tatarskii VI. Introduction to Statistical Radiophysics. In: Rytov SM, editor. Moscow: Nauka; 1978. p. 464

[3] Ishimaru A. Wave Propagation and Scattering in Random Media. New York: Wiley-IEEE Press; 1999. p. 600

[4] Oleschko K, Korvin G, Balankin AS, Khachaturov RV, Flores L, Figueroa B, Urrutia J, Brambila F. Fractal scattering of microwaves from soils. Physical Review Letters. 2002;**89** (18):188501-1–188501-4

[5] Potapov AA. Earth Covering Image Synthesis in Optic and Millimeter Wave Bands [dissertation]. Moscow: IREE RAS; 1994. p. 436

[6] Potapov AA. Fractals in Radio Physics and Radar. Moscow: Logos; 2002. p. 664

[7] Potapov AA. Fractals in Radio Physics and Radar: Topology of Sample. Moscow: University Library; 2005. p. 848

[8] Potapov AA. Fractals and chaos as basis of new challenging technologies in modern radio systems. In: Cronover R, editor. Introduction to Fractal and Chaos. Moscow: Technosphere; 2006. pp. 374-479

[9] Podosenov SA, Potapov AA, Sokolov AA. Pulse Electrodynamics of Wide-Band Radio Systems. In: Potapov AA, editor. Moscow: Radiotekhnika; 2003. p. 720

[10] Bunkin BV, Reutov AP, Potapov AA, et al. Aspects of Prospective Radar. Moscow: Radiotekhnika; 2003. p. 512

[11] Bistrov RP, Potapov AA, Sokolov AV. Millimeter Band Radars with Fractal Processing. Moscow: Radiotekhnika; 2005. p. 512

[12] Nonlinear Radar: Collection of Articles (Library of Journal "Nonlinear World": Scientific Series "Fractals. Chaos. Probability"). Gorbachev AA, Koldanov AP, Potapov AA, Chigin EP, editors. Moscow: Radiotekhnika; 2005. p. 96

[13] Nonlinear Radar: Collection of Articles (Library of Journal "Nonlinear World": Scientific Series "Fractals. Chaos. Probability"). Gorbachev AA, Koldanov AP, Potapov AA, Chigin EP, editors. Moscow: Radiotekhnika; 2006. p. 168

[14] Nonlinear Radar: Collection of Articles (Library of Journal "Nonlinear World": Scientific Series "Fractals. Chaos. Probability"). Gorbachev AA, Koldanov AP, Potapov AA, Chigin EP, editors. Moscow: Radiotekhnika; 2007. p. 128

[15] Potapov AA. The textures, fractal, scaling effects and fractional operators as a basis of new methods of information processing and fractal radio systems designing. Proceedings of SPIE. 2009;**73740**:73740E-1–73740E-14

[16] Potapov AA, German VA. Detection of artificial objects with fractal signatures. Pattern Recognition and Image Analysis. 1998;**8**(2):226-229

[17] Potapov AA, German VA. Processing of optic and radar images of the earth surface by fractal methods. Journal of Communications Technology and Electronics. 2000;**45**(8):853-860

[18] Potapov AA. Fractals in radio physics and radio location: Fractal analysis of signals. Journal of Communications Technology and Electronics. 2001;**46**(3):237-246

[19] Potapov AA, German VA. Effects of deterministic chaos and strange attractor in the radar of dynamics systems of the vegetative cover type. Technical Physics Letters. 2002;**28** (7):586-588

[20] Potapov AA. Fractals in radiophysics and radar: Fundamental theory of wave scattering by a fractal surface. Journal of Communications Technology and Electronics. 2002;**47** (5):461-487

[21] Potapov AA. The fractal analysis in modern problems of radars and radiophysics. Radiotekhnika (Moscow). 2003;(8):55-66

[22] Potapov AA. New information technology in radar detection of low-contrast targets based on probabilistic texture and fractal features. Journal of Communications Technology and Electronics. 2003;**48**(9):1012-1029

[23] Reutov AP, Potapov AA, German VA. Strange attractors and fractals as a basis of new dynamics model of radar signals, scattering by vegetative cover. Nonlinear World (Moscow). 2003;**1**(1-2):12-27

[24] Potapov AA. Topology of sample. Nonlinear World (Moscow). 2004;**2**(1):4-13

[25] Potapov AA, German VA. Methods of measuring the fractal dimension and fractal signatures of a multidimensional stochastic signal. Journal of Communications Technology and Electronics. 2004;**49**(12):1370-1391

[26] Gorbachev AA, Potapov AA, Tarakankov SP. Distant diagnostic of dynamical systems based on electromagnetic waves nonlinear scattering. Nonlinear World (Moscow). 2004;**2** (5-6):310-314

[27] Potapov AA, Bulavkin VV, German VA, Vyacheslavova OF. The researching of micro structure surfaces with methods of fractal signatures. Technical Physics. 2005;**50**:560-575

[28] Bulavkin VV, Potapov AA, German VA, Vyacheslavova OF. The theory of fractals in problem of estimation surfaces. Russian Journal of Heavy Machinery. 2005;(6):19-25

[29] Potapov AA. The appliance of synergetic principles of nonlinear dynamic and fractals to development of the new information technologies for modern radio systems. Radiotekhnika (Moscow). 2005;(8):73-88

[30] Gulyaev YV, Nikitov SA, Potapov AA, Davydov AG. Design of fractal radio systems: Numerical analysis of electromagnetic properties of the Sierpinski fractal antenna. Journal of Communications Technology and Electronics. 2005;**50**(9):988-993

[31] Potapov AA, Il'yin EM, Chigin EP, German VA. Development and structure of the first etalon dictionary of fractal properties of target classes. Electromagnetic Phenomena. 2005;**5**(2(15)):107-142

[32] Opalenov YuV, Potapov AA. Fractal radar radonometry: Endomorphism of the radon's conception. Electromagnetic Phenomena. 2005;**5**(2(15)):143-159

[33] Kolesov VV, Potapov AA. Information technologies on dynamic chaos for telecommunication, radar and navigation systems. Electromagnetic Phenomena. 2005;**5**(2(15)):91-106

[34] Potapov AA, Potapov AA Jr, Potapov VA. The fractal capacitor, fractal operators, and fractal impedances. Nelineinyi Mir. 2006;**4**(4-5):172-187

[35] German VA, Potapov AA. Example of use of a hypothesis of self – similarity in a problem of detection of an acoustic signal. Nonlinear World (Moscow). 2006;**4**(4-5):204-207

[36] Potapov AA, German VA. The methods of fractal digital processing of weak signal and low contrast images. Avtometriya (Novosibirsk, Russia). 2006;**42**(5):3-25

[37] Potapov AA, German VA. The fractal non – parametric detector of radio signals. Radiotekhnika (Moscow). 2006;(5):30-36

[38] Gulyaev YV, Nikitov SA, Potapov AA, German VA. Concepts of scaling and fractal dimension in the design of a fractal detector of radio signals. Journal of Communications Technology and Electronics. 2006;**51**(8):909-916

[39] Potapov AA, Laktyunkin AV. Microwaves scattering on fractal surfaces as a new line of investigations. In: Proceedings of the Second European Conference on Antennas and

Propagation EuCAP 2007; 11-16 November 2007; The EICC, Edinburgh, UK. Edinburgh: The Institution of Engineering and Technology & EurAAP AISBL; 2007. p. MoPP.016. pdf. 6 pp

[40] Potapov AA. The theory of functionals of stochastic backscattered fields. Journal of Communications Technology and Electronics. 2007;**52**(3):245-292

[41] Laktyunkin AV, Potapov AA. Waves scattering dependence on the statistical parameters of classical and fractal rough surfaces. In: Proceedings of XXIX URSI General Assembly; 7-16 August 2008; USA, Chicago, Illinois. Chicago: University of Illinois at Chicago; 2008. p. BP16.1(228). pdf. 4 pp

[42] Potapov AA, Laktyunkin AV. Theory of the wave scattering by anisotropic fractal surface. Nonlinear World (Moscow). 2008;**6**(1):3-36

[43] Laktyunkin AV, Potapov AA. Radio waves scattering dependence on the statistical parameters of classical and fractal rough surfaces. In: Program 3rd European Conference on Antennas and Propagation EuCAP 2009; 23-27 March 2009; Berlin, Germany. Berlin: EurAAP; 2009. p. 24

[44] Laktyunkin AV, Potapov AA. Frequency and spatial features of waves scattering on fractals. In: Book of Abstracts 2nd International Conference (CHAOS' 2009) on Chaotic Modeling, Simulation and Applications; 01-05 June 2009; Chania, Crete, Greece. Chania: National and Kapodistrian University; 2009. p. 40

[45] Potapov AA. Can we build an adaptive fractal radio system? In: Proceedings of PIERS 2009 in Moscow "Progress in Electromagnetics Research Symposium"; 18-21 August, 2009; Moscow, Russia. Cambridge, MA: Electromagnetics Academy; 2009. pp. 1798-1802

[46] German VA, Potapov AA, Sykhonin EV. Fractal characteristics of radio thermal radiation of a different layer of atmosphere in a range of millimeter waves. In: Fractal Characteristics of Radio Thermal Radiation of a Different Layer of Atmosphere in a Range of Millimeter Waves; 18-21 August, 2009; Moscow, Russia. Cambridge, MA: Electromagnetics Academy; 2009. pp. 1813-1817

[47] Laktyunkin AV, Potapov AA. On frequency and spatial correlation of waves scattered by fractal surface. In: Program PIERS 2009 in Moscow "Progress in Electromagnetics Research Symposium"; 18-21 August, 2009; Moscow, Russia. Cambridge, MA: Electromagnetics Academy; 2009. p. 59

[48] Potapov AA. Fractals, scaling and fractal operators for radar problems: Fractal radio systems designing. In: Rohling H, editor. Proceedings of International Radar Symposium (IRS - 2009); 09-11 September, 2009; Hamburg, Germany. Hamburg: Technical University Hamburg-Harburg and German Institute of Navigation (TUHH and DGON); 2009. pp. 667-670

[49] Dyson F, Montroll E, Kas M, Fisher M. Stability and Phase Transitions. Moscow: Mir; 1973. p. 376

[50] Berry MV. Diffractals. Journal of Physics A. 1979;**12**(6):781-797

[51] Berry MV, Blackwell ZV. Diffractal echoes. Journal of Physics A. 1981;**14**(11):3101-3110

[52] Jaggard DL, Kim Y. Diffraction by band-limited fractal screens. Journal of Optical Society of America A. 1987;**4**:1055-1062

[53] Jaggard DL, Sun X. Scattering from fractally corrugated surfaces. Journal of Optical Society of America A. 1990;**7**:1131-1139

[54] Jakeman E. Fraunhofer scattering by a sub-fractal diffuser. Optica Acta. 1983;**30**:1207-1212

[55] Bertoni HL, Felsen LB, editors. Directions in Electromagnetic Wave Modeling. New York: Plenum Press; 1991. p. 560

[56] Jordan L, Hollins RC, Jakeman E. Experimental measurements of non-Gaussian scattering by a fractal diffuser. Applied Physics B. 1983;**31**:179-186

[57] Ball R, Blunt M. A fractal model for the impedance of a rough surface. Journal of Physics A. 1988;**21**:197-204

[58] Bourrely C, Chiappetta P, Torresani B. Light scattering by particles of arbitrary shape: A fractal approach. Journal of Optical Society of America A. 1986;**3**(2):250-255

[59] Sinha SK. Scattering from fractal structures. Physica D. 1989;**38**:310-314

[60] Jaggard DL, Sun X. Fractal surface scattering: A generalized Rayleigh solution. Journal of Applied Physics. 1990;**68**:5456-5462

[61] Lin N, Lee HP, Lim SP, Lee KS. Wave scattering from fractal surfaces. Journal of Modern Optics. 1995;**42**(1):225-241

[62] Potapov AA, Laktyun'kin AV. Frequency coherence function of a space-time radar channel forming images of an anisotropic fractal surface and fractal objects. Journal of Communications Technology and Electronics. 2015;**60**(9):962-969

[63] Alexander AP. Chaos theory, fractals and scaling in the radar: A look from 2015. In: Skiadas C, editor. The Foundations of Chaos Revisited: From Poincaré to Recent Advancements. Switzerland, Basel: Springer International Publishing; 2016. pp. 195-218

[64] Beckmann P, Spizzichino A. The Scattering of Electromagnetic Waves from Rough Surfaces. New York: Pergamon Press; 1963. p. 503

[65] Ogilvy JA. Theory of Wave Scattering from Random Rough Surfaces. Bristol: Adam Hilger; 1991. p. 292

[66] Jaggard DL. On Fractal Electrodynamics. In: Kritikos HN, Jaggard DL, editors. Recent Advances in Electromagnetic Theory. New York: Springer-Verlag; 1990. pp. 183-223

[67] Jaggard DL. Fractal electrodynamics: Wave interaction with discretely self – similar structures. In: Baum C, Kritikos HN, editors. Symmetry in Electrodynamics. London: Taylor & Francis; 1995. pp. 231-281

[68] Potapov AA. The global fractal method and the fractal paradigm in fundamental radar problems. In: Book of Abstracts International Conference on "Dynamics, Bifurcations and Strange Attractors" Dedicated to the Memory of L.P. Shil'nikov (1934–2011); July 1–5, 2013; Nizhni Novgorod, Russia. Nizhni Novgorod: Lobachevsky State University of Nizhni Novgorod; 2013. p. 98

[69] Potapov AA. The base of fractal antenna theory and applications: utilizing in electronic devices. In: Proceedings of the 2013 IX International Conference on Antenna Theory and Techniques; 16–20 September, 2013; Odessa, Ukraine. Odessa: Odessa National O.S. Popov Academy of Telecommunications; 2013. pp. 62-67

[70] Potapov AA. The global fractal method, fractal paradigm and the fractional derivatives method in fundamental radar problems and designing of revolutionary radio signals detectors. In: Zbornik Radova Konferencije MIT - Matematicke i informacione tehnologije; 5–14 September 2014; Vrnjackoj Banji od 5. do 9. septembra i u Becicima od 10. do 14. septembra 2013. godine. Kosovska Mitrovica: Prirodno-matematicki fakultet Ulverziteta u Pristini (Serbia); 2014. pp. 539-552

[71] Potapov Alexander A. Fractal radar: Towards 1980 – 2015. In: Proceedings of CHAOS 2015 International Conference; 26-29 May 2015; France. Paris: Henri Poincaré Institute; 2015. pp. 559-573

[72] Potapov AA, Rassadin AE. Feynman integrals, fractal paradigm and new point of view on hydroacoustics. Eurasian Physical Technical Journal. 2015;**12**(1(23)):3-13

[73] Podosenov SA, Potapov AA, Foukzon J, Menkova ER. Nonholonomic, fractal and linked structures in relativistic continuous media, electrodynamics, quantum mechanics and cosmology. In: Potapov AA, editor. 3 volumes. Moscow: Lenand; 2015. p. 1128

[74] Potapov AA. New conception of fractal radio device with fractal antennas and fractal detectors in the MIMO systems. In: Book of Abstracts 9th International Conference (CHAOS' 2016) on chaotic modeling, simulation and applications; 23-26 May 2016; Senate House, University of London, London, UK. London: University of London; 2016. p. 85

[75] Potapov AA, Laktyun'kin AV. Frequency coherence function of a radar channel forming images of a fractal surface and fractal objects. In: Book of Abstracts 9th International Conference (CHAOS' 2016) on Chaotic Modeling, Simulation and Applications; 23-26 May 2016; Senate House, University of London, London, UK. London: University of London; 2016. pp. 85-86

[76] Potapov AA, Potapov VA. Scaling of the fractals antennas. In: Book of Abstracts 9th International Conference (CHAOS' 2016) on Chaotic Modeling, Simulation and Applications; 23–26 May 2016; Senate House, University of London, London, UK. London: University of London; 2016. pp. 86-87.

[77] Potapov AA. The fractal - scaling radiolocation: Formation history 1980–2015. Chaotic Modeling and Simulation (CMSIM). 2016;(3):317-331

[78] Andrianov MN, Potapov AA. Increase efficiency of system of data transmission in fading channels using MIMO Systems. In: Shunjun W, editor. Proceedings of the 2016 CIE International Conference on Radar "Radar 2016"; October 10-12, 2016; China, Guangzhou. Beijing: Chinese Institute of Electronics (CIE); 2016. pp. 119-123

[79] Potapov AA, Zhang W. Simulation of new ultra-wide band fractal antennas based on fractal labyrinths. In: Shunjun W, editor. Proceedings of 2016 CIE International Conference on Radar "Radar 2016"; October 10-12, 2016; China, Guangzhou. Beijing: Chinese Institute of Electronics (CIE); 2016. pp. 319-323

[80] Potapov AA, German VA, Pahomov AA. Processing of images obtained from unmanned aerial vehicles in the regime of flight over inhomogeneous terrain with fractal-scaling and integral method. In: Shunjun W, editor. Proceedings of 2016 CIE International Conference on Radar "Radar 2016"; October 10-12, 2016; China, Guangzhou. Beijing: Chinese Institute of Electronics (CIE); 2016. pp. 585-587

[81] Potapov AA, Lazko FF. Gradients distribution matrices of and lacunarity in the capacity of texture measure of SAR and UAVs images. In: Shunjun W, editor. Proceedings of 2016 CIE International Conference on Radar "Radar 2016"; October 10-12, 2016; China, Guangzhou. Beijing: Chinese Institute of Electronics (CIE); 2016. pp. 605-608

[82] Potapov AA, German VA. A local dispersion method of measuring a fractal dimension and fractal signatures. In: Shunjun W, editor. Proceedings of 2016 CIE International Conference on Radar "Radar 2016"; October 10–12, 2016; China, Guangzhou. Beijing: Chinese Institute of Electronics (CIE); 2016. pp. 799-803

[83] Potapov AA. Strategic directions in synthesis of new topological radar detectors of low-contrast targets against the background of high-intensity noise from the ground, sea and precipitations. In: Shunjun W, editor. Proceedings of 2016 CIE International Conference on Radar "Radar 2016"; October 10-12, 2016; China, Guangzhou. Beijing: Chinese Institute of Electronics (CIE); 2016. pp. 692-696

Chapter 10

Fractal Geometry and Porosity

Oluranti Agboola, Maurice Steven Onyango,
Patricia Popoola and Opeyemi Alice Oyewo

Additional information is available at the end of the chapter

http://dx.doi.org/10.5772/intechopen.68201

Abstract

A fractal is an object or a structure that is self-similar in all length scales. Fractal geometry is an excellent mathematical tool used in the study of irregular geometric objects. The concept of the fractal dimension, *D*, as a measure of complexity is defined. The concept of fractal geometry is closely linked to scale invariance, and it provides a framework for the analysis of natural phenomena in various scientific and engineering domains. The relevance of the power law scaling relationships is discussed. Fractal characteristics of porous media and the characteristic method of the porous media are also discussed. Different methods of analysis on the permeability of porous media are discussed in this chapter.

Keywords: fractal geometry, fractal structure, fractal dimension, porous media, permeability

1. Introduction

Fractal is one of the subjects, which recently attracted attention in natural science and social science. A fractal is defined as a geometric object whose fractal dimension is larger than its topological dimension. Many fractals also have a property of self-similarity; within the fractal lies another copy of the same fractal, smaller but complete. Mandelbrot [1] referred to fractals as structures consisting of parts that, in some sense, are similar to integers; fractals are of a fine (non-integer) dimension (*D*) that is always smaller than the topological dimension. In the past 40 decades, fractal theory has significantly contributed to the characterization of the distribution of physical or other quantities with a geometric support. In science and engineering, fractal geometry provides a wide range and powerful theoretical framework that is used to describe complex systems, which have been successfully applied to the quantitative description of microstructures such as surface roughness and amorphous metal structure [2].

Typically, microstructure elements can be explained using the Euclidean dimension (d). With respect to point defects (e.g., vacancies and interstitial atoms), $d = 0$; with respect to linear defects (dislocations), $d = 1$; with respect to planar defects (twins), $d = 2$; and with respect to three-dimensional (3-D) formations, $d = 3$. Nonetheless, the Euclidean dimension cannot be used to illustrate structural elements differing from standard ones (e.g., points or straight lines). Thus, a well-known grain boundary, being the most significant element of the microstructure, is curvilinear, and this form can be described by the fractal dimension (D) correlating to $1 \leq D \leq 2$. Surface defects may also be illustrated using the fine dimension that will commensurate to the range $2 \leq D \leq 3$ [3]. Therefore, fractal theory introduces a new quantitative parameter-fractal dimension for illustrating structures, which, because of its universal nature, is appropriate for illustrating structures in systems types. With a system such as a deformed solid, the fractal concept provides the possibility of quantitatively illustrating the elements of the initial microstructure (e.g., phases, grain, boundaries, etc.) and the structures formed during deformation [3]. Fractal theory thus provides a new and effective method for characterizing complex structure of the engineering materials. The theory of fractals is considered a basis for quantitative description by means of the fractal dimension of various structures.

An extremely disordered morphology, such as surface roughness and porous media having the self-similarity property, is scrutinized by fractal geometry. This implies that the morphology stays similar in magnification over a broad range. Another significant attribute of natural fracture is that their formation needs the supply of a large amount of energy externally [3]. If microstructure formation is preferentially caused by a phenomena taking place outside of thermodynamic equilibrium, they are also characterized by fractal property. This implies that the description of highly disordered microstructures on the basis of conventional approach is not sufficient [3]. Thus, most of the objects that occur in nature are disordered and irregular, and they do not follow the Euclidean illustration because of the scale-dependent measures of length, area, and volume [4, 5]. Examples of such objects are the surfaces of mountains, coastlines, microstructure of metals, and so on. These objects are termed fractals and are illustrated by a non-integral dimension known as fractal dimension [1]. The fractal property is a physical property expressed at the super-molecular level, at a microscopic scale, and at a macroscopic scale.

The phenomenon of fractal is ubiquitous in a wide array of materials, such as the fracture of nanoparticle composites [6–8], the growth of crystal [9–12], the quasicrystal structure [13], the fracture of martensite morphology [14, 15], the porous materials [16–19], and the deposited film [20–25]. These materials are of uncommon class of disordered materials and usually show complex microstructures. Fractal theory has been widely used in many fields of modern science since it was presented by Mandelbrot [1] in 1982. It has been used in studying permeability of porous media [17, 26–28], dual-porosity medium [29, 30], evaluating dislocation structure [31], simulating the failure of concrete [32–35], analyzing fracture surfaces or network [36, 37], and thermal conductivity performance [38]. Fractal has also been used to establish the morphology of highly irregular objects imbedded onto two- and three-dimensional spaces and is defined as two- and three-dimensional fractal dimensions [39].

2. Fractal structure

Fractal structure is a structure that is characterized with self-similarity, that is, it is composed of such fragments whose structural motif is repeated if the scale changes. Fractal structure outlined the degree of occupancy of a structure in a space (dimension), which is not an integer value. Therefore, n-dimensional fractal occupies an intermediate position that lies between the n-dimensional and $(n + 1)$-dimensional objects. Recursive functions are used to construct a fractal object. An important characteristic of fractal structure is the scale independence [40]. Thus, fractal structures do not have a single length scale, and fractal processes (e.g., time series) cannot be characterized by a single time scale [41]. Fractal structures are associated to rough or fragmented geometric structures [42]. The complexity of a fractal structure is described by its fractal dimension; this is greater than the topological dimension. It is much easier to obtain fractal dimension from datasets by using fractal analysis, for example, digital images, obtained from the investigation of natural phenomena, and from theoretical models. Different techniques to perform fractal analysis include box-counting, lacunarity analysis, multifractal analysis, and mass methods. An interesting application of fractal analysis is the description of fractured surfaces [43]. Mandelbrot et al. [42] have shown that fractured surfaces are fractal. Zhang [44] reported a quadratic polynomial relationship between the rock burst tendency and fractal dimension of fracture surface. A fractal dimension threshold of $\hat{d}_f$ was found, and there was a positive correlativity between the rock burst tendency index and the fractal dimension when $d_f \leq \hat{d}_f$, an inverse correlativity when $d_f \leq \hat{d}_f$. In the investigation of fractured surfaces, Liang and Wu investigated the relationship between the fracture surface fractal dimension and the impact strength of polypropylene nanocomposites. A strong correlation was observed, and it indicated that the fracture surface of the composites was fractal, and the relationship between the impact strength and fractal dimension of the composites obeyed roughly exponential function [7]. Lung et al. have also demonstrated that there is a relation between the roughness and the fractal dimension of the surface [45].

3. Fractal analysis

Fractal analysis is defined as a contemporary method of applying non-traditional mathematics to patterns that defy understanding with traditional Euclidean concepts. It means assessing the fractal description of data, and it is a common technique to study a variety of problems. It consists of different methods assigned to a fractal dimension and other fractal characteristics to a dataset. It, in essence, measures complexity using fractal dimension. In fractal analysis, other different parameters can also be assessed [43], for instance, lacunarity and succolarity, and can be used to classify and segment images [46]. Whatever type of fractal analysis has to be done, it always rests on some kind of fractal dimension. In fractal analysis, complexity is a change in detail with change in scale. The simplest form of fractal dimension is described using the relation in Eq. (1).

$$N = \frac{1}{S^D} \tag{1}$$

where N is the number of self-similar "pieces," S is the linear scaling factor (sizes) of the pieces to the whole, D is the dimension that characterizes the (invariant) relationship between size and number. Rearranging the elements in Eq. (1), one can solve for D.

$$D = -\left(\frac{\log N}{\log S}\right) \tag{2}$$

D is an algebraic equation, that is, Eq. (1) can give a dimension, which is the concept of geometry, not algebra. Let's say, one-dimensional line is cut into pieces, each of which is a fraction (S) of the original line, like making $S = 1/4$. For example, one-dimensional line can be cut into pieces such that each one-fourth will be the size of the original line, then N will be equal to four little lines. Then one can say that Eq. (1) gives the line a fractal dimension $D = 1$, because $N = 1/\left(\frac{1}{4}\right)^1$. If a two-dimensional square area is cut into pieces, the side of which is one-fourth the size of the original square, then $N = 16$ little squares. Eq. (1) will then tell that the area of the square has a fractal dimension $D = 2$, because $N = 1/\left(\frac{1}{4}\right)^2$. If a three-dimensional cube volume is cut into pieces, such that the side of which is one-fourth the size of the original cube, then N will be equal to 64 little cubes. Eq. (1) then tells that the volume of the cube has a fractal dimension of $D = 3$, because $N = 1/\left(\frac{1}{4}\right)^3$. No matter the value of S, N will still be found as $1/S^D$pieces when one-, two-, and three-dimensional objects have been cut into pieces. Thus, Eq. (1) gives the correct fractal dimension for one-dimensional line, two-dimensional area, and three-dimensional volume [47].

3.1. Concepts of the fractal dimension

The ratio that gives statistical index of intricacy and compares how detailed a shape (fractal pattern) changes with the scale at which it is measured is called fractal dimension. It is sometimes identified by a measure of the space filling volume of a pattern that states how a fractal scale is different from the space it is rooted in; a fractal dimension is not always an integer [48–50]. There are several different concepts of the fractal dimension of a geometrical configuration [5].

There are several ways of measuring length-related fractal dimensions. Mandelbrot [51] first proposed the concept of a fractal dimension to describe structures, which look the same at all length scales. His concept takes into account the measuring of the perimeter of an object with several lengths of rulers (spans or calipers) (using a trace method). For a fractal object, the plot of the log of the perimeter against the log of the ruler lengths will give a straight line with a negative slope S. This plot will then result to $D = 1 - S$ [52]. Although this is mainly mathematical concept, many examples in nature that can be closely approximated to fractal objects are available for only over a particular range of scale. The likes of these objects are generally named self-similar in order to indicate their scale invariant structure. The common attribute of such objects is that their length (for a curve object, otherwise it could be the area or volume) mainly rests on the length scale used for measuring it, and the fractal dimension provides the exact nature of this reliance [53]. Fractal dimensions (D) are numbers used to quantify these properties [5]. In fractal geometry [1], the fractal dimension, D, is given as:

$$D = \lim_{r \to 0} \frac{\log(N_r)}{\log(1/r)} \tag{3}$$

This is a statistical quantity that shows how a fractal totally fills the space when viewed at finer scales.

The second concept was proven by Pentland on the basis that the image of a fractal object is also a fractal [54], which has made scientific investigations on the methods of estimating the fractal dimensions of images. Many researchers have put great efforts into this field of fractal geometry, and many methods for estimating fractal dimensions of certain objects have been proposed since the establishment of fractal geometry theory. Typical methods of this concept involve the use of spectral analysis and box-counting. Usually, spectral analysis method applies fast Fourier transformation (FFT) to image in order to obtain the coefficients and mean spectral energy density. The fractal dimension can be evaluated by analyzing the power law reliance of spectral energy density and the square size [55]. The box-counting method is the widely used method for calculating fractal dimensions in the natural sciences; this is called box-counting dimension. It is a method based on the concept of "covering" the border, it is also known as the grid method. Sets of square boxes (i.e., grids) are used here in order to cover the border. Each set is represented by a box size. The number of boxes essential to cover the border is considered a function of the box size. **Figure 1** is an example of the log of the number of covering boxes of each size times the length of a box edge plotted against the log of the length of a box edge. Furthermore, a straight line with slope S which is equal to the dimension D will be obtained [52]. The slope is defined as the amount of change along the Y-axis, divided by the amount of change along the X-axis. Any result with a steeper slope shows that the object is more "fractal," which means it gains in complexity as the box size reduces. Any result with a lower-valued slope shows that the object is closer to a straight line, less "fractal," and that the amount of detail does not grow as quickly with an increase in magnification. Again, the 3-D space containing a specific object, partitioned into boxes of a certain size and how many boxes could fill up the object, is also accounted for. With the use of ratio r in Eq. (1), in order to decide the box size, the box-counting method will account for the total number of boxes (i.e., Nr of Eq. (1)) that are needed to form the object. The fractal dimension D of Eq. (1) can then be estimated from the least square linear fit of log(Nr) versus log($1/r$) by counting Nr for different scaling ratio r [56].

Several traditional box-counting methods have been used for the calculation of the fractal dimensions of images, this includes differential box-counting (DBC) method [57], Chen et al.'s approach [58], the reticular cell counting method [59], Feng's method [60], and so on. DBC method has been proved to have better performance than other methods [61]. Many analyses have been done in order to improve the DBC method [62–64].

A third concept was developed by Flook [65], and the method of this concept is called the dilation method. Dilation, in this case, means a widening and smoothing of the border. This

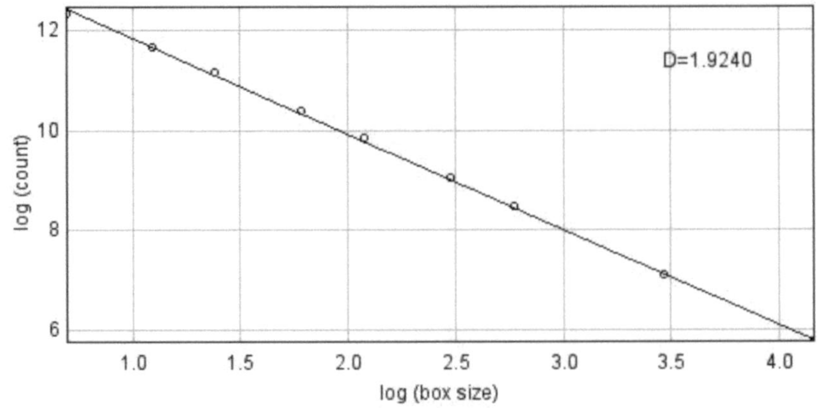

Figure 1. Example of fractal dimension of a material.

can be accomplished by convolution operation with a binary disk, that is, all the non-zero components of all the convolution kernels have a (Boolean) unitary value. The result is a thickened, but grey-scale border. All non-zero pixels are thresholded to a Boolean one when returning this border to Boolean one values. The speed at which the total surface area of the border raises as a function of the diameter of the convolution kernel relies on the dimension *D*. The log of each resulting area, divided by the kernel diameter, is plotted against the log of the kernel diameter [52]. A straight line results with a negative slope *S*, and *D* can be further estimated.

3.2. Power law scaling relationship

A functional relationship between two quantities is known as power law. This relationship takes place when a relative change in one quantity results in a proportional change in the other quantity and independent of the initial size of those quantities; thus, one quantity varies as power law to another. The characteristic of fractals is known as power law scaling. Therefore, a relationship, which yields a straight line on log-log coordinates, can often identify an object or phenomena as fractal. Although not all power law relationships are due to fractals, an observer needs to consider the existence of such relationship in order to know if the system is self-similar [66]. Self-similarities indicate the existence of scaling relationship which implies the type of a relationship called "power law." Thus, self-similarities give rise to the power law scaling. The power law scalings are shown as a straight line when the logarithm of the measurement is plotted against the logarithm of the scale at which it is measured. Fractal dimension is based on self-similarities; thus, power law scaling can be used to determine the fractal dimension. For a set to achieve the complexity and irregularity of a fractal, the number of self-similar pieces must be related to their size by power law [47]. The power law scaling describes how the property $L(r)$ of the system depends on the scale r at which it is measured using the relation in Eq. (4).

$$L(r) = Ar^{\alpha} \quad (4)$$

The fractal dimension describes how the number of pieces of a system depends on the scale r, using the relation in Eq. (5).

$$N(r) = Br^{-D} \tag{5}$$

where B is a constant. The similarity between Eqs. (4) and (5) means that one can determine the fractal dimension D from the scaling exponent α if one knows how the measure property $L(r)$ depends on the number of pieces $N(r)$. For example, for each little square of sides of an object with two-dimensional area, the surface area is proportional to r^2. Thus, one can determine the fractal dimension of the exterior of such an object by showing that the scaling relationship of the surface area depends on the scale r. For example, to determine the fractal dimension D from the scaling exponent is to derive the function of the dimension $f(D)$, such that the property measured is proportional to $r^{f(D)}$ [66]. If the experimentally determined scaling of the measured property is proportional to r^{α}, then the power of the scale r can be equated to the relation in Eq. (6):

$$f(D) = \alpha \tag{6}$$

Then, one can solve for D.

4. Fractal characteristics of porous media

Porous media include many manmade as well as natural materials. All solid substances are in fact porous either to some degree or at some length scale [67]. A porous medium is a randomly multi-connected medium with channels randomly obstructed. The quantity that measures how "holed" the medium is due to the presence of these channels, and it is called the porosity of the medium. A pore network description can represent the porous medium as an ensemble of pores and throats of different geometries and sizes that can take values from appropriate distributions [68]. Therefore, fractal theory gives a favorable layer of structures of different models that will address the complexity of the disordered, heterogeneous, and hierarchical porous media like soil, materials with fracture networks. Theoretically, Yu et al. [69] provided an overview of the physical properties of ideal fractal porous media and explained how natural heterogeneous materials can exhibit both mass- and pore-fractal scaling. Cihan et al. [70] reported new analytical models for predicting the saturated hydraulic conductivity based on the Menger sponge mass fractal. They tested their model predictions against lattice Boltzmann simulations of flow performed in different configurations of the Menger sponge.

Fractal models have been used to describe the solid volume, the pore volume, or the interface between the two phases of porous media. In the past three decades, fractal models of pore space were developed and used in the petroleum physics with application in hydraulic system and in engineering communities with the application electrical conductivity [71, 72]. Turcotte [73] proposed a fractal fragmentation model, which identified a physical basis for the existence of fractal soils in the scale invariance of the fragmentation of soil particles. Hence, his model elucidated a mechanism in which scale-independent fragmentation processes could form fractal distribution of particles, giving theoretical legitimacy to the study of fractal models on porous media. Fragmentation can be viewed as the chief mechanism of physical weathering [67].

4.1. Characteristic method of porous media

Hunt [67] stated that model characteristics are defined so that the porosity and water retention functions are identical to those of the discrete and explicit fractal model of Rieu and Sposito [74] (called hereafter the RS model). They began with a description of virtual pore size fractions in a porous medium that permits a facile foundation that conceptualized the fractal of solid matrix and pore space. These concepts resulted in equations used in solving the porosity and bulk density of both the size fractions and the porous medium in terms of a characteristic fractal dimension, *D*.

A porous medium with a porosity from a broad range of pore sizes was considered, the porosity decreases in mean (or median) diameter from $p_0 - p_{m-1} (m \geq 1)$. A bulk element of the porous medium has the volume V_0, massive enough to contain all sizes of pore; it has porosity ϕ and the dry bulk density σ_0. They divided the pore-volume distribution of V_0 mathematically into m virtual pore size fractions, with the ith virtual size fraction defined by:

$$P_i \equiv V_i - V_{i+1} \; (i = 0, \ldots, m - 1) \tag{7}$$

where P_i represents the volume of pores entirely made of size p_i contained in V_i which is the ith partial volume of the porous medium, which itself has all pores of size $\leq p_i$. The partial volume V_{i+1} is therefore incorporated in V_i and the partial volume V_{m-1} is then incorporated in the smallest pore-size fraction P_{m-1}, together with the residual solid volume symbolized by V_m. They stated that the solid material whose volume is V_m will not be chemically or mineralogically homogeneous. Its mass density, symbolized by σ_m, represents an average "primary particle" density. Eq. (7) gives the bulk volume of the porous medium which can mathematically be represented as the summation of m increments of the basic pore-size fraction $P_0 - P_{m-1}$ added to a residual solid volume V_m:

$$V_0 = \sum_{i=1}^{m-1} P_i + V_m \tag{8}$$

The porosity of the medium can then be given as [74]:

$$\phi \equiv \frac{(V_0 - N^m V_m)}{V_0} \tag{9}$$

$$= 1 - (1 - \Gamma)^m \tag{10}$$

Going by the fractal dimension of Eq. (3), proposed by Mandelbrot [1], the fractal dimension is related closely to the pore coefficient, Γ [74].

$$\Gamma = 1 - Nr^3 \tag{11}$$

which, with Eq. (3), results to:

http://dx.doi.org/10.5772/intechopen.68201

$$\Gamma = 1 - r^{3-D} \ (\Gamma < 1, \ r < 1) \tag{12}$$

It was shown from Eq. (12) that in a fractal porous medium where pore sizes are scaled by the same ratio $r < 1$, the fractal dimension increases with the decrease in the magnitude of the pore coefficient [74]. Thus, the relation between the porosity and the fractal dimension from Eqs. (10) and (12) gives:

$$\phi = 1 - (r^{3-D})^m \tag{13}$$

For a given value of the exponent m, the porosity of a fractal porous medium decreases as the fractal dimension increases. Further, Eq. (13) shows that the fractal dimension of a porous medium must be < 3.

Moreover, integration over the continuous pore size distribution between qr and r, where q is the ratio of radii of successive pore classes in fractal soil, r is the pore radius, $q < 1$ is an arbitrary factor, yields the contribution to the porosity from each size class obtained by RS model. The distribution of pore sizes is defined by the following probability density function [75]:

$$W(r) = \frac{3 - D_p}{r_m^{3-D_p}} r^{-1-D_p} r_0 \leq r \leq r_m \tag{14}$$

where r_0 and r_m refer explicitly to the minimum and maximum pore radii, respectively. The power law distribution of pore sizes is bounded by a maximum radius, r_m, and truncated at the minimum radius, r_0. Eq. (14), as written, is compatible with a volume, r^3, for a pore of radius r and D_p describes the pore space. The result for the total porosity derived from Eq. (14) is given in [75] as:

$$\phi = \frac{3 - D_p}{r_m^{3-D_p}} \int_{r_0}^{r_m} r^3 r^{-1-D_p} dr = 1 - \left(\frac{r_0}{r_m}\right)^{3-D_p} \tag{15}$$

Eq. (15) is exactly as in RS (Eq. (13)). If a particular geometry for the pore shape is envisioned, it is possible to change the normalization factor to maintain the result for the porosity and also maintain the correspondence to RS [67].

4.2. Fractal analysis on the permeability of porous media

The fluid flow through porous media is governed by geometrical properties, such as porosity, properties of the flowing fluid, the connection and the tortuosity of the pore space.

The transport phenomena in porous media, that include single-phase and multiphase fluid flow through porous media, electrical and acoustical transport in porous media, and heat transfer in porous media, are focused on common interests and have emerged as a separate field of study [76–79]. A matrix of a porous medium combined with fractured networks is called the dual-porosity medium. In the dual-porosity media, fracture and matrix are generally considered as different media, each with its own property. Thus, gas flow through these dual-porosity media could consist of two physically distinguished migration processes: one is associated to the movement of gas through the larger-scale fractures, that is, a permeability flow, which can be described by Darcy law, the other is related to the movement of gas inside matrix blocks, that is, diffusion processes, which may be involved in several different mechanisms, subject to the pore size [30, 74, 80].

In reality, surfaces of capillaries are rough and have great impact on fluid flow behavior and permeability of a porous medium. Analytically, permeability expression is a function of the relative roughness, the tortuosity fractal dimensions, capillaries sizes, and surface roughness, together with the microstructural parameters (such as the characteristic length, the maximum and minimum pore diameters, and the fractal dimensions) [19].

4.3. Methods of fractal analysis on the permeability of porous media

Fractal, multifractal, Gaussian, and log-normal models have been initiated, perhaps in all scale range. The validation of unchanging theoretical framework used in calculating transport properties, at least at some scales, has the capacity of eliminating much confusion regarding the appropriate theoretical approaches used and the appropriate model to choose [67]. Investigation on gas flow through a dual-porosity medium, for example, a flow domain made up of matrix blocks (with low permeability) implanted in a network of fractures, is not common. Physical and computer modeling are commonly used for permeability of porous media. Different methods of analysis on the permeability of porous media will be discussed in this section.

Zheng and Yu [30] studied the permeability of a gas with the use of matrix porous media embedded with randomly distributed fractal-like tree networks. The scientific expression for gas permeability in dual-porosity media was obtained based on the pore size of matrix and the mother channel diameter of embedded fractal-like tree networks having fractal distribution. It was discovered that gas permeability was a function of structural parameters, which includes the fractal dimensions for pore area and tortuous capillaries, porosity and the maximum diameter of matrix, the length ratio, the diameter ratio, the branching levels, and angle of the embedded networks used for dual-porosity media. The model that was initiated does not contain any empirical constant. The model predictions were validated with the available experimental data and simulating results, a fair agreement among them was found. An analysis of the influences of geometrical parameters on the gas permeability in the media was done.

Khlaifat et al. [80] experimentally studied a single-phase gas flow through fractured porous media of tight sand formation of Travis Peak Formation under different operation conditions. Their study enhanced gas recovery from low permeability reservoirs by the creation of a single fracture perpendicular to the flow direction. The porous medium sample that was taken into account was a slot-pore-type tight sand from the Travis Peak Formation with permeability in the microdarcy range and a porosity of 7%. A number of single-phase experiments that include water and gas were performed at different pressure drops conditions ranging from 100 to 600 psig and at overburden pressures of 2000, 3000, and 4000 psig, respectively. It was shown from their results that the sample used was very sensitive to overburden pressure. Again, it was shown from the experimental data that the presence of a fracture in a low permeability porous media is the main factor responsible for reinforcing the gas recovery from tight gas reservoirs. The presence of a fracture reinforces the gas flow, due to the increase in overall permeability and the creation of different flow patterns, which locally shifted the two-phase flow away from capillary force domination region. Furthermore, the fracture aperture played a significant role in enhancing flow due to both reconfigurations of connecting pores and joining of the non-connecting pores to the flow network.

A well-testing technique for Devonian shale gas reservoirs characterized by a low storage and high flow-capacity natural fractures fed by a high storage, low flow-capacity rock matrix was developed by Kucuk and Sawye [81] by using analytical methods and numerical simulator. They developed analytical solutions in order to analyze the basic fractured reservoir measurable factors that influence well productivity. These measurable factors are fracture system porosity and permeability, matrix porosity and permeability, and matrix size. They found that the traditional way of testing the well does not usually work for fractured Devonian shale gas reservoirs. Most of the time, the two parallel straight lines with a vertical separation are not shown in the semi-log plot of the drawdown and build-up data. They further found that the inter-porosity flow parameter is not the only parameter, which characterizes the nature of semi-log straight line.

A permeability model assumed to be comprised of a bundle of tortuous capillaries whose size distribution and roughness of surfaces follow the fractal scaling laws has been derived for porous media [19]. The proposed model includes the effects of the fractal dimensions for size distributions of capillaries, for tortuosity of tortuous capillaries, and for roughness of surfaces on the permeability. The proposed model is given by Eq. (16):

$$K_R = \frac{\pi L_0^{1-D_T} D_f \lambda_{\max}^{3+D_T}}{128A(3 + D_T - D_f)}(1-\varepsilon)^4 \tag{16}$$

where K_R denotes the permeability for flow in porous media with roughened surfaces. Eq. (16) indicates that the permeability is a function of the relative roughness ε, the fractal dimensions D_T (the fractal dimension for tortuosity of tortuous capillaries) and D_f (the fractal dimension for pore space), as well as the structural parameters A (cross-sectional area), L_0 (the representative length or straight line along the flow direction of a capillary), and $\lambda_{\max}$ (maximum capillary diameter). Eq. (16) also shows that the higher the relative roughness, the lower the permeability value; this can be explained by saying that the flow resistance is increased with the increase in roughness. This is consistent with a physical situation [19].

The proposed Eq. (16) was found to be a function of the relative roughness ε, the fractal dimension D_T for tortuosity of tortuous capillaries, and structural parameters A, L_0, and $\lambda_{\max}$. The ratio of the permeability for rough capillaries to that for smooth capillaries follows the quadruplicate power law of $(1 - \varepsilon)$ given by Eq. (17). That is, Eq. (17) indicated that the decrease of permeability for porous media with roughened surfaces in capillaries follows the quadruplicate power law of $(1 - \varepsilon)$. The authors concluded that the permeability of porous media with roughened capillaries will be drastically decreased with the increase in relative roughness, and the proposed model can reveal more mechanisms that affect the flow resistance in porous media than conventional models [19]. K in Eq. (17) is the permeability of porous media with smooth capillaries.

$$\eta = K_R/K = (1 - \varepsilon)^4 \tag{17}$$

Zinovik and Poulikakos [82] evaluated the relationships between porosity and permeability for a set of fractal models with porosity approaching unity and a finite permeability. Prefractal tube bundles generated by finite iterations of the corresponding geometric fractals can be used as a model porous medium where permeability-porosity relationships are derived analytically as explicit algebraic equations. Their investigation showed that the tube bundles generated by finite iterations of the corresponding geometric fractals can be used to model porous media where the permeability-porosity relationships are derived analytically. It was further shown that the model of prefractal tube bundles can be used to obtain fitting curves of the permeability of high porosity metal foams and to provide insight on permeability-porosity correlations of the capillary model of porous media.

All the methods discussed here have shown that the permeability of a porous media is strongly affected by its local geometry and connectivity, the matrix size of the material, and the pores available for flow. All the methods gave concept and knowledge of fractal geometry in relation to the characterization of the porous structure with respect to the permeability of the porous media.

5. Conclusion

Fractal is considered a self-similar system. It has been confirmed that the fractal technique is a powerful technique that has been successfully used in the characterization of the geometric and structural properties of fractal surfaces and pore structures of porous materials. It gives an understanding on how the geometry affects the physical and chemical properties of systems since their complex patterns are better described in terms of fractal geometry if the self-similarity is satisfied. It also builds a bridge between micro-morphology and macro performance. This chapter shows that the structural and functional characters of porous materials depend on the pore structure, which can be described effectively by the fractal theory.

Author details

Oluranti Agboola[1,2]*, Maurice Steven Onyango[2], Patricia Popoola[2] and Opeyemi Alice Oyewo[2]

*Address all correspondence to: funmi2406@gmail.com

1 Department of Chemical Engineering, Covenant University, Nigeria

2 Department of Chemical, Metallurgical, and Materials Engineering, Tshwane University of Technology, South Africa

References

[1] Mandelbrot BB. The fractal geometry of nature. 1st ed. W.H. Freeman and Company: San Francisco; 1982. 34p. ISBN-13: 978-0716711865

[2] Li A, Geng H, Zhou Y, Ding L. Fabrication of nanoporous copper ribbons by dealloying of $Mn_{70}Cu_{30}$ alloy and fractal characterization of their porosity. Rev. Adv. Mater. Sci. 2013;**33**:50–60.

[3] Ivanova VS, Bunin IJ, Nosenko VI. Fractal material science: a new direction in materials science. JOM. 1998;**5(1)**:52–54. DOI: 10.1007/s11837-998-0068-1

[4] Gao J. Fractal characteristics for goose down assemblies as porous media. J. Comput. 2012;**7(2)**:2884–2890. DOI:10.4304/jcp.7.12.2884-2890

[5] Stănescu V, Olteanu M, lorea-Spiroiu M, Rusu Rusu M. Using fractal analysis to describe collagen-chitosan matrices. Analele UniversitãŃii din Bucuresti – Chimie, Anul XVII (serie nouă). 2008;**2**:47–51.

[6] dos Santos SF, Rodrigues J. de. A. Correlation between fracture toughness, work of fracture and fractal dimensions of alumina-mullite-zirconia composites. Mater. Res. 2003;**6(2)**:219–226. DOI.org/10.1590/S1516-14392003000200017

[7] Liang JZ, Wu CB. Fractal characterization of impact fracture surface of polypropylene nanocomposites. Adv. Polym. Technol. 2012;**31(1)**:71–81. DOI 10.1002/adv.20237

[8] Pramanik B, Tadepalli T, Mantena PR. Surface fractal analysis for estimating the fracture energy absorption of nanoparticle reinforced composites. Materials. 2012;**5**:922–936. DOI:10.3390/ma5050922

[9] Marangoni AG, Rousseau D. Is plastic fat rheology governed by the fractal geometry of the fat crystal network? J. Am. Oil Chem. Soc. 1996;**73**:991–994. DOI: 10.1007/BF02523406

[10] Ohta S, Honjo H. Growth probability distribution in irregular fractal-like crystal growth of ammonium chloride. Phys. Rev. Lett. 1998;**60(7)**:611–614. DOI: 10.1103/PhysRevLett.60.611

[11] Batte HD, Marangoni AG. Fractal growth of milk fat crystals is unaffected by microstructural confinement. Cryst. Growth Des. 2005;**5**:1703–1705. DOI: 10.1021/cg050144v

[12] Kulkarni AS, Beaucage G. Quantification of branching in disordered materials. J. Polym. Sci: Part B: Polym. Phys. 2006;**44**:1395–1405. DOI: 10.1002/polb.20794

[13] Bandres MA, Rechtsman MC, Segev M. Topological photonic quasicrystals: fractal topological spectrum and protected transport. Phys. Rev. 2016;**6**:1–12. DOI:10.1103/PhysRevX.6.011016

[14] Riosa PR, Guimarães JRC. Microstructural path analysis of martensite burst. Mater. Res. 2010;**13(1)**:119–124.

[15] Skrotzki B. The course of the volume fraction of martensite VS. temperature function mx (T). J. Physique IV. 1991;**1**:367–372. DOI:10.1051/jp4:1991455

[16] Starzewski MO. Continuum mechanics models of fractal porous media: integral relations and extremum principles. J. Mech. Mater. Struct. 2009;**4(5)**:901–912.

[17] Zhao L, Wu GC, He JH. Fractal approach to flow through porous material. Int. J. Nonlinear Sci. Numer. Simul. 2009;**10(7)**:897–901. DOI:10.1515/IJNSNS.2009.10.7.897

[18] Xu P, Cai J, Sasmito AP, Jangam SV, Yu M. Transport phenomena in porous media and fractal geometry transport phenomena in porous media and fractal geometry. J. Chem. 2015;**2015**:1–2. DOI:10.1155/2015/486501

[19] Yang S, Liang M, Yu B, Zho M. Permeability model for fractal porous media with rough surfaces. Microfluid. Nanofluid. 2015;**18**:1085–1093. DOI 10.1007/s10404-014-1500-1

[20] Lapsker I, Azoulay J, Rubnov M, Regev Z, Peled RC, Peled A. Image analysis of structural changes in laser irradiated thin films of photodeposited a-Se. Appl. Surf. Sci. 1996;**106**:316–320. DOI: 10.1016/S0169-4332(96)00390-X

[21] Fleury V. Branched fractal patterns in non-equilibrium electrochemical deposition from oscillatory nucleation and growth. Nature. 1997;**390**:145–148. DOI:10.1038/36522

[22] Zhou JG, He Z, Gou J. Fractal growth modeling of electrochemical deposition in solid freeform fabrication. Proceedings of the Tenth Solid Freeform Fabrication Symposium; August 1999.

[23] Caricato AP, Martino M, Romano F, Mirchin N, Peled A. Pulsed laser photodeposition of a-Se nanofilms by ArF laser. Appl. Surf. Sci. 2007;**253**:6517–6521. DOI:10.1016/j.apsusc.2007.01.036

[24] De Nicola F, Castrucci P, Scarselli M, Nanni F, Cacciotti I, De Crescenzi M. Multi-fractal hierarchy of single-walled carbon nanotube hydrophobic coatings. Sci. Rep. 2015;**5**:1–9. DOI: 10.1038/srep08583

[25] Yadav RP, Kumar M, Mittal AK, Pandey AC. Fractal and multifractal characteristics of swift heavy ion induced self-affine nanostructured BaF_2 thin film surfaces. Chaos. 2015;**25**:1–9. DOI: 10.1063/1.4928695

[26] Yu B, Cheng P. A fractal permeability model for bi-dispersed porous media. Int. J. Heat Mass Transfer. 2002;**45**:2983–2993.

[27] Tan XH, Li XP, Liu JY, Zhang GD, Zhang LH. Analysis of permeability for transient two-phase flow in fractal porous media. J. Appl. Phys. 2014;**115(11)**:113502–113507. DOI: 10.1063/1.4868596

[28] Wang JZ, Ma J, Ao OB, Zhi H, Tang HP. Review on fractal analysis of porous metal materials. J. Chem. 2015;1–6. DOI: 10.1155/2015/427297

[29] Hamm SY, Bidaux P. Dual-porosity fractal models for transient flow analysis in fissured rocks. Water Resour. Res. 1996;**32(9)**:2733–2745. DOI: 10.1029/96WR01464

[30] Zheng Q, Yu B. A fractal permeability model for gas flow through dual-porosity media. J. Appl. Phys. 2012;**111(2)**:024316. DOI: 10.1063/1.3679070

[31] Vinogradov A, Yasnikov IS, Estrin Y. Stochastic dislocation kinetics and fractal structures in deforming metals probed by acoustic emission and surface topography measurements. J. Appl. Phys. 2014;**115(23)**:233506–233515. DOI: 10.1063/1.4884682

[32] Mechtcherine V, Mtiller HS. Fractological investigations on the fracture in concrete. Fracture Mechanics of Concrete Structures, de Borst et al (eds), Swets & Zeitlinger, Usse, 2001;81–88.

[33] Carpinteri A, Lacidogna G, Pugno N. A fractal approach for damage detection in concrete and masonry structures by the acoustic emission technique. Acoust. Technol. 2004;**38**:31–37.

[34] Bazant ZP, Yavari A. Is the cause of size effect on structural strength fractal or energetic–statistical? Eng. Fract. Mech. 2005;**72**:1–3. DOI: 10.1016/j.engfracmech.2004.03.004

[35] Zhang H, Wei D. Fracture and Damage Behaviors of Concrete in the Fractal Space. J. Mod. Phys. 2010;**1**:48–58. DOI:10.4236/jmp.2010.11006

[36] Yuan WJ, Zhu QY, Deng CJ, Zhu HX. Fractal analysis of fracture surfaces in refractories. China's Refractories. 2014;**1**:27–31.

[37] Gao MZ, Jin WC, Dai ZX, Xie J. Relevance between abutment pressure and fractal dimension of crack network induced by mining. Int. J. Min. Sci. Technol. 2013;**23(6)**:925–930. DOI: 10.1016/j.ijmst.2013.11.008

[38] Zhu FL, Cui SZ, Gu BH. Fractal analysis for effective thermal conductivity of random fibrous porous materials. Phys. Lett. A. 2010;**374(43)**:4411–4414. DOI:10.1016/j.physleta.2010.08.075

[39] Lee C, Kramer TA. Prediction of three-dimensional fractal dimensions using the two-dimensional properties of fractal aggregates. Adv. Colloid Interface Sci. 2004;**112**:49–57. DOI: 10.1016/j.cis.2004.07.001

[40] Lei X, Kusunose K. Fractal structure and characteristics scale in the distributions of earthquake epicentres, active fault and rivers in Japan. Geophys. J. Int. 1999;**139**:754–762. DOI: 10.1046/j.1365-246x.1999.00977.x

[41] Li BL. Fractal dimension in Encyclopedia of Environmetrics. Edited by Abdel H. El-Shaarawi and Walter W. Piegorsch, John Wiley & Sons, Ltd; 2002. 821–825 p. DOI: 10.1002/9780470057339.vaf015

[42] Mandelbrot BB, Passoja DE, Paulay AJ. Fractal character of fracture surfaces of metals. Nature. 1984;**308**:721–722. DOI:10.1038/308721a0

[43] Corrêa RD, Meireles JB, Huguenin JAO, Caetano DP, da Silva L. Fractal structure of digital speckle patterns produced by rough surfaces. Physica A. 2013;**392**:869–874. DOI: 10.1016/j.physa.2012.10.023

[44] Zhang ZZ. Fractal dimension of fracture surface in rock material after high temperature. Adv. Mater. Sci. Eng. 2015;1–6. Article ID 468370.

[45] Lung CW, Jiang J, Tian EK, Zhang CH. The relation between fractal dimension and roughness index for fractal surface. Phys. Rev. E. 1999;**60**:5121–5130.

[46] de Mello RHC, Vieira EA, Conci A. Characterizing the lacunarity of objects and image sets and its use as a technique for the analysis of textural patterns. In: ACIVS'06 Proceedings of the 8th international conference on Advanced Concepts for Intelligent Vision Systems; 18–21 September 2006; Antwerp, Belgium pp. 208–219.

[47] Brown C, Liebovitch L. Fractal analysis: Quantitative application in social sciences. Washington DC, USA: SAGE Publication Inc; 2010. pp. 6–16.

[48] Falconer K. Fractal Geometry: Mathematical foundation and application. 2nd ed. New York: Wiley; 2003. p. 308.

[49] Hans S. Space-filling curves. Berlin: Springer-Verlag; 1994. p. 156. 10.1007/978-1-4612-0871-6

[50] Tamás V. Fractal growth phenomena. 2nd Ed. Singapore New Jersey: World Scientific; 1992. p. 10.

[51] Mandelbrot BB. How long is the coast of Britain? Statistical self-similarity and fractional dimension. Science. 1967;**156**:636–638.

[52] Smith Jr TG, Lange GD, Marks WB. Fractal methods and results in cellular morphology—dimensions, lacunarity and multifractals: review article. J. Neurosci. Methods. 1996;**69**:123–136. DOI: 10.1016/S0165-0270(96)00080-5

[53] Babinec P, Kučera M, Babincová M. Global characterization of time series using fractal dimension of corresponding recurrence plots: from dynamical systems to heart physiology. HarFA. 2005;87–93.

[54] Pentland AP. Fractal-based description of nature scenes. IEEE Trans. Pattern Anal. Mach. Intell. 1984;**6(6)**:315–326.

[55] Huang J, Turcotte DL. Fractal image analysis: application to the topography of Oregon and synthetic images. J. Opt. Soc. Am. A-Opt. Image Sci. Vis. 1990;**7(6)**:1124–1130.

[56] Long M, Peng F. A box-counting method with adaptable box height for measuring the fractal feature of images. Radio Eng. 2013;**22(1)**:208–213.

[57] Sarkar N, Chaudhuri BB. An efficient differential box counting approach to compute fractal dimension of image. IEEE Trans. Syst. Man Cybern. 1994;**24(1)**:115–120. DOI: 10.1109/21.259692

[58] Chen SS, Keller JM, Crownover RM. On the calculation of fractal features from images. IEEE Trans. Pattern Anal. Mach. Intell. 1993;**15(10)**:1087–1090. DOI: 10.1109/34.254066

[59] Gangepain J, Roques-Carmes C. Fractal approach to two dimensional and three dimensional surface roughness. Wear. 1986;**109**:119–126. DOI: 10.1016/0043-1648(86)90257-7

[60] Feng J, Jin WC, Chen CT. Fractional box-counting approach to fractal dimension estimation. In: Proceedings of International Conference on Pattern Recognition, 25–29 August, 1996; Vienna, Austria: IEEE; pp. 854–858. DOI: 10.1109/ICPR.1996.547197

[61] Xie W, Xie W. Fractal-based analysis of time series data and features extraction. Signal Process. 1997;**13(2)**:98–104.

[62] Li J, Sun C, Du Q. A new box-counting method for estimation of image fractal dimension. In: Proceedings of IEEE International Conference on Image Processing, 8–11 October, 2006; USA: pp. 3029–3022,

[63] Xu S, Weng Y. A new approach to estimate fractal dimensions of corrosion image. Pattern Recognit. Lett. 2006;**27**:1942–1947. DOI: 10.1016/j.patrec.2006.05.005

[64] Li J, Du Q, Sun C. An improved box-counting method for image fractal dimension estimation. Pattern Recognit. 2009;**42**:2460–2469 DOI:. 10.1016/j.patcog.2009.03.001

[65] Flook AG. The use of dilation logic on the quantimet to achieve fractal dimension characterisation of textured and structured profiles. Powder Technol. 1978;**21**:295–298. DOI: 10.1016/0032-5910(78)80099-0

[66] Bassingthwaighte JB, Liebovitch LS, West BJ. Fractal physiology. New York: Springer; 1994. pp. 16–20.

[67] Hunt AG. Basic transport properties in natural porous media: continuum percolation theory and fractal model. Complexity. 2005;**10(3)**:22–37. DOI: 10.1002/cplx.20067

[68] Yiotis AG, Tsimpanogiannis IN, Stubos AK. Fractal characteristics and scaling of the drying front in porous media: a pore network study. Drying Technol. 2010;**28**:981–990. DOI: 10.1080/07373937.2010.497087

[69] Yu B, Cai J, Zou M. On the physical properties of apparent two phase fractal porous media. Vadose Zone J. 2009;**8**:177–186. DOI: 10.2136/vzj2008.0015

[70] Cihan A, Sukop M, Tyner JS, Perfect E, Huang H. Analytical predictions and lattice Boltzmann simulations of intrinsic permeability for mass fractal porous media. media. Vadose Zone J. 2009;**8**:187–196. DOI: 10.2136/vzj2008.0003

[71] De Gennes PG. Partial filling of a fractal structure, by a wetting fluid. In: Physics of disordered materials. Adler D, Frizsche H, Ovshinsky SR. Editors, New York: Plenum Press; 1985; pp. 227–241.

[72] Katz AJ, Thompson AH. Fractal sandstone pores: Implications for conductivity and pore formation. Phys. Rev. Lett. 1985;**54**:1325–1328. DOI:https://doi.org/10.1103/PhysRevLett.54.1325

[73] Turcotte DL. Fractals and fragmentation. J. Geophys. Res. 1986;**91**:1921–1926.

[74] Rieu M, Sposito G. Fractal fragmentation, soil porosity, and soil water properties I. Theory. Soil Sci. Soc. Am. J. 1991;**55**:1231.

[75] Hunt AG, Gee GW. Application of critical path analysis to fractal porous media: comparison with examples from the Hanford site. Adv. Water Res. 2002;**25**:129–146. DOI: 10.1016/S0309-1708(01)00057-4

[76] Cai J, Yu B, Zou M, Luo L. Fractal characterization of spontaneous co-current imbibition in porous media. Energy Fuels, 2010;**24(3)**:1860–1867. DOI: 10.1021/ef901413p

[77] Xu M, Dehghanpour H. Advances in understanding wettability of gas shales. Energy Fuels. 2014;**28(7)**:4362–4375. DOI: 10.1021/ef500428y

[78] Mirzaei-Paiaman A, Masihi M. Scaling of recovery by co-current spontaneous imbibition in fractured petroleum reservoirs. Energy Technol. 2014;**2(2)**:166–175. DOI: 10.1002/ente.201300155

[79] Kerunwa A, Onyekonwu MO, Olafuyi AO, Anyadiegwu CIC. Fractal model to characterize spontaneous imbibition in porous media. IJERA. 2016;**1(6)**:46–54.

[80] Khlaifat B, Abdelaziz K, Hani Q, Hamid A. Influence of a Single Fracture and Its Aperture on Gas Production from a Tight Reservoir, AAPG Search and Discovery Article, 2011;#**40732**:1–9.

[81] Kucuk F, Sawyer WK. Transient flow in naturally fractured reservoirs and its application to devonian gas shales. Presented at SPE Annual Technical Conference and Exhibition; 21–24 September 1980; Dallas. Texas; SPE-9397-MS.

[82] Zinovik I, Poulikakos D. On the permeability of fractal tube bundles. Transport Porous Med. 2011;**94**:747–757. DOI: 10.1007/s11242-012-0022-0

Chapter 11

Analysis and Application of Decaying Turbulence with Initial Fractal Geometry

Hiroki Suzuki, Shinsuke Mochizuki,
Yasuhiko Sakai and Koji Nagata

Additional information is available at the end of the chapter

http://dx.doi.org/10.5772/67942

Abstract

In this paper, we address high-Schmidt-number (Sc) scalar turbulent mixing that results from grid-generated turbulence using the initial fractal geometry of the velocity profile. More specifically, as was proposed in our recent study, we adopt an initial flow field generated by a fractal grid and apply it to a water channel experiment based on a high-Sc-number scalar-mixing layer in order to create grid-generated turbulence, and thus solve our current research problem. The high-Sc-number scalar and velocity fields of the grid-generated turbulence are then measured using planar laser-induced fluorescence (PLIF) and particle image velocimetry (PIV), respectively. By means of fractal analysis, this study specifically addresses the turbulent mixing phenomena in which the fractal dimension of the mixing interface of an observed high-Sc-number scalar field is calculated. Additionally, we discuss the efficiency of using fractal grids as devices for enhancing high-Sc-number scalar turbulent mixing by observing turbulent intensities and dissipation by PIV.

Keywords: fractal analysis, fractal dimension, turbulent mixing, grid-generated turbulence, initial value problem

1. Introduction

Fractals are widely found in nature. In fluid mechanics, the fractals found in turbulence have been investigated as a means of understanding and modeling fundamental phenomena, one of which is the non-Gaussian nature found in small-scale fluctuations [1]. For instance, since the skewness factor of a longitudinal velocity derivative that relates to turbulence vortex stretching is approximately −1 to −0.2 rather than zero, small-scale turbulence fields, in which the probability density distribution of fluctuations is described by a non-Gaussian function, have been discussed from the perspective of fractals. In such turbulent flows, kinetic energy is first injected into a large-scale region, after which it is transferred to a smaller-scale region, before dissipating into the smallest-scale region. In such cases, it is possible to formulate the transfer of kinetic energy in the inertial subrange. For structure function modeling, which also characterizes the nature of turbulent flows in physical space, the fractal perspective, in which the fractal dimension is a characteristic quantity, is applied [2]. In engineering applications, turbulent flows are often considered to be mixers, and the flows produced are thus referred to as turbulent mixing [3]. The characteristics of the mixing interface between two mixtures, which increases in area with time due to turbulent mixing, can also be quantified by using the fractal method.

Turbulence decays temporally when the mean flow does not provide kinetic energy to the turbulence. In decaying turbulence, kinetic energy follows the decay law (the parameters of which are the decay coefficient and decay exponent) and has been examined in previous studies (e.g., [4–6]). The decay coefficient magnitude relates to the drag coefficient of a turbulence-generating grid, while the decay exponent magnitude is of the order of unity and could be used to characterize the decay characteristics of decaying homogeneous turbulence. Recent studies (e.g., [7, 8]) have examined the effects of the initial conditions on the turbulence decay characteristics. Specifically, it has been determined that, within the range of moderate Reynolds numbers, decay exponent values depend on the initial turbulence decay conditions. Additionally, since the initial fractal geometry of the velocity profile may affect grid-generated turbulence decay characteristics, thereby resulting in turbulence decay, the effects of the initial velocity profile fractal geometry on decaying turbulence have also been investigated in several previous studies (e.g., [9–23]). These studies investigated turbulence produced by a fractal geometry turbulence-generating grid (referred to hereafter as a fractal grid (FG)) and most such studies used turbulence created by a square-type fractal grid (referred to hereafter as fractal grid turbulence) when conducting wind tunnel experiments (e.g., [10–14]), water channel experiments (e.g., [15–17]), and numerical simulations (e.g., [18–23]).

Since the observed diffusion resulting from turbulence has been determined to be much larger than that which can be produced by viscous diffusion, one of the engineering applications of turbulence is turbulent mixing. In addition, since rapid turbulence mixing of heat and mass due to turbulence has obvious engineering applications, there have been previous studies that have attempted to enhance such mixing (e.g., [24, 25]). However, few works have applied water channel experiments to the study of high-Schmidt-number (Sc) scalar mixing caused by fractal grid turbulence [26]. Since the diffusion coefficient of a high-Sc-number scalar is significantly smaller than that of heat in air and water, the contribution of turbulent diffusion is dominant in high-Sc-number scalar turbulent mixing. Therefore, this study will address turbulent mixing by focusing on the fractal geometry of the initial velocity profile, which is set using a fractal grid. More specifically, high-Sc-number scalar turbulent mixing resulting from turbulent diffusion in fractal grid turbulence will be examined, and mixing interfaces will be discussed in terms of fractals to clarify the effects of turbulent mixing resulting from fractal grid turbulence.

In this section, the effects of the initial fractal geometry of the velocity profile on the turbulent mixing characteristics, and the high-Sc-number scalar-mixing layer in grid-generated turbulence, are presented. As in our previous study, a flow with an initial fractal geometry is generated using a fractal grid, but the present experiment is based on a water channel experiment and utilizes both planar laser-induced fluorescence (PLIF) (e.g., [27, 28]) and particle image velocimetry (PIV) (e.g., [29]) to investigate the high-Sc and fluctuating velocity fields. Note that the PLIF and PIV apparatuses used herein were constructed in-house and are validated in the present work.

The fractal dimensions of the mixing interface (e.g., [30, 31]) as well as the turbulent statistics in the high-Sc-number scalar-mixing layer, where the fractal dimension is calculated using the box counting method, are shown. Additionally, we discuss the efficiency of fractal grids as devices for enhancing turbulent mixing by using the turbulent intensities and viscous dissipations measured via PIV, and then show a periodic box approach to further address the effects of the initial fractal geometry of the velocity profile on the decay characteristics and turbulent mixing. The Fourier spectral method used for the periodic box approach simulations is also described and verified.

2. Experimental analysis

2.1. Flow and mass transfer fields

In this section, we discuss flow and mass transfer fields in regular grid and fractal grid turbulences. **Figure 1** shows schematic diagrams of the experimental apparatus used. The size of the test section in the present water channel experiment is 1500 × 100 × 100 [mm^3] for the streamwise (x), transverse (y), and span-wise (z) directions, respectively. The coordinate system origin is set at the center of the turbulence-generating grid. A head tank is used to provide the free stream, and a splitter plate is placed in the contraction upstream of the test section to produce the scalar-mixing layer.

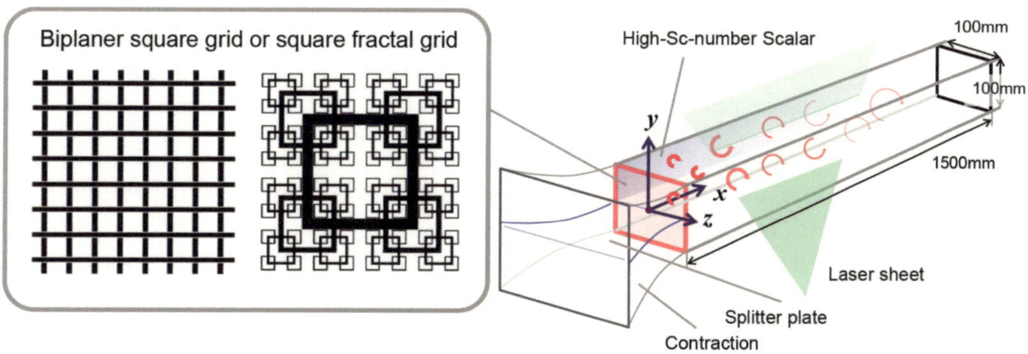

Figure 1. Water-channel experiment schematics. The origin of the coordinate system is placed at the center of the turbulence-generating grid. In this experiment, a biplane square grid or a square fractal grid is set at the entrance of the test section.

An aqueous solution of Rhodamine B, which is considered to be a high-Sc-number scalar, is set in the lower layer of the flow leading into the test section. Here, the flow of the upper layer does not include fluorescent dye because the concentration field of a high-Sc-number scalar is measured via the PLIF technique. Since PIV is used to measure the grid-generated turbulence flow field in this experiment, the laser system and high-speed camera used for PLIF and PIV observations were also employed in the present experimental apparatus.

The present study used both square grids and a fractal grid [10] for turbulence generation. In the first experiment, two square grids were used to produce grid-generated turbulence. The first square grid (SG1), which generates conventional grid turbulence (CGT) and was used to validate the PIV and PLIF experiments, consisted of round bars and was the same one used in our previous water-channel experiment [2, 4]. The second square grid (SG2) also generates conventional grid turbulence of which the characteristics were compared with those of fractal grid turbulence. The present fractal grid was based on that proposed by that used in the wind tunnel experiment conducted in the previous studies.

The fractal dimensions D_f of all grids were set at two. The thickness ratio t_r, which is defined as the ratio of thickness between the largest and smallest bars of the grid [10], was set at 1, 1, and 9.76 for SG1, SG2, and FG1, respectively. The grid mesh size M_{eff} was defined as $M_{eff} = (4D^2/P_M)\ (1 - \sigma)^{1/2}$, where D^2 is the cross-sectional area of the channel, P_M is the fractal perimeter length of the grid [10], and $Re_M = U_o M_{eff}/\nu$ is the mesh Reynolds number, where U_o and ν are the cross-sectionally averaged mean velocity and kinematic viscosity, respectively. The mesh sizes of SG1, SG2, and FG1 are 20, 10, and 5.68 [mm], respectively. The mesh Reynolds number Re is defined as $Re = U_o M_{eff}/\nu$ is 2500 for the three turbulence-generating grids. This Reynolds number value is comparable to those used in our previous experiment [15, 24]. Values of U_o for SG1, SG2, and FG1, which are set by using electromagnetic flow meters, are 0.126, 0.251, and 0.442 [m/s], respectively.

2.2. PLIF technique

In the present experiment, the PLIF technique is used to measure the high-Sc-number scalar field. Here, it should be noted that PLIF does not affect the flow field of grid-generated turbulence, and that the spatial resolution of PLIF is sufficiently accurate for scalar field measurements. Therefore, we are confident that the high-Sc-number scalar field in the grid-generated turbulence could be determined accurately.

Rhodamine B was used in the PLIF experiment because it is considered to be a high-Sc-number scalar. The fluorescent dye contained in this solution is excited by a 532-nm continuous wave (CW) laser with a maximum power of approximately 4 W. The measurement region $L_x \times L_y$ is 50 × 100 [mm^2], where L_x and L_y are the width of the measurement region for x and y directions, respectively. The number of grid points $N_x \times N_y$ is 1800 × 3600, where N_x and N_y are the number of pixel points for the x and y directions, respectively.

In this study, the magnitude of fluorescence in the observed high-Sc-number scalar was measured using a Nikon D700 single-lens reflex (SLR) camera, which was deemed suitable for PLIF due to its large pixel number and high sensitivity. Furthermore, the 36.0 × 23.9 [mm^2]

solid-state image-sensing area of the camera corresponds roughly to the image area of a 35-mm film camera. The camera was equipped with a Nikkor 50 mm F/1.2S lens with an optical filter mounted to cutoff dye fluorescence at wavelengths above 540 nm. The F-stop set was set to 2.0, the ISO sensitivity was set to 400, and the depth of bit in the captured RAW images was set to 14 because those settings were deemed appropriate for our PLIF measurements.

In PLIF experiments, a reference image that is captured with a constant concentration under pre-experiment conditions is often used. However, PLIF experiments contained sources of error that could not be eliminated using a reference image alone [27]. Specifically, the observed fluorescence intensity depends on the experimental conditions, which varies continuously even if the dye concentration is constant because the quantum yield of the dye is a function of time. Additionally, since camera gain also depends on time, the excitation laser intensity may experience temporary fluctuations that affect the observed fluorescence intensity. Furthermore, the intensity of the excitation laser is reduced by the presence of fluorescent dye along the laser path, which may produce spatial variations in the fluorescence intensity even if the dye concentration is spatially constant. Accordingly, in this experiment, a post-processing scheme was used to correct for these sources of error. Specifically, the following processing equations were applied: $\beta I_{j+1} \phi = \beta I_j \phi \exp[\varepsilon C'_{ref} (1 - C_j)\, dy_j]$ and $V_{j+1}/C_{j+1} = (V_j/C_j) \exp[\varepsilon C'_{ref} (1 - C_j)\, dy_j]$, where a value in the initial state, $j = 0$, is given as follows: $C_0 = V_0/(\beta I_0 \phi)$. Here, the factors, C, V, β, ϕ, and I, are dye concentration, brightness value, camera gain, quantum yield, and laser intensity for a measured image of fluorescence intensity, respectively, which are normalized by those of the reference image. Also, ε and C'_{ref} are the absorption coefficient and the actual concentration for a reference image. **Figure 2** presents schematic diagrams showing our correction method.

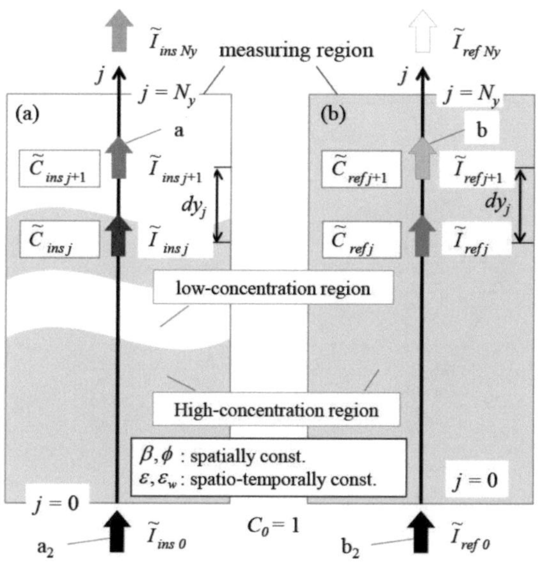

Figure 2. Schematics of observed and reference scalar fields, shown by (a) and (b), used for the present data-processing method in the PLIF. Laser attenuation could also be corrected via the present processing method.

We validated our PLIF technique and examined the effectiveness of the above correction method under the three experimental conditions. In Case 1, the influences of the factors that vary and fluctuate with time and absorption of the excitation laser were reduced; in Case 2, only the influence of the varying factors was reduced; in Case 3, neither the influence of the varying factors nor the laser excitation was reduced. In the validation and examination stage, a scalar-mixing layer in the grid-generated turbulence produced by SG1 was measured. **Figure 3(a)** shows transverse profiles of the mean concentration used for validation and examination. As can be seen in the figure, correcting for factor variations produced a significant influence. By contrast, laser absorption had little effect on the mean concentration profile. The mean concentration profiles of Cases 1 and 2 agree with our previous experiment [24], which used the same experimental conditions as the current experiment. **Figure 3(b)** shows transverse profiles of the variance of concentration fluctuations. As can be seen in the figure, the observed variance of concentration fluctuations is affected by the absorption of the excitation laser as well as the temporally varying/fluctuating factors.

2.3. PIV

In this experiment, PIV was applied to flow field measurements. However, in order to visualize a flow field for measurement, fine polyester particles, the diameter and specific gravity of which were 50 μm and 1.03, respectively, were added to the solution. These particles provided sufficiently high response to flow field fluctuations. In these experiments, particle response to flow field fluctuations was examined using the classical theoretical formula, in which the frequency response and phase lag were calculated. For frequencies up to 1 kHz, the magnitude of both frequency response and phase lag was found to be smaller than 1%. For PIV visualization, we employed the same CW laser that was used in the PLIF experiment, and the high-speed camera was used to capture brightness values for the particles illuminated by the laser. Here, the measurement area $L_x \times L_y$ was 7.5 × 40 [mm²].

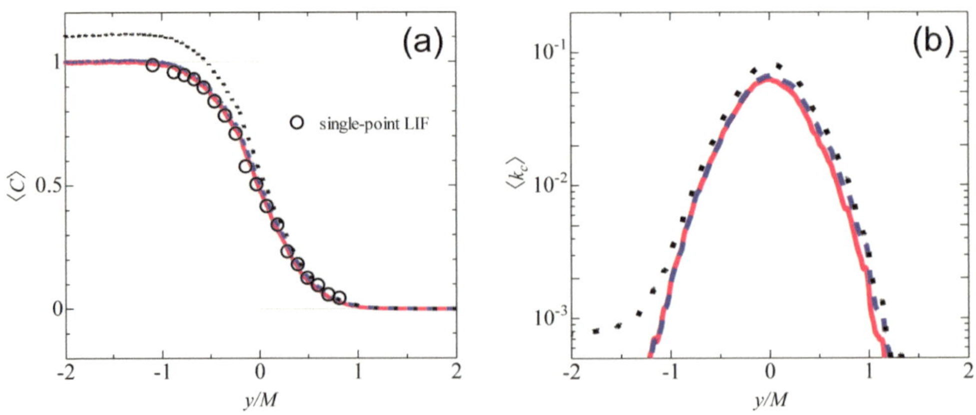

Figure 3. Validation of the present PLIF measurement for mean scalar (a) and scalar fluctuation variance (b). Red-solid, blue-dashed, and black-dotted lines show results of ⟨C⟩ and ⟨kc⟩ for Cases 1, 2, and 3, respectively. In the mean scalar, the results of Case 3, in which neither the influence of the factors nor laser attenuation is corrected, are shown. In the scalar fluctuation variance results, laser attenuation affects may be significant.

http://dx.doi.org/10.5772/67942

In this experiment, custom-designed software that calculates instantaneous velocity vectors from the set of brightness values for the illuminated fine particles [15] was used. The specific process was as follows. First, the temporal field of brightness value was calculated from the captured images by using Open Source Computer Vision (Open CV), which eliminated the need to conduct preprocedures. The main framework of our PIV is based on the recursive local-correlation method with a 50% overlap. Here, the interrogation windows with the offset determined were applied to the PIV software using the iterative procedure. For subpixel analysis, the gradient method was used instead of Gaussian fitting because the latter requires large interrogation windows to reduce the PIV experimental uncertainty, and because the spatial resolution of the PIV could be improved sufficiently via the gradient method.

We validated our PIV using grid-generated turbulence measurements that were conducted before the PIV measurement results were obtained. The grid-generated turbulence used for the validation of the present PIV was the same as that used for PLIF measurements. **Figure 4(a)** shows the intensity of streamwise velocity fluctuations along the centerline. As shown in the figure, the streamwise profile of the velocity intensity could be approximated by a power law. The agreement of the intensity of the velocity fluctuations with the power law validates the PIV measurement. In the present study, the decay exponent of the power law, where the decay exponent magnitude is on the order of unity, was also measured. As shown in our previous works, decay exponent values can be larger than unity. The decay exponent value measured in this study was 1.2, which agrees with the result of previous experiments (e.g., [4–6]). The power spectrum of the streamwise velocity fluctuations was also used for PIV validation. As shown in **Figure 4**, the power spectrum observed by our PIV technique agrees well with the laser-Doppler velocimetry measurements of our previous experiment [24], thereby indicating that the present PIV could accurately measure velocity fluctuations in grid-generated turbulence.

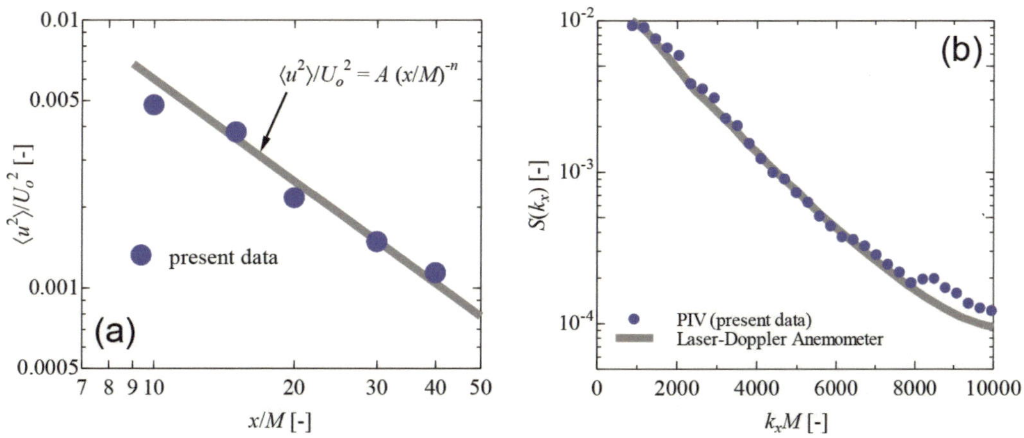

Figure 4. Validation of the present PIV measurement in grid-generated turbulence, where (a) and (b) show the normalized intensity of streamwise velocity fluctuation and power spectra of the streamwise velocity fluctuation at $x/M = 18$, where the laser-Doppler anemometer data are measured by Ito et al. (2002) [24]. The agreement shown here validates our current PIV measurements.

3. Results and discussion

3.1. Turbulent mixing and fractal dimensions

Here, we address the turbulent mixing of the high-Sc-number scalar in the grid-generated turbulence of SG2 and FG. **Figure 5(a)** and **(c)** shows the instantaneous scalar of the fractal grid turbulence and classical grid turbulence in which the instantaneous scalar is normalized by the initial scalar difference. As can be seen in the figure, the high-Sc-number scalar is mixed by the grid-generated turbulence. Furthermore, as also shown in the figure, the yellow region in the fractal grid turbulence is wider than that in the grid-generated turbulence. Note that the yellow color indicates a scalar value at the mixing interface, the normalized value of which is 0.5.

Figure 5(b) shows a transverse profile of the mean scalar at this streamwise distance. As can be seen in the figure, the magnitude of the transverse gradient at the centerline of the fractal grid turbulence is smaller than that in the grid-generated turbulence. **Figure 6(a)** and **(c)** shows instantaneous scalar fluctuations of the scalar-mixing layer around $x/M_{eff} = 80$. As can be seen in the figure, similar to the instantaneous scalar results, the wider region in which the scalar fluctuation exists is also found in the case of the fractal grid turbulence. **Figure 6(b)** shows transverse profiles of the scalar fluctuation intensity. Here, it can be seen that the intensity profile width in the fractal grid turbulence is larger than that in the grid-generated turbulence.

Since enhancing scalar turbulent mixing increases the scalar-mixing layer width in grid-generated turbulence, the intensity of turbulent mixing is quantified by using a length scale that characterizes scalar-mixing layer width. **Figure 7** shows streamwise profiles of the mean scalar

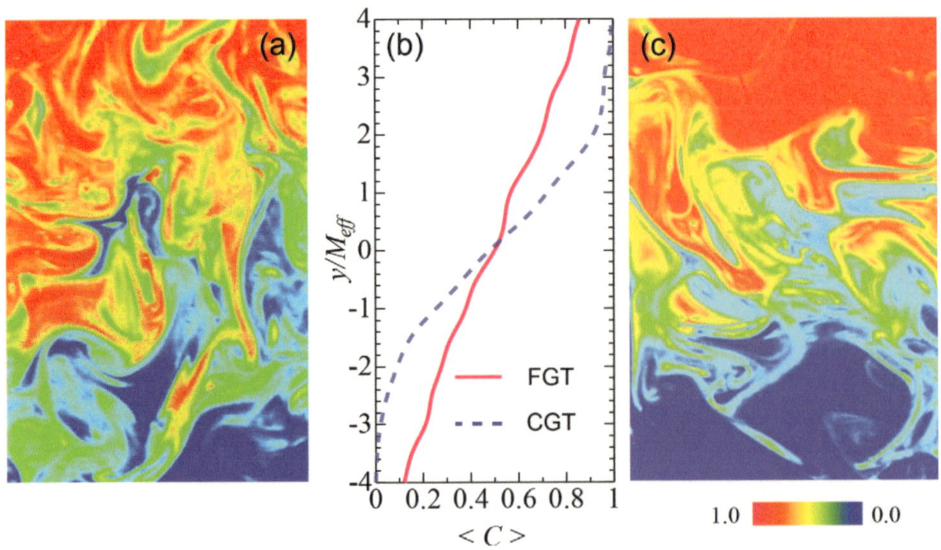

Figure 5. Instantaneous scalar field in the fractal-grid turbulence (FGT) (a) and conventional-grid turbulence (CGT) (c) around x/Meff = 80. The mean scalar profile at $x/M_{eff} = 80$ is also shown in (b). Here red, yellow, and blue indicate C = 1, C = 0.5, and C = 0, respectively.

http://dx.doi.org/10.5772/67942

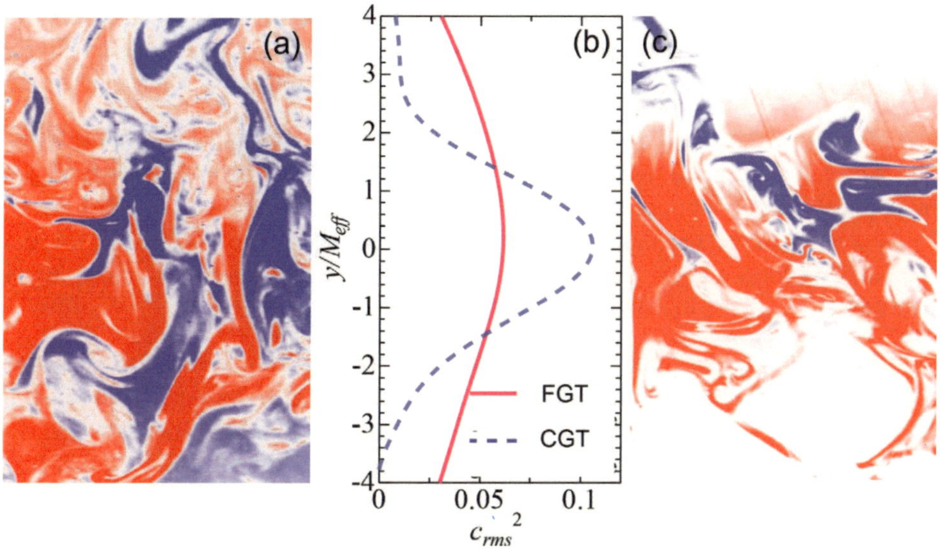

Figure 6. Instantaneous scalar fluctuation in the fractal-grid turbulence (FGT) (a) and conventional-grid turbulence (CGT) (c) around $x/M_{eff} = 80$. Scalar fluctuation intensity profiles at $x/M_{eff} = 80$ are also shown in (b).

profile half width. As can be seen in the figure, the mean scalar profile half widths in grid-generated turbulence follow a function of the square root of x/U, where x/U [s] is the time dimension and is referred to as convection time. In fractal grid turbulence, the half-width streamwise profile includes the mean scalar half-width profile, which is also characterized by the square root of x/U. Note that fractal grid turbulence half-width values are larger than those in grid-generated turbulence, thereby indicating that turbulent mixing is enhanced by using the fractal grid. The scalar fluctuation intensity width is also characterized by its half width. As with the mean scalar

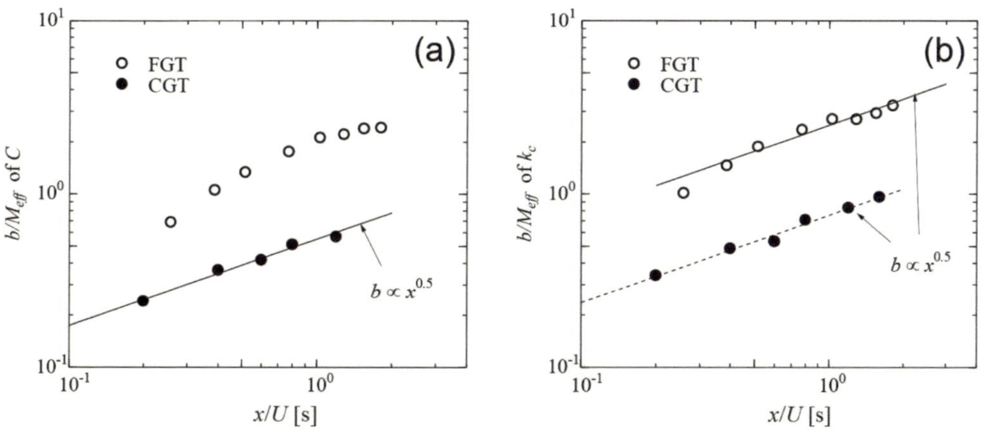

Figure 7. Profiles of the mean scalar half width and scalar variance as functions of the convection time x/U [s]. Here profiles of mean scalar half width and scalar variance half width are shown in (a) and (b), respectively. In the mean scalar and scalar variance results, the half width for the fractal-grid turbulence is larger than that for the conventional-grid turbulence.

half-width results, the larger half width of the scalar fluctuation intensity is found in the scalar-mixing layer due to fractal grid turbulence. These larger half-width values confirm the high-Sc-number scalar-mixing enhancement achieved by fractal grid application.

Next, we discuss the enhanced turbulent mixing of the high-Sc-number scalar in the fractal grid turbulence by calculating the fractal dimension using a box-counting method (e.g., [30, 31]). In the box-counting method, boxes that follow the relationship are counted: $|C - C_{th}| \approx 0$. Here, C and C_{th} are a value of the normalized scalar in the mixing layer and a threshold, respectively. The counted boxes satisfy the following relation: $N(\delta) = \alpha\ \delta^{-D}$, where δ is the characteristic length of the counted boxes and the area of the counted box is equal to δ^2.

Figure 8(a) shows a counted box value, which is a function of the scale δ, for the scalar-mixing layer in the fractal grid turbulence, where $x/M_{eff} = 40$ and $C_{th} = 0.5$. As shown in the figure, the counted box values decrease as the scale δ increases. Furthermore, the observed counted box value follows the basic relationship of the box-counting method, $N(\delta) = \alpha\ \delta^{-D}$. It should be noted that the value of D also depends on the scale δ, and that the scales δ in the large and small values of d show large- and small-scale scalar fractal dimension values. A value of D for the large δ, which is found to be constant, is different from that for the small δ. Here, the large- and small-scale fractal dimension values are within the range of 1–2.

A fractal dimension value, which is considered to differ between larger and smaller scales, is calculated for the fractal grid and classical grid turbulence. **Figure 8(b)** shows the observed values of the fractal dimension, where the convection time for the fractal grid turbulence is nearly equal to that for the classical grid turbulence. As the figure shows, scalar field fractal dimension values in fractal grid turbulence are larger than those found in the classical grid turbulence. These larger fractal dimension values are found in both large- and small-scale scalar fields. A large-scale fractal dimension value does not depend on the threshold value,

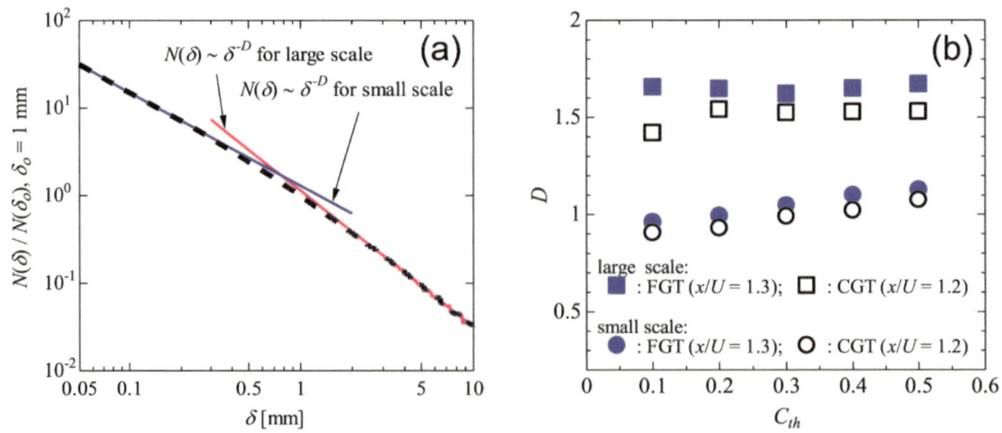

Figure 8. Counted boxes of the scalar field in the fractal-grid turbulence around $x/M_{eff} = 40$ and $C_{th} = 0.5$ (a), which are given by the box-counting method. Profiles of the counted boxes satisfy $N(\delta) \sim \delta^{-D}$. In (b), fractal dimension of the large- and small-scale scalar fields in fractal-grid and conventional-grid turbulence around $x/U = 1.25$ [s] are shown. The fractal dimension for the fractal-grid turbulence is larger than that for the conventional-grid turbulence.

http://dx.doi.org/10.5772/67942

although a small-scale fractal dimension value in the scalar field may depend slightly on the threshold value. The fractal dimension of a scalar field for which turbulent mixing does not occur is unity. Fractal dimension values approach two as turbulent mixing is enhanced because our PLIF measurement technique uses a two-dimensional measurement for the scalar field. The larger fractal dimension value found in turbulent mixing due to fractal grid turbulence indicates the enhancement of turbulent mixing achieved by using a fractal grid.

Figure 9 shows streamwise fractal dimension evolution as a function of convection time. Here, it can be seen that the fractal dimension of large-scale turbulent mixing resulting from fractal grid turbulence increases with convection time. The fractal dimension of the small-scale field for fractal grid turbulence is also larger than that for the classical grid turbulence, thereby indicating that fractal grid application could enhance both small- and large-scale turbulent mixing. Since the small-scale fractal dimension difference found between the two turbulence-generating grids would not be large, further experiments may be needed to determine the incremental fractal dimension increase achieved by applying the fractal grid.

3.2. Turbulent mixing device efficiency

Next, we discuss the enhancement of turbulent mixing achieved via fractal grid using PIV measurement results, starting by examining velocity fluctuation intensity. **Figure 10(a)** shows streamwise profiles of (3/2) $\langle u^2 \rangle/(1/2)\ U_o^2$, referred to hereafter as turbulence intensity. In the classical and fractal grid turbulence, the large-scale anisotropy is constant for time and approximates unity. Therefore, the quantity of the turbulence intensity (3/2) $\langle u^2 \rangle/(1/2)\ U_o^2$ would correspond to turbulent kinetic energy. The longitudinal direction is the convection time, which is normalized by the unit time [s].

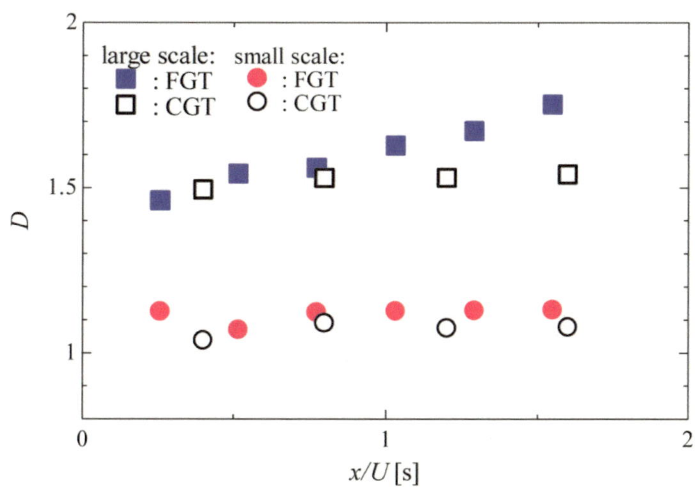

Figure 9. Streamwise profile of the fractal dimensions. The square and circle denote fractal dimension for large- and small-scale scalar fields, respectively. The fractal dimension for fractal-grid turbulence is larger than that for conventional-grid turbulence in large- and small-scale scalar fields.

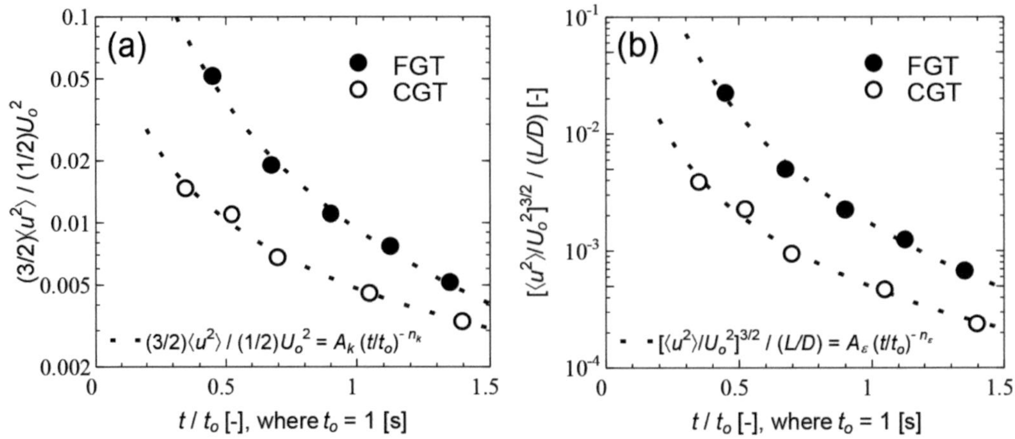

Figure 10. Streamwise profiles of the turbulent intensity (a) and the dissipation (b) for fractal-grid turbulence and conventional-grid turbulence, where the longitudinal direction is the normalized convection velocity. These profiles satisfy each power law with different decay exponent and decay coefficient values.

As shown in the figure, the classical and fractal grid turbulence intensities decrease as the convection time increases. Also, the fractal grid could be considered to generate decaying turbulence with a turbulence intensity that is larger in the same bulk energy magnitude because it is normalized by the bulk flow kinetic energy. The figure also shows that the classical grid turbulence intensity satisfies the following power law: $(3/2)\ \langle u^2\rangle/(1/2)\ U_0^{\ 2} = A_k\ (t/t_0)^{-nk}$, where A_k and n_k are the decay coefficient and decay exponent, respectively. Fractal grid turbulence intensity also follows a power law with the different parameter values.

Viscous dissipation is also examined in this study because the measurement allows potentially serious result uncertainties. Accordingly, observed viscous dissipation ε, which is referred to as dissipation, is calculated and defined as follows: $\varepsilon = [\langle u^2\rangle/U_0^{\ 2}]^{3/2}/(L/D)$. Here, L is the integral scale of the streamwise velocity fluctuation. As shown in **Figure 10(b)**, streamwise dissipation profiles in the classical and fractal grid turbulence also follow each power law: $[\langle u^2\rangle/U_0^{\ 2}]^{3/2}/(L/D) = A_\varepsilon\ (t/t_0)^{-n\varepsilon}$. Here, A_ε and n_ε are the decay coefficient and decay exponent of the dissipation, respectively.

Next, the decay exponents and decay coefficients of the fractal and classical grid turbulence are measured. **Figure 11(a)** shows the classical and fractal grid turbulence intensity decay exponents. As can be seen in the figure, the turbulence intensity decay exponents are measured as 2.1 for fractal grid turbulence and 1.1 for the classical grid, and the fractal grid turbulence decay exponent is larger than that for conventional grid turbulence. This larger fractal grid turbulence value implies that it includes dynamics other than the viscous dissipation, such as turbulent diffusion (e.g., [12]). **Figure 11(b)** shows the decay coefficients of fractal grid turbulence normalized by that of conventional grid turbulence, where the turbulence intensity decay coefficients and dissipation are shown.

As shown in the figure, the normalized decay coefficient value for fractal grid turbulence intensity is about two. Since a decay coefficient is directly related to the drag coefficient of a turbulence-generating grid, a larger decay coefficient value for turbulence intensity would have a larger drag coefficient in the grid. Furthermore, since fractal grid turbulence size characteristics depend primarily on the largest and smallest grids, larger grids can be expected to produce larger grid drag coefficients. The dissipation decay coefficient of fractal grid turbulence normalized by the conventional grid turbulence is also shown in the figure, and is about 3.5. In turbulent diffusion, which is modeled by two-particle diffusion, the turbulent diffusion coefficient is proportional to viscous dissipation. Therefore, the decay coefficient increment of the dissipation indicates that of the turbulent diffusion coefficient. However, it should be noted that the dissipation decay coefficient increment is considerably larger than that of the turbulence intensity. Qualitatively, turbulent-mixing enhancement increases as the drag coefficient is increased. As can be seen in the figure, applying the fractal grid produces a drag coefficient-related dissipation increment that is related to turbulent diffusion coefficient and is larger than the turbulence intensity increment. Therefore, **Figure 11(b)** confirms the efficiency of a fractal grid as a device for enhancing turbulence mixing.

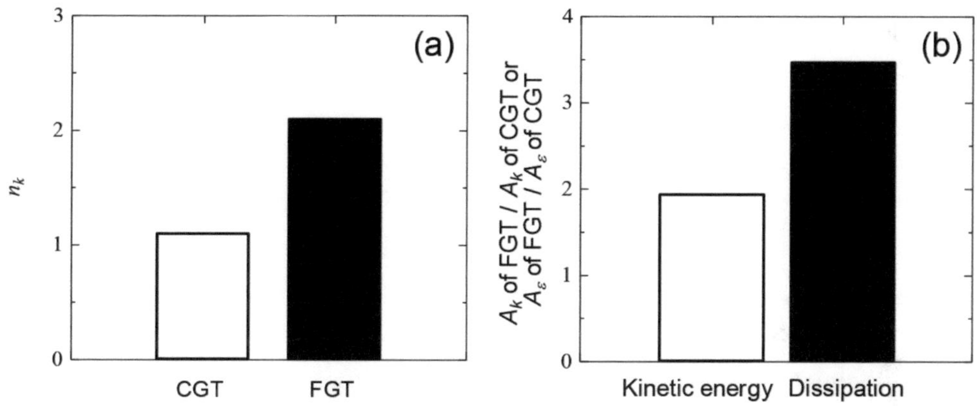

Figure 11. Decay exponents of the turbulent intensity (a) and the ratio of the decay coefficients between fractal-grid and conventional-grid turbulence (b). The dissipation decay coefficient increase, which relates to the turbulent diffusion coefficient, is larger than that of the turbulent intensity, which relates to the grid-drag coefficient.

4. Conclusion

This study has addressed high-Sc-number scalar turbulent mixing in decaying turbulence through experiments in which decaying turbulence was generated using a grid placed inside a uniform flow. In recent years, fractal grids, which are new turbulence-generating grids with shape-based fractal perspectives, have been proposed, and high-magnitude turbulent intensities have been found in the decaying turbulence produced by such grids. The present study applies the decaying turbulence produced by a fractal grid to high-Sc-number scalar turbulent mixing in a water channel experiment measured by PLIF and PIV in order to clarify fractal grid turbulence mixing. Conventional grid turbulence, which is produced via a biplane conventional grid, was also measured. Fractal dimensions, which were calculated using the box-counting method as well as relevant statistics, were used to clarify high-Sc-number scalar turbulent mixing in the grid turbulences examined.

We started by validating the PLIF and PIV measurement techniques used and showed that our correction for laser attenuation effects caused by the dye used definitely improved scalar fluctuation measurement accuracy. Next, the PIV was validated by using the decay characteristics and spectra of conventional grid turbulence. As shown in the PLIF visualization results provided herein, the use of a fractal grid increased the scalar-mixing layer width, and the scalar-mixing layer width increment could be quantitatively observed by using the characteristics length of the mean scalar profile and scalar fluctuation intensity.

Our fractal dimension investigation, which was calculated in order to discuss high-Sc-number scalar turbulent mixing in fractal grid turbulence, showed that the use of a fractal grid could enhance both large- and small-scale turbulent mixing. The decay exponent, turbulence intensity decay coefficient, and dissipation were then calculated from PIV measurements, the results of which showed a larger decay exponent magnitude in fractal grid turbulence. Next, we discussed using the decay coefficient to quantify the efficiency of a fractal grid as a device for enhancing turbulent mixing. The results showed that when utilizing a fractal grid, the turbulent diffusion coefficient increase in the generated grid turbulence was found to be larger than the grid drag coefficient.

In our future works, the dynamics that are expected to be found in enhanced turbulent mixing due to fractal grid turbulence will be studied. By further investigating the phenomena from the perspective of dynamics, a new turbulent-mixing device may further be developed.

Acknowledgements

The authors gratefully acknowledge the assistance of Professor J. Christos Vassilicos of Imperial College London and Dr. Takashi Kubo of Meijo University, Nagoya, Japan. A portion of this study was carried out under the Research Cooperative Program between the Japan Society for the Promotion of Science and The Royal Society. This study was also supported in part by Grants-in-Aid (nos. 25289030, 25289031, and 15K17970) from the Japanese Ministry of Education, Culture, Sports, Science and Technology.

Author details

Hiroki Suzuki[1]*, Shinsuke Mochizuki[2], Yasuhiko Sakai[3] and Koji Nagata[3]

*Address all correspondence to: h.suzuki@yamaguchi-u.ac.jp

1 Graduate School of Sciences and Technology for Innovation, Yamaguchi University, Ube, Japan

2 Department of Mechanical Science and Engineering, Nagoya University, Nagoya, Japan

3 Department of Aerospace Engineering, Nagoya University, Nagoya, Japan

References

[1] Sreenivasan KR and Antonia RA: The phenomenology of small-scale turbulence. Annual Review of Fluid Mechanics. 1997;**29**:435–472. DOI: 10.1146/annurev.fluid.29.1.435

[2] Sreenivasan KR: Fractals and multifractals in fluid turbulence. Annual Review of Fluid Mechanics. 1991;**23**:539–604. DOI: 10.1146/annurev.fl.23.010191.002543

[3] Dimotakis PE: Turbulent mixing. Annual Review of Fluid Mechanics. 2005;**37**:329–356. DOI: 10.1146/annurev.fluid.36.050802.122015

[4] Comte-Bellot G and Corrsin S: The use of a contraction to improve the isotropy of grid-generated turbulence. Journal of Fluid Mechanics. 1966;**25**:657–682. DOI: https://doi.org/10.1017/S0022112066000338

[5] Mohamed MS and LaRue JC: The decay power law in grid-generated turbulence. Journal of Fluid Mechanics. 1990;**219**:195–214. DOI: https://doi.org/10.1017/S0022112090002919

[6] Suzuki H, Mochizuki S, and Hasegawa Y: Validation scheme for small effect of wind tunnel blockage on decaying grid-generated turbulence. Journal of Fluid Science and Technology. 2016;**11**:JFST0012. DOI: http://doi.org/10.1299/jfst.2016jfst0012

[7] Lavoie P, Djenidi L, and Antonia RA: Effects of initial conditions in decaying turbulence generated by passive grids. Journal of Fluid Mechanics. 2007;**585**:395–420. DOI: https://doi.org/10.1017/S0022112007006763

[8] Sinhuber M, Bodenschatz E, and Bewley GP: Decay of turbulence at high Reynolds numbers. Physical Review Letters. 2015;**114**:034501. DOI: https://doi.org/10.1103/PhysRevLett.114.034501

[9] Vassilicos JC: Dissipation in turbulent flows. Annual Review of Fluid Mechanics. 2015;**47**:95–114. DOI: 10.1146/annurev-fluid-010814-014637

[10] Hurst D. and Vassilicos JC: Scalings and decay of fractal-generated turbulence. Physics of Fluids. 2007;**19**:035103. DOI: http://dx.doi.org/10.1063/1.2676448

[11] Valente PC and Vassilicos JC: The decay of turbulence generated by a class of multi-scale grids. Journal of Fluid Mechanics. 2011;**687**:300–340. DOI: https://doi.org/10.1017/jfm.2011.353

[12] Nagata K, Sakai Y, Inaba T, Suzuki H, Terashima O, and Suzuki H: Turbulence structure and turbulence kinetic energy transport in multiscale/fractal-generated turbulence. Physics of Fluids. 2013;**25**:065102. DOI: http://dx.doi.org/10.1063/1.4811402

[13] Hearst RJ and Lavoie P: Decay of turbulence generated by a square-fractal-element grid. Journal of Fluid Mechanics. 2014;**741**:567–584. DOI: https://doi.org/10.1017/jfm.2013.684

[14] Nagata K, Saiki T, Sakai Y, Ito Y, and Iwano K: Effects of grid geometry on non-equilibrium dissipation in grid turbulence. Physics of Fluids. 2017;29:015102. DOI: http://dx.doi.org/10.1063/1.4973416

[15] Suzuki H, Nagata K, Sakai Y, and Ukai R: High-Schmidt-number scalar transfer in regular and fractal grid turbulence. Physica Scripta. 2010;**T142**:014069. DOI: http://dx.doi.org/10.1088/0031-8949/2010/T142/014069

[16] Gomes-Fernandes R, Ganapathisubramani B, and Vassilicos JC: Particle image velocimetry study of fractal-generated turbulence. Journal of Fluid Mechanics. 2012;**711**:306–336. DOI: https://doi.org/10.1017/jfm.2012.394

[17] Gomes-Fernandes R, Ganapathisubramani B, and Vassilicos JC: The energy cascade in near-field non-homogeneous non-isotropic turbulence. Journal of Fluid Mechanics. 2015;**771**: 676–705. DOI: https://doi.org/10.1017/jfm.2015.201

[18] Nagata K, Suzuki H, Sakai Y, Hayase T, and Kubo T: Direct numerical simulation of turbulent mixing in grid-generated turbulence. Physica Scripta. 2008;T132:014054. DOI: http://dx.doi.org/10.1088/0031-8949/2008/T132/014054

[19] Laizet S and Vassilicos JC: Multiscale generation of turbulence. Journal of Multiscale Modelling. 2009;**1**:177–196. DOI: http://dx.doi.org/10.1142/S1756973709000098

[20] Suzuki H, Nagata K, Sakai Y, and Hayase T: Direct numerical simulation of turbulent mixing in regular and fractal grid turbulence. Physica Scripta. 2010;**T142**:014065. DOI: http://dx.doi.org/10.1088/0031-8949/2010/T142/014065

[21] Laizet S and Vassilicos JC: DNS of fractal-generated turbulence. Flow, Turbulence and Combustion. 2011;**87**:673–705. DOI: 10.1007/s10494-011-9351-2

[22] Suzuki H, Nagata K, Sakai Y, Hayase T, Hasegawa Y, and Ushijima T: Direct numerical simulation of fractal-generated turbulence. Fluid Dynamics Research. 2013;**45**:061409. DOI: http://dx.doi.org/10.1088/0169-5983/45/6/061409

[23] Zhou Y, Nagata K, Sakai Y, Suzuki H, Ito Y, Terashima O, and Hayase T: Relevance of turbulence behind the single square grid to turbulence generated by regular- and multiscale-grids. Physics of Fluids. 2014;**26**:075105. DOI: http://dx.doi.org/10.1063/1.4890746

[24] Ito Y, Nagata K, and Komori S: The effects of high-frequency ultrasound on turbulent liquid mixing with a rapid chemical reaction. Physics of Fluids. 2002;**14**:4362–4371. DOI: http://dx.doi.org/10.1063/1.1518508

[25] Sponfeldner T, Soulopoulos N, Beyrau F, Hardalupas Y, Taylor AMKP, and Vassilicos JC: The structure of turbulent flames in fractal- and regular-grid-generated turbulence. Combustion and Flame. 2015;**162**:3379–3393. DOI: http://dx.doi.org/10.1016/j.combustflame.2015.06.004

[26] Sakai Y, Nagata K, Suzuki H, Ito Y: Mixing and diffusion in regular/fractal grid turbulence. In: Sakai Y, Vassilicos, JC, editors. Fractal Flow Design: How to Design Bespoke Turbulence and Why: Springer International Publishing; 2016. pp. 17–73. DOI: 10.1007/978-3-319-33310-6

[27] Crimaldi JP: Planar laser induced fluorescence in aqueous flows. Experiments in Fluids. 2009;**44**:851–863. DOI: 10.1007/s00348-008-0496-2

[28] Suzuki H, Nagata K, and Sakai Y: Quantitative visualization of high-Schmidt-number turbulent mixing in grid turbulence by means of PLIF. Journal of Visualization. 2012;**15**:109–117. DOI: 10.1007/s12650-011-0121-y

[29] Raffel M, Willert CE, Wereley S, and Kompenhans J: Particle Image Velocimetry: A Practical Guide. Springer-Verlag Berlin Heidelberg; 2013. DOI: 10.1007/978-3-540-72308-0

[30] Villermaux E and Innocenti C: On the geometry of turbulent mixing. Journal of Fluid Mechanics. 1999;**393**:123–147. DOI: https://doi.org/10.1017/S0022112099005674

[31] Suzuki H, Nagata K, Sakai Y, and Hasegawa Y: Fractal analysis of turbulent mixing in fractal-generated turbulence by planar laser-induced fluorescence. Physica Scripta. 2015;**T155**:014062. DOI: http://dx.doi.org/10.1088/0031-8949/2013/T155/014062